Images of the Wildman in Southeast Asia

The book examines 'wildmen', images of hairy humanlike creatures known to rural villagers and other local people in Southeast Asia and elsewhere. Sometimes described in considerable detail, the creatures are reported as still living or as having survived until recent times. The aim of the book is to discover the source of these representations and their status in local systems of knowledge, partly in relation to distinct categories of spiritual beings, known animals and other human groups. Although largely focused on Indonesia, it explores images of the wildman from various parts of Southeast Asia and beyond, including the Asian mainland, Africa, North America, Australia and Oceania.

Images of the Wildman in Southeast Asia reveals how, in Southeast Asia and elsewhere, 'wildmen' cannot readily be explained as imaginary constructs rooted in cultural values and social institutions, nor as simply another kind of 'spirit'. Also critically examined is a view of such figures as fundamentally similar expressions of a pan-human 'archetype'. Forth concludes that many Asian and African figures are grounded in experience or memories of anthropoid apes supplemented by encounters with ethnic others. Representations developed among European immigrants (including the North American 'sasquatch') are, in part, similarly traceable to an indirect knowledge of primates, informed by long-standing European representations of hairy humans that have coloured Western views of non-Western peoples and which may themselves originate in ancient experience of apes. At the same time, the book demonstrates how Indonesian and other Malayo-Polynesian images cannot be explained in the same way and explores the possibility of these reflecting an ancient experience of non-sapiens hominins.

Gregory Forth is Professor in the Department of Anthropology at the University of Alberta, Canada.

Images of the Wildman in Southeast Asia

An anthropological perspective

Gregory Forth

Routledge
Taylor & Francis Group
LONDON AND NEW YORK

First published 2008 by Routledge
2 Park Square, Milton Park, Abingdon, Oxon OX14 4RN

Simultaneously published in the USA and Canada
by Routledge
711 Third Avenue, New York, NY 10017

Routledge is an imprint of the Taylor & Francis Group, an informa business

First issued in paperback 2012

Reprinted 2009

Typeset in 10/12pt Times New Roman by Graphicraft Limited, Hong Kong

British Library Cataloguing in Publication Data
A catalogue record for this book is available from the British Library

Library of Congress Cataloging in Publication Data
Forth, Gregory L.
Images of the Wildman in Southeast Asia: social values, archetypes and existing creatures / Gregory Forth.
p. cm.
Includes bibliographical references and index.
1. Wild men—Southeast Asia. 2. Fossil hominids—Southeast Asia. 3. Legends—Southeast Asia. 4. Southeast Asia—Folklore. I. Title.
GR308.F67 2008
398.20959—dc22

2008024912

ISBN13: 978-0-7103-1354-6 (hbk)
ISBN13: 978-0-415-53348-5 (pbk)
ISBN13: 978-0-203-88624-0 (ebk)

Contents

List of illustrations

Maps

Plates

Tables

Preface

The idea of a comparative study of wildman images first came to me in 1984 when I began ethnographic research among the Nage people, who reside in the central part of the eastern Indonesian island of Flores. It was not long after I arrived in Bo'a Wae, the centre of the Nage region, that I became involved in a typically animated discussion with Emiel Waso Ea (1941–1994), a son of the last Nage rajah and my principal Florenese host. Describing my previous research on the neighbouring island of Sumba (1975–76), I happened to mention—in a conversation about the kinds of spiritual beings recognized by followers of eastern Indonesian indigenous religion—the hairy hominoids Sumbanese call 'mili mongga' (see Chapter 4). Emiel responded immediately that the Nage knew beings similar to these called 'ebu gogo'. He added that the ebu gogo were not spirits but creatures that used to reside in the vicinity and had been exterminated by humans long ago.

It was also in the 1980s that certain hominoidal figures, including the North American sasquatch and the Chinese wildman, were gaining renewed attention, even to a measured extent among some anthropologists (see, for example, M. Ames and M. Halpin 1980). The anthropological interest was not sustained. In some ways, this is not surprising, but it is regrettable nonetheless. Images of wildmen—creatures bearing a resemblance, sometimes striking, to models of pre-sapiens hominins—are recognizable in many of the world's cultures and if nothing else it is as cultural images, or representations, that they should command the attention of social and cultural anthropologists. More particularly, images of any kind should, following a time-honoured anthropological method, be considered as phenomena hypothetically informed by social values and institutions and connected with other images and ideas that compose a social system or culture.

As a general position, this is one approach that has shaped the present book. But it is not the only approach. Indeed, a major concern is to explore ways in which representations of wildmen, sometimes better described as reports or rumours, might reflect something other than imaginary cultural constructs. Like my Nage friend, many people consider scientifically unrecognized hominoids to be real and at the same time quite different from spiritual or supernatural beings. By contrast, a large majority of anthropologists would no doubt consider them unreal—which is to say, unconnected with anything empirical.

Owing in part to this general disciplinary tendency to regard empirically unattested phenomena as entirely or mostly imaginary and even representations of empirical phenomena as primarily socially constructed, the topic of this book presents challenges for anyone trying to keep a reasonably open mind about the object of study. Yet a further difficulty lies in the ambiguity of the figure of the wildman itself, or rather the variety of figures that can be accommodated under this rubric. Scientific zoology tells us that such creatures are certainly, or almost certainly, fictional; scholarship in the humanities, which has been almost entirely concerned with the European wildman (which I discuss in Chapter 8), simply assumes it. As anthropologists, however, we need to be mindful of the ontological and epistemological distinctions recognized by other cultures and, in a non-trivial sense, to respect what ordinary folk tell us. With regard to that segment of our ridiculously diverse discipline that maintains connections to evolutionary biology, palaeontology and other branches of physical science, we also need to be aware of the natural, including zoological, contexts in which human cultural traditions develop and operate. Presumptions of empirical unreality can therefore be tricky. All anthropologists will be familiar with creatures whose local representation sounds utterly fantastic but which nevertheless turn out to be undeniably empirical species (the pangolin as conceived by the Lele of Central Africa may be the first example that comes to mind). In the same vein, everyone will know that fully attested and locally well-known animals (tigers and pythons, to cite just two Southeast Asian examples) are often credited with extraordinary powers of a sort usually ascribed to spirits.

Nevertheless, it is probably true that anyone investigating 'hairy hominoids' is likely to be dismissed as a 'believer'. I do not consider myself a believer; but neither am I a convinced 'atheist'. It has sometimes been suggested that the topic of wildmen is appealing because people want them to exist. Some people evidently do. But my impression is that many do not and reasons are not difficult to find. People like the Nage are certainly glad the ebu gogo no longer exist, for reasons that will become clear in Chapter 2. As for westerners, quite apart from the challenge they would pose to prevailing anthropological theories and methodology, the prospect, as improbable as it may seem, of things that sound like non-sapiens hominins actually turning up somewhere would present tremendous philosophical challenges bearing on the perennial question of the human–animal boundary. More practical but equally extraordinary questions would also be raised concerning how to treat creatures falling just a little bit shy of the divide (presuming the divide itself were to survive). Things would be much easier, then, if wildmen did not exist—and for their sake, at least, one should probably hope that they do not! Yet philosophical and practical difficulties raised by the existence of something do not of course rule out its possibility or probability. And in any case, an equally fascinating anthropological question is why so many non-western people should think wildmen do exist, or did exist until recently. The curiosity is underlined when one considers that non-western cultures are as diverse as they are different from western cultures.

As indicated, I first started thinking about this book some twenty-five years ago. Owing to other commitments, it has taken a long time to come to fruition. That it has finally done so is due largely to two developments in 2004. In that year, I was awarded a McCalla Professorship by the University of Alberta, tenable in the academic year 2004–05. After having conducted ethnographic research bearing on the topic for some twenty years, I felt I was ready to begin writing comparatively on wildmen and the professorship allowed me the time to do so. A trip to the Netherlands in June 2004 provided the opportunity to extend my reading concerning images reported from Indonesia and elsewhere in Southeast Asia. A sabbatical during the following year (2005–06) facilitated further research, especially during my tenure as Senior Fellow at the International Institute for Asian Studies (IIAS) at Leiden University and as British Academy Visiting Professor at the University of Kent. In both Leiden and Kent I was able to present ideas discussed here in special seminars and the theme of Southeast Asian wildman images was the topic of a 'Masterclass' conference I convened under the auspices of IIAS in February 2006.

But back to 2004. It was also in June of that year that I met a member of the discovery team which, as I then learned, had just the previous year been excavating in a cave in western Flores and had found the sub-fossil remains that were to be interpreted as a new hominin species, *Homo floresiensis*. Details of the discovery were made public in late October 2004. Dating to well within the period modern humans inhabited Indonesia, the find relates to my topic in ways that will soon become clear. Suffice it to say that speculation connecting the new hominin with the figure locally known as 'ebu gogo' (not to mention the published remark by one well-known British biologist that someone should start searching Flores for surviving specimens of *floresiensis*) lent a new impetus to my project. It also underlined the need for someone, preferably an anthropologist, to address the more general issue of possible connections between palaeontological reconstructions of non-*sapiens* hominins and the hairy hominoids described in a variety of non-western traditions.

Acknowledgements

This study has benefited from the considerable assistance of a variety of scholars, partly in the form of information and opinions provided by way of personal communications. Particular help is acknowledged in the text or in endnotes. In a more general vein, however, I should like to give special thanks to several individuals, listed below mostly with respect to the countries of their current academic affiliations.

In the United Kingdom, my gratitude extends to Roy Ellen, C.W. Watson and Roger Just (all of the University of Kent); David Chivers and Martin Walsh (both of Cambridge University), Michael Hitchcock (of London Metropolitan University); Stephen Oppenheimer (Oxford University) and Chris Stringer and Robert Kruszynski (both of the Department of Palaeontology, Natural History Museum, London). I am particularly grateful to David Chivers and also to Deborah Martyr, Achmad Yanuar, Barney Long and Jeremy Holden for allowing me to quote freely from unpublished reports written in connection with field research concerning the Sumatran 'orang pendek' (Chapter 5).

In the Netherlands, special thanks are due to Annekieke Rintjema, Tom van den Berge, Raymond Corbey, Jet Bakels, John de Vos, Gert van den Bergh, Herman Rijksen, Juara Ginting and Han Vermeulen. Drs Rintjema, who has written a master's thesis on the topic, was particularly helpful in providing advice on materials relating to Sumatran hominoids; Dr Corbey and Dr Bakels both read papers at the previously mentioned 'Masterclass' conference and in various ways helped me think clearly and critically about a number of relevant issues. To Professors Wim Stokhof and Reimar Schefold I am most grateful for supporting my Leiden Fellowship.

Australia looms large as the work place of numerous Southeast Asianists, many of whom I have been in correspondence with over a number of years in connection with this project. Among these are David Bulbeck and Peter Bellwood (both of the Australian National University), Alexander Adelaar and E. Douglas Lewis (both of the University of Melbourne), Hans Brunner and Graham Joyner. Thanks are due to all. Dr Bulbeck, an archaeologist and physical anthropologist, was another valuable contributor to the Leiden 'Masterclass' conference and since then I have regularly benefited from his knowledge of human evolution and prehistory in Southeast Asia and adjacent regions. Dr Robert Cribb of the Australian National University was kind enough to provide me with electronic copies he had made of unpublished documents relating to the Sumatran orang pendek, a topic in which he also has a scholarly interest, while Dr Richard (Bert) Roberts of the University of Wollongong was especially helpful in making available copies of Peter Schouten's painting of *Homo floresiensis*.

Mostly in a more general vein, I am similarly indebted to several scholars in the United States and Canada. These include: Adrienne Mayor (of Stanford University), Genese Sodikoff (Rutgers University) and Owen Beattie (University of Alberta).

Among fellow ethnographers of Flores who have provided information and advice, I am especially grateful to Maribeth Erb, Takashi Sugishima, Eriko Aoki, Andrea Molnar and Father Philipus Tule. Father Tule, himself a Florenese and now Rector of a major Catholic seminary on the island, has also generously served as my sponsor during two visits to eastern Indonesia. My visit to Sumba in 2005 was greatly facilitated by research assistance provided by Mr Bernardus Retang Wohangara, a lecturer in English at Soegijapranata Catholic University in Semarang, Java, as well as by the generous support of his Sumbanese family in Lambanapu. Kevin Tan of Singapore, who recently completed his doctorate at the University of Alberta, very kindly located for me copies of newspaper articles relating to putative hominoids reported from the Malay Peninsula in 1953–54.

In regard to institutional support, I should like to express my gratitude to the British Academy, the University of Kent and the International Institute of Asian Studies in Leiden for providing financial assistance and allowing me use of their excellent facilities during my sabbatical year. Although I have drawn on the resources of a number of libraries and archives, special thanks are due to the KITLV (Royal Institute of Linguistics and Anthropology) in Leiden and especially to the Institute's chief librarian, Drs Sirtjo Koolhof, whose interest in my project and enthusiastic assistance have been much appreciated. Special thanks are also due to John de Vos, of the Natural History Museum in Leiden, for making available a series of newspaper clippings concerning the 'orang pendek', compiled by Eugene Dubois and forming part of the Dubois Collection held by the museum. In the same vein, I should like to acknowledge the University of Alberta libraries. Not only have I regularly been impressed by the university's own extensive collections, but where items were not available an invaluable resource has been the libraries' interlibrary loan service. My thanks are also due to The Indonesian Institute of Sciences (LIPI), Nusa Cendana and Artha Wacana Universities in Kupang and St. Paul's Major Seminary in Ledalero (Flores) for, at various times and in a variety of ways, sponsoring and facilitating my research in Indonesia. Over a number of years, my Indonesian researches have received generous financial support from the Social Sciences and Humanities Research Council of Canada and several research funds within the University of Alberta, for which I am equally grateful.

A very special debt is owed to numerous Indonesian informants and assistants from the islands of Flores and Sumba, many of whom are more accurately described as friends. Proverbially, these are too many to mention by name and I shall not endeavour to do so. In respect of substantial field assistance, however, provided in several cases over many years, I should like to acknowledge Cornelis Kodhi Léjo, Fidelis Laja Ga'e, Nico Nua, Petrus Lape Ga'e and Endy Paji—all of Flores.

Finally, I should like to thank my wife Christine, herself an anthropologist familiar with Indonesia, who in so many ways has helped me bring this project to completion.

Gregory Forth
Department of Anthropology
University of Alberta
October 2007

Map 1. Southeast Asia.

1 Introduction

Among hairy, bipedal humanlike creatures reputed to exist in remote places, most people will have heard of the Himalayan 'yeti' and the North American 'sasquatch' (or 'bigfoot'). Far less well known is the occurrence of such images—'wildmen' according to a definition I articulate below—in many of the world's cultures. Rather than colonialist distortions as some historians have supposed, or figures constructed from European travellers' tales (Dudley and Novak 1972; Gouda 1995), even the better-known figures have their origins in representations maintained by members of small-scale, non-western societies. Non-western folk, moreover, frequently describe the creatures prosaically, in much the same way as they describe animals known to western science.

Often specified as smaller than local human beings, among the most intriguing wildman images are ones reported from parts of Southeast Asia. Recently, the region has received much attention owing to the discovery, on the eastern Indonesian island of Flores, of the remains of a tiny hominin. Interpreted as a new species distinct from yet contemporary with our own, this creature, designated as *Homo floresiensis*, has been hypothetically linked by some members of the discovery team with legendary hominoids called 'ebu gogo' (Morwood and Van Oosterzee 2007: 154–55). 'Ebu gogo' denotes a category of reputedly extinct hominoids recognized by the Nage people of central Flores, a population of about 50,000 cultivators, raisers of livestock and occasional hunters. Since the middle of the last century, the majority of Nage have converted to Roman Catholicism, yet they retain much of their indigenous cosmology, religion and oral tradition. Nage culture and social organization have been described by the present author in a series of publications. Among these is a book entitled *Beneath the Volcano* (Forth 1998a) which contains the first published mention of the ebu gogo. By contrast, the present study has a more comparative objective. The major aim is to locate the ebu gogo in relation to comparable images found in other parts of Southeast Asia and the world. Another is to explore ways in which these images may be connected with both humans and non-human species that are known to exist in local environments or, from the reconstructions of palaeoanthropology, are known to have existed in the past.

At the risk of the reader's early alienation, I must proceed directly to definitions. Somewhat arbitrarily, I have selected 'wildman' as a general term for hairy, bipedal creatures of generally human form, typically leading a cultureless existence in desolate places far from human settlements. There are two further specifications: the creatures are described as more humanlike than known apes and their existence remains unrecognized by modern science. 'Wildman' thus does not refer to attested primate species, though in some respects the term could quite accurately be applied to anthropoid apes. Nor do I employ it for 'feral' humans, 'primitive' tribes, outlaws, social deviants, or 'men with tails'—despite previous applications of 'wild man' and 'wild men' to all of these.[1] Some distinction is supplied by my use throughout of the compound ('wildman' rather than 'wild man'). At the same time, I have been unable to find a neutral plural. 'Wild people' and 'wild folk' are too suggestive of human beings and do not permit compounding. It may therefore be necessary to emphasize that 'wildman' covers both males and females. And in case anyone should take offence at the adjective 'wild', I might add that it is intended to reflect local cultural evaluations of wildmen, especially as creatures lacking in culture and often articulate language and not, in itself, as any kind of anthropological interpretation.

As the foregoing definition might suggest, I could just as well designate the subject of this book as 'hairy hominoids'. Apart from its quaint ring, the expression is perhaps less definite than 'wildman'. However, as I do employ it from time to time, I should stress that by 'hominoid' I refer simply to any humanlike being. Although the term can be applied to anthropoid apes, it is not meant to include attested non-human primates. At the same time, the label does not imply any prejudgement as to whether possible empirical referents are human or not human. 'Hominoid' is not to be confused with 'hominid' or 'hominin', terms with quite precise meanings in scientific taxonomy, both denoting zoological categories that include the species *Homo sapiens*. Throughout the present work I follow newer taxonomic usage and employ 'hominin' to refer to anatomically modern humans, extinct species (or 'chrono-species') of the genus *Homo* and the extinct genera *Australopithecus* and *Paranthropus*. In an older usage, this grouping was called 'hominids', a usage I avoid partly because of its similarity to 'hominoid'.[2]

As a qualifier of 'hominoid', the adjective 'hairy' is crucial. The local categories that form my subjects typically comprise creatures described as hairy-bodied. Since the focus of this book is moreover on hominoids regarded as *smaller* than local humans, if these were not specified as hairy (that is, hairier than normal humans) there would in many cases be little to distinguish them, physically at least, from the host of generally glabrous fairies, dwarfs and other diminutive figures typically attributed with supernatural powers that populate folk cosmologies the world over. Occasionally, I discuss creatures otherwise resembling hairy hominoids which are not hirsute, or are not unequivocally so. I treat these, however, only where relevant to my overall comparative purpose.

The difficulty of finding a single term for the subject of this study is telling. A related circumstance is the lack of anthropological interest in the topic, despite the widespread occurrence of ethnographic instances. Most attention to wildmen has come from humanists—historians, philosophers, folklorists and the like—and these have focused squarely on the European wildman (or 'wild man'), a figure of late mediaeval and Renaissance art and literature. As I discuss in Chapter 8, this figure has less in common than might be supposed with the mostly Asian images otherwise treated in this book. The historical character and provenance of the European representation may partly explain anthropological inattention to the topic. As a reference, for example, to non-European images such as the 'Chinese wildman' (Chapter 7), 'wildman' has latterly gained some currency in the recently emerged and academically marginal field of cryptozoology, the study of animals not recognized by scientific zoology. But this association, too, has hardly endeared the subject to anthropologists; indeed, probably more than claims on the topic by historians and folklorists, it illuminates what could reasonably be construed as an anthropological avoidance of the wildman.

To the extent that anthropologists have touched upon ethnographically documented categories of wildmen, they have usually treated them, implicitly and uncritically, as varieties of 'spirits', or (which is to say much the same thing) as entirely imaginary, socially constructed beings. In regard to this last qualification, an instructive contrast is provided by the 'witch', an image apparently even more widespread than the wildman. Since the inception of professional anthropology, witches and witchcraft have been investigated in a great variety of cultures and accounting for them has been the object of a range of theoretical writings. The difference would appear to be connected with the impossibility of witchcraft for the modern western mind, combined with the invariable identification of witches, by definition, with particular humans beings and often with specific social categories (women, lower-ranking people, ethnic others).[3] More than anything else, this last feature has rendered witches and witch-craft singularly susceptible to a sociological interpretation. As this should imply, wildmen, especially when locally described in a way reminiscent of palaeoanthropological reconstructions, are less readily dismissed as fantastic beings; they do not lead the 'double life' of witches; and partly for this reason are far less easily explained as social imaginaries. Implicitly, then, the anthro-pological response has been one of avoidance and denial of the wildman as a distinct analytical category, amounting to a virtual taboo.

How far local wildman categories might be explicable as imaginary social or cultural constructs is a matter to which I attend throughout this work. The fact is, however, that I have found relatively little in particular social systems that could account for locally posited hominoids or features specifically attributed to them. Nor do their features obviously vary in accordance with more general cultural differences—although in some cases particular wildman images do suggest regional features coinciding with ethno-linguistic boundaries (as I show in Chapter 9). This does not mean that some such features are not obviously imaginary. But whether the categories themselves are entirely without empirical basis is another matter. The ubiquity of wildman images and the concomitant difficulty of linking them with specific cultural traditions raise the question of how far these may arise naturally in human cognition—a question necessarily left for the concluding chapter. Otherwise, my approach is broadly 'ethnozoological', viewing the representations, heuristically, as possible components of local environments (or historical environments) and considering categories of unattested hominoids as part of a general scheme of folk zoological (or, in some cases, folk palaeontological) knowledge. My principal concern, therefore, is a series of named local categories which, although they have yet to be firmly identified with known zoological species—or indeed, simply, as instances of *Homo sapiens*—are also not unequivocally varieties of spirits or other non-empirical beings.

Research in both ethnozoology and cognitive anthropology has shown animal categories to be subject to two sorts of representation and components of two varieties of knowledge. One is an intuitive 'general purpose' zoological knowledge grounded in universal perception (Berlin 1992). Where this pertains to classification, its product is an ethnotaxonomy. The other kind is a symbolic knowledge, prospectively contributing to a 'symbolic classification' (Forth 2004a: 63) and comprising counter-intuitive mythical and religious ideas specially learned and often deliberately transmitted in the course of enculturation (Atran 1990: 219). Accordingly, people can have a detailed empirical knowledge of an animal kind and still ascribe to it all sorts of fantastic qualities. While symbolic representations, too, can ultimately be grounded in experience, they also need not affect the way people ordinarily behave towards the animal, especially where practical or material interests are concerned. By the same token, symbolic knowledge is more likely to be expressive of particular cultural and ideological interests than is empirical knowledge.[4]

In view of these contrasts, the question becomes how far wildmen are intelligible as articles of one or the other kind of knowledge. As will become apparent, categories of wildmen reported from different parts of the world vary considerably in zoogeographical credibility. Whereas many appear to lack any local zoological basis, others can plausibly be linked with experience of some natural species. This is not to suggest that from ethnological study of local categories one can definitely conclude that any reflects undiscovered species—let alone non-sapiens hominins. The main problem for any such thesis, of course, is the 'lack of a body', which is to say, authenticated evidence of an actual specimen. So long as this remains the case, zoologists who claim that wildmen (or unrecognized species locally identified as such) do not exist will always be right. On the other hand, since many are described as creatures long or recently extinct, or otherwise rare and decreasing in number, the possibility of some basis in cultural memory of former species is worthy of consideration.

Current social theory typically treats human memory as a social construct: an artefact of present interests and values rather than an accurate reflection of actual events. To a degree, this view is of course unexceptionable. Yet there is evidence for people collectively preserving long-term memory of entities or events, some verifiable and doing so moreover in the absence of apparent support from social, political, or economic interests—a matter I take up again in the concluding chapter. There are other grounds for considering possible zoological bases for legends of hairy hominoids. Even if no hominins other than *Homo sapiens* survive to the present (something I, like most people, find highly probable), there is no reason in principle—or according to current biological theory—why they should not do so. We certainly know that *Homo neanderthalensis* was for a long time contemporaneous with modern humans. Australopithecines were evidently contemporary with early forms of *Homo* for even longer. And insofar as *Homo floresiensis* is confirmed as a new species, this hominin will have shared the earth with *Homo sapiens* until a considerably more recent date (to at least 12,000 years ago). Modern palaeoanthropology makes such contemporaneity of different species of the genus *Homo* rather more likely than it would have seemed even a few decades ago. I refer to a move away from a strict unilinear view of human evolution (in which one chrono-species is replaced by another), as well as a newfound possibility of hominin evolution in Asia commencing before the development of *Homo* (Dennell and Roebroeks 2005). In addition, ever since 1812, when Cuvier declared that there were no more large animals left to discover, a good number have come to light, including many in the last decades of the twentieth century. Southeast Asia has had its fair share of these, some of which were well known to local people—the real folk zoologists—long before their discovery by western scientists. In eastern Indonesia, perhaps the most famous is the giant lizard *Varanus komodoensis*, the so-called Komodo dragon of which inhabitants of Komodo, Flores and Sumbawa were certainly aware long before 1912, when a Dutch herpetologist first described the reptiles (see Auffenberg 1981; Verheijen 1982).

Comparison of indigenous and scientific representations of what are palpably the same creatures is one kind of interest which the present book hopes, in a limited way, to advance. Another, more directly relevant to the main topic, concerns local representations of scientifically unattested animals hypothetically linked with recently extinct species. The latter include subfossil primates of Madagascar, an island whose Malayo-Polynesian-speaking inhabitants—thus linguistic relatives of Florenese and other Indonesians—also recognize a variety of hairy wildman (Chapter 8).

With the discovery of a possible new hominin on Flores, there is arguably new reason to consider whether wildman images and legends could ultimately reflect some basis in human experience of other species, including scientifically undiscovered species, especially in Asia. In a global perspective, however, I am not seeking to advance any single kind of explanation; indeed, different cases suggest explanations of quite different kinds. Invoking a broad anthropological perspective, my purpose is rather to review possibilities of explanation and interpretation and thus to contribute to the development of questions which others might care to address. One possibility, of course, is that wildmen everywhere are entirely imaginary. But if this is so, the images cannot be solely explained as functions of particular cultural traditions, for their occurrence in a great range of societies suggests instead something like an 'archetype' of the human imagination, much as Needham (1978) proposed for the figure of the witch. The witch and the wildman may therefore have more in common than their unequal anthropological treatment would suggest. Yet what is especially intriguing about wildmen and especially those described by villagers in eastern Indonesia and other parts of Asia, is their resemblance to palaeoanthropological reconstructions of extinct hominins. (We are back, then, to comparisons between the Nage figure of ebu gogo and *Homo floresiensis*.)

Apart from the variability of wildman categories from different parts of the world in regard to zoological or zoogeographical plausibility, a general finding of this study concerns the size of legendary hominoids. Contrary to the giant-sized 'bigfoot' (or 'sasquatch') and the 'abominable snowman' ('yeti') of popular Western imagination, wildmen are frequently conceived as smaller than human beings, or at least as no larger—indeed, often about the size of the Florenese ebu gogo or the type specimen of *Homo floresiensis*. Another general conclusion concerns the marginality of these figures in the cultures in which they occur. Seemingly consistent with their putative rarity or extinction, very often the hominoids do not occupy a prominent place in local religion, ritual, or narrative traditions (mythology, legend, or folktale). As an experienced ethnographer of the Lio people told me with regard to the Florenese category named 'lae ho'a' (Chapter 3), these do not appear 'culturally important' because local people do not often speak of them or know much about them (Takashi Sugishima, pers. comm., May 2004). This, of course, is one respect in which wildmen obviously differ from what are usually described as 'spiritual beings'. Sometimes local people attribute supernatural powers to putative hominoids. But, as already pointed out, this does not distinguish them from representations of many attested animal kinds. Also, the evidence does not reveal any marked or distinctive tendency to 'spiritualize' wildmen—an inclination possibly exemplified, rather ironically, by the recent media designation of *Homo floresiensis* as a creature of fantasy: a 'hobbit'.

On the other hand, ethnographic information relating to single cultural contexts does often reveal a relative distinction, previously remarked by the primatologist John Napier (1972) and cryptozoologist Bernard Heuvelmans (1980), between what may be called 'mythical' and 'naturalistic' representations of wildmen. One context in which the contrast is apparent is where one and the same category, described in everyday discourse in a thoroughly matter-of-fact way, is given a more fantastic representation in myth or folktale. As I later show in regard to the ebu gogo, the same opposition is discernible in other discursive contexts. The co-existence of 'naturalistic' and 'mythical' representations is partly explicable with reference to the cognitive anthropological distinction remarked just earlier between empirical (or 'encyclopaedic' in the terminology of Sperber 1975) and symbolic knowledge. But it also presents a puzzle. As suggestive as it may be, the simultaneous presence of prosaic description and fabulous depiction does not of course prove that the former necessarily and accurately records a real animal. Yet if both were equally imaginary, we should then have a sort of double fantasy, a rather unparsimonious interpretation that seems to cry out for simpler explanation.

Organization and sources

Beginning in eastern Indonesia, where my interest in the topic of wildmen originated some twenty-five years ago, my exploration of comparative contexts for Florenese images of hairy hominoids follows a circular course. The more particular point of departure is the ebu gogo; and, since much of the final chapter is also devoted to this category, the Florenese hominoids provide the terminus as well. Chapter 2 is concerned solely with the ebu gogo and is based entirely on my own fieldwork in the Nage region of central Flores, conducted at various times between 1984 and 2005. Chapter 3 also draws mostly on my eastern Indonesian ethnography, as does that part of Chapter 4 dealing with the mili mongga of Sumba Island, where I carried out fieldwork for two years in 1975 and 1976 and for shorter periods in 1983 and 2005.[5]

Turning to parts of western Indonesia and Mainland Southeast Asia, figures described in Chapter 5 and 6 are largely informed by older reports which, while obviously valuable, are less detailed than my own eastern Indonesian ethnography. A partial exception is the orang pendek of Sumatra, described in Chapter 5, which has been the subject of numerous published accounts and commentaries, mostly from the 1920s and 1930s and of a series of unpublished reports of primatological fieldwork from the 1990s. Dealing with wildman images from other parts of Asia, Europe, North America, Africa, Australia and Oceania, Chapters 7–9 are also for the most part based on older publications. Noteworthy exceptions include some of the literature on Madagascar, where one hominoid category has drawn the attention of modern ethnographers and recent cryptozoological discussion of the Australian 'yahoo' and North American 'sasquatch'. Discussion of Madagascar at the end of Chapter 8 returns us to Malayo-Polynesian-speaking peoples, thus facilitating a transition to Melanesia, Polynesia and other parts of Oceania in Chapter 9 and completing the circle that leads back to Flores in the final chapter. (For reasons that should become clear, it has proven more convenient to treat the Philippines with Oceania and Taiwan than with other parts of Southeast Asia.)

In some degree, the diverse character and quality of the evidence reflect the scant attention so far given to wildmen, not just by anthropologists but also by other serious investigators of natural environments and cultural traditions outside of Europe. In regard to Indonesia and adjacent parts of Southeast Asia, in the mid-1990s I surveyed a number of anthropological colleagues who had conducted ethnographic research in the region, asking whether they had ever come across local beliefs in beings like the Nage ebu gogo. My enquiries drew few results. This might reasonably be taken to mean that representations of wildmen are indeed infrequently encountered in Indonesia. But there are other possible reasons they rarely show up on the anthropological radar. One is their generally marginal status in local cultures, already mentioned with reference to the view of one Flores ethnographer on the local unimportance of one example. As for the ebu gogo, it is also germane that older European Catholic missionaries I questioned in the 1980s and 1990s, who had worked for years in the Nage region, were unfamiliar with the representation. Based in eastern Sumba, the Protestant missionary D.K. Wielenga was certainly aware of the mili mongga (Chapter 4). This, however, is explained by Wielenga's diligence in recording traditional narratives, in which the creatures occur quite regularly, and indeed much more frequently than do other Indonesian hominoids. My own knowledge, acquired during fieldwork in 1975–76 and 2005, it is important to add, stemmed initially from direct questioning about the Sumbanese figure as described by Wielenga.[6]

The fact that hairy hominoids cannot definitely be described as a regular representation in a region as large as Indonesia prospectively qualifies the idea of the wildman as a universal property of human imagination, a hypothesis I critically discuss in Chapter 10. Any such thesis, of course, requires close attention to the degree of resemblance between hypothetical instances, which of course demands detailed examination of particular cases. This task commences, with the ebu gogo, in the following chapter.

Plate 1.1 Hendrikus Lako (deceased 2004), a Nage man who in 1984 first provided the author with detailed information on the ebu gogo.
Source: G. Forth.

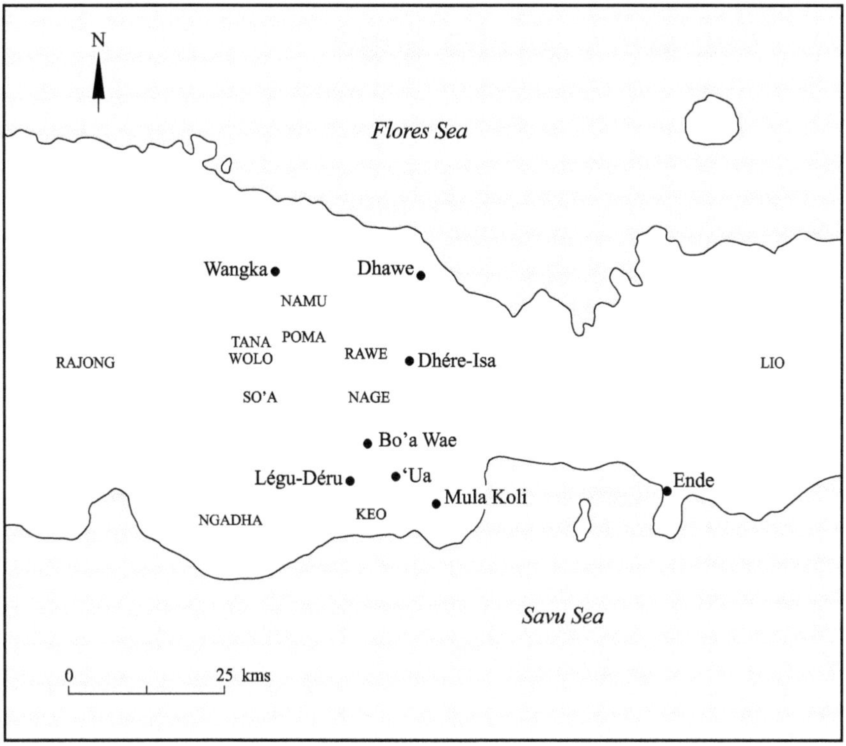

Map 2. Central Flores.

2 The story of ebu gogo

Named 'ebu gogo', the wildmen of Nage are reputed once to have inhabited a cave located high on the northern slope of Ebu Lobo, a large and active volcano in south central Flores. People of the Nage region of 'Ua (see Map 2) encountered ebu gogo several generations ago, when they moved from Wolo 'Ua, an area further to the west. Before this, Nage say, no one lived in the present territory of 'Ua, nor anywhere near the wildmen's cave. Although associated with 'Ua, stories of ebu gogo are known elsewhere in the Nage region, including the vicinity of Bo'a Wae, the main Nage village located about 3 kilometres to the west of 'Ua.[1]

In this chapter, I review local accounts of ebu gogo as related by more than forty informants during a dozen visits I made to Flores between 1984 and 2005. Most of the informants were middle-aged or elderly individuals, many from 'Ua and neighbouring settlements. The bulk of this information, I should emphasize, was recorded before 2004 and thus before the publication of verbal and visual reconstructions of *Homo floresiensis*. Information on ebu gogo I collected in 2005 did not contradict earlier descriptions. Nor did anything I heard that year suggest that palaeoanthropological interpretations of *floresiensis* reported in Indonesian media were widely known among Nage, or had had any effect on the 'Ua tradition (Forth 2006). Of the many accounts I recorded, one of the clearest and most comprehensive was given in 1996 by Julius Poi, an 'Ua man born in 1922 and thus at the time in his seventies (see Plate 2.1). In 2005, I was able to speak again to Old Julius and what he told me then matched his earlier statements to a remarkable degree. Julius Poi is a respected and knowledgeable elder, but his account is typical rather than exemplary. The following summarizes his 1996 account:

> Ebu gogo were like humans: they walked erect and did not have tails, but were covered in hair and their faces resembled a monkey or an orang-utan, with large canine teeth. The females had long breasts—so long that they carried them over their shoulders and suckled their infants from behind. Ebu gogo were shorter than people, just over one metre, but they were exceptionally strong and very powerful runners; humans were unable to catch them.

> There were perhaps as many as fifty ebu gogo living in Lia Ula, a cave about a kilometre upslope of the old village of 'Ua; the cave was very large and contained a passage, perhaps a kilometre in length, that led to an eastern exit. Ebu gogo knew no tools or weapons; nor did they use fire. They ate their food raw. They never bathed and therefore smelled very bad; people could detect them by their smell. Ebu gogo stole crops from the gardens and field huts of 'Ua people. They would sometimes appear at feasts held in 'Ua village and in smaller subsidiary settlements. People would then give them food, which they would swallow whole. Being greedy and uncultivated, they also swallowed vessels made of gourd and coconut shells. When attending these feasts, the ebu gogo would perform a kind of circle dance, by themselves and separately from their hosts. They were able to speak, but their speech was different from Nage and difficult to understand.
>
> Once a female ebu gogo was caught stealing from the fields of an 'Ua ancestor named Huma Léli. Before beginning her foray, she left her infant in Huma's field hut; and when he later returned and discovered the small creature, he promptly stabbed it to death with a palm rib. The mother then came back to collect her baby, but was set upon by Huma Léli's dogs. Ebu gogo were scared of dogs; they were also frightened of bamboo head combs. Addressing Huma by name, the female ebu gogo exclaimed that his dogs had bitten her child to death. She then fled with the dead infant.
>
> Because of their thieving ways, the 'Ua people later decided to exterminate the ebu gogo. Once, after the creatures had attended a feast, 'Ua men waited until the ebu gogo had all returned to their cave. They then sealed the cave's eastern exit and tossed a large quantity of arenga palm fibre, 500 bales in all, into the entrance at Lia Ula. This, they told the ebu gogo, was for them to use as sleeping mats; but being stupid, the creatures mistook the fibre for clothing. In the last bale of fibre the 'Ua men inserted a firebrand, thus causing a conflagration that killed the ebu gogo inside the cave. Only a male and female pair, who had been out searching for food, were able to escape destruction. They fled to Mount Ua, in the region of Tana Wolo. After the ebu gogo were exterminated a great mass of maggots emerged from Lia Ula and crawled for half a kilometre, before dying in the heat of the sun.

This last detail is one of several evidently fantastic components of the legend. But before exploring these, I propose to 'suspend disbelief' and evaluate local descriptions of ebu gogo as reflections of creatures and events possibly grounded in some empirical reality. To this end I employ details from various informant accounts to reconstruct a composite image of the ebu gogo to which various questions may then be directed. The foregoing sample account paints quite a detailed picture of the ebu gogo, especially the creatures' physical appearance and behaviour. This is all the more striking when one considers that the creatures are supposed to have been extinct in the Nage region for some two hundred years. With the exception of very simple and stylized masks, Nage do not produce visual images of anything they call 'ebu gogo'. Nevertheless, many Nage were able to describe the wildmen with the same level of detail as they described extant local mammals. Whenever I asked people how they could ascribe particular physical features to the ebu gogo without ever having seen one, they always replied that they had learned about them from their parents and grandparents, who would have been closer to the phenomenon.

What the ebu gogo looked like

Although somewhat various, Nage reports of ebu gogo morphology and behaviour generally produce a consistent image of a coherent perceptual entity, but not one so stereotypical as, for example, representations of categories of spiritual beings. Invariably, the impression they conveyed was one either of a primitive form of humanity (a member of the genus *Homo*) or a relatively large, bipedal non-human primate.

Most informants described ebu gogo as shorter than local humans, although some indicated a height not significantly different from modern Nage. Since Nage men usually range between 1.6 and 1.7 metres, for the ebu gogo to be considered unusually 'short' they could have stood no more than 1.4 to 1.5 metres. By the same token, some Nage said ebu gogo were not significantly shorter than themselves, so heights above 1.5 metres would also be possible.[2] Just five people described ebu gogo as 'tall', that is, bigger than Nage men. All, however, were people resident outside 'Ua, in three cases at some considerable distance. Also in three instances, it became clear that large size was likely something informants had inferred from other features, such as the creatures' reputed ability to devour large objects whole. As one elderly man put it, they must have been large since 'a small-bodied (creature) would have been incapable (of this)'.

If Nage ideas about ebu gogo's height reveal broad agreement, views on their breadth were rather more mixed. Although few informants provided specific details, most described the wildmen as robust or stout and no one characterized them as thin. A robust build accords with attributes mentioned more often: considerable physical strength and a mostly vegetarian diet. On the other hand, other comments suggested they were not significantly more robust than local humans, who are mostly slim.

My records include few comments on ebu gogo limbs, though two people described them as having long fingernails. Responses to questions about arm and leg length indicated these were comparable to human limbs, in proportion to the smaller body size. In other words, there is no indication that the ebu gogo had long arms and short legs like apes. According to Julius Poi, the wildmen's feet were small and flat; another man also described them as small, but furthermore as somewhat distinctively shaped and immediately recognizable. These were the only people who mentioned foot size and the only ones to refer to footprints. Both described the prints as a former means (in addition to a distinctive odour) of detecting ebu gogo; according to the more general report, the creatures themselves were rarely encountered visually. Everyone I spoke with was certain that ebu gogo moved bipedally like humans and no one described them as possessing tails. Their posture was erect and not bent or stooped and they were strong walkers and fast runners. People frequently remarked how they must have been strong and skilful climbers to manage the steep climb to their cave.

Nage often described ebu gogo looking like monkeys, especially in regard to the head and face. Others depicted them more as extremely 'ugly' ('e'e', dialectal 're'e') humans. As regards Julius Poi's comparing them with orang-utans (Malay, Indonesian 'mawas'), it should be emphasized that Flores has no apes; nor is there palaeontological evidence for apes east of Wallace's Line (see Map 4) during the period of human occupation. Nage mention of orang-utans in relation to ebu gogo reflects modern print appearances of the primates and—since even in recent decades Nage have rarely had access to books, magazines, or newspapers—particularly the depiction, since 1992, of the red ape on 500 rupiah Indonesian banknotes.

Ebu gogo facial features were often described in some detail. Their noses were mostly characterized as flat and wide or simply as simian. Speaking in 1988 (thus before the banknotes were issued), one man specified ebu gogo noses as 'broad and flat and rounded at the end'. A few informants described the eyes and ears as large, but otherwise of human appearance. By contrast, Julius Poi, who was very much inclined to compare the ebu gogo with simians, suggested that their eyes were small, 'like a monkey's'. Another man said he had heard the eyes were 'reddish'. With some regularity, Nage referred to ebu gogo as having broad faces and wide or large mouths and a couple of reports indicated low foreheads and receding chins. Describing the chin and lower jaw as monkey-like, Old Julius remarked that stones thrown at the heads of ebu gogo would have had no effect, as their skulls were thick. I seem not to have heard any specific description of the brows or orbital ridges. In view of the general notion that ebu gogo resembled monkeys—by virtue of their reputed lack of fire, clothing and technology, as well as their hairiness—Nage may well have inferred some details of morphology from the appearance of local monkeys (long-tailed macaques, *Macaca fascicularis*), the only known non-human primate on Flores and neighbouring eastern Indonesian islands. On the other hand, one man linked ebu gogo's gross facial features to their reputed ability to swallow large objects whole.

Consistent with reports of relatively large mouths, several people described the wildmen's teeth as large or very large ('the size of an adult's thumb nail' or 'like a water buffalo's teeth'). Such descriptions mostly referred to the front teeth. Others mentioned large canine teeth, with a few informants comparing these to macaque canines. Only Julius Poi spoke of small teeth, which he added, were 'turned inwards, like a monkey's'.

With the possible exception of long, pendulous female breasts, the physical feature of ebu gogo that Nage most often mentioned spontaneously and indeed almost invariably, was their hairiness. There was general agreement that female as well as male ebu gogo had hairy bodies. But while most people spoke of 'long', 'thick', or profuse hair extensively covering the body, others described ebu gogo as not significantly hairier than the most hirsute local humans. Although Nage usually depicted ebu gogo bodies as covered entirely in hair ('toto weki', 'covering the entire body'), two men described them as hairiest on the chest. Only one man, who seemed to envisage the wildmen as veritable balls of fur, implicitly specified the hair as also covering the female breasts. Another two individuals compared the body hair to that of a wild pig—that is, thick (in regard to individual strands), coarse and stiff. It goes without saying that terms like 'long', 'short' and 'thick' are vague and subjective and it was not always clear how far these judgements derived solely from comparison with humans. Nevertheless, one gets a fairly definite impression of a creature hairier than a human being but probably not as hairy as a monkey or other non-human animal.[3]

Only a few people said ebu gogo had hairy faces. Others mentioned faces mostly free of hair, while another described male ebu gogo as possessing short beards. Most informants spoke of long head hair. Commonly described as unkempt and dishevelled, the wildman's head hair provides Nage with a standard simile; someone whose hair is uncombed and untidy and nowadays left to grow long, is thus compared to an ebu gogo ('kau fu bhia ebu gogo', 'you have hair like ebu gogo'). Nage usually described both the head and body hair as 'black' or 'dark', the virtually uniform hair colour of Nage themselves. Still, it should be noted that Nage 'mite' ('black') denotes a range of darker colours, including dark brown and dark grey; accordingly, the colour was occasionally compared to that of macaque pelage—a greyish brown or brownish grey. Whenever ebu gogo's skin colour was mentioned, it was always characterized as 'black', which is how Nage describe their own complexions.

How the ebu gogo behaved

Nage invariably described ebu gogo as lacking a material culture. They had no tools, nor did they know the use of fire. Indeed, they were as much afraid of fire as they were of dogs. Thus, while stories portray the wildmen as engaging in circle dancing when attending 'Ua feasts, unlike Nage they would not dance around a fire for fear that their hair might be set alight. The wildman's reputed technological inadequacy partly illuminates a general view of the creatures as having been 'stupid' ('dhozo') or unintelligent, an evaluation commonly conveyed in connection with standard narratives relating how they would swallow eating vessels and the like—mistaking these for food—and how they fatally mistook palm fibre for clothing. Despite the ebu gogo's assessment of palm fibre, they are always described as going naked. As the wildmen reputedly never bathed, they possessed an unpleasant odour. One man speculated that they smelled like wild pigs; another more definitely compared their odour to the nest of the Flores Giant Rat (*Papagomys armandvillei*), a scent hunters use to locate the animals.

Nage depict ebu gogo diet as mostly vegetarian, consisting largely of fruits and vegetables stolen from people's fields and orchards. Specific cultivars included eggplants, maize, tubers, gourds, bananas, papaya and the arenga palm (in the form of tapped juice). Interestingly, cucumbers seem to be mentioned more than other foods, perhaps because they figure in the well-known story of Huma Léli (referred to in Julius Poi's account). Some Nage said ebu gogo ate no meat; but citing their attendance at 'Ua feasts, others explained that they would consume cooked meat provided by humans. Less often mentioned is a habit of swallowing whole young dogs, pigs and even human infants. Moreover, this reputation, seems to refer specifically to events related in standard narratives, while swallowing puppies appears inconsistent with the wildmen's great fear of dogs. Otherwise, Nage never described ebu gogo as consuming raw meat; nor are they credited with any kind of hunting ability.

Accounts of ebu gogo always describe a single population inhabiting the cave at Lia Ula. Estimates of their numbers ranged from fifty to more than a hundred individuals. According to one report, ebu gogo only ever travelled in pairs, consisting either of mates or a mother and infant. (Presumably, an exception was when the creatures appeared at 'Ua feasts, since it is implicit that these events were attended by the entire population of Lia Ula.) The same informant further stated that the ebu gogo were rarely encountered in daylight, that they were largely nocturnal and usually stole from fields at night. Others denied this nocturnalism, pointing out that the encounter with the ancestor Huma Léli took place during the daytime.

Two other behavioural proclivities of the ebu gogo were mentioned with some regularity: abduction of human children and anthropophagy. A few Nage spoke of kidnapping as a general practice, but more often the reputed abduction was specified as occurring on a single occasion. A story told by 'Ua villagers thus describes how the wildmen of Lia 'Ua abducted two small children named Sedho and Lilo. According to one variant, as the children grew tired of eating only the raw food given them by the ebu gogo, they returned home to request fire; however, their distraught parents naturally kept them and they never did return to the wildmen's cave. According to another version, the ebu gogo captured the children specifically as a way of acquiring fire for themselves, thereafter sending the abductees to request a fire-brand from the nearest human habitation. Unfortunately for the wildmen, the villagers recognized them as youngsters who had gone missing, detained them and then sent them home to their parents. The second variant has an obvious allegorical quality: as Nage themselves articulated, it was owing to their stupidity that the ebu gogo were never able to obtain fire (and, one might further infer, were incapable of attaining the cultural condition, partly symbolized by fire, that confers complete humanity). Yet another variant, recorded in a Nage village well to the west of 'Ua, described the wildmen abducting a child to teach them how to cook; the wily youngster, however, heated stones instead of food and when the dull-witted wildmen swallowed these, they were killed and the child escaped.[4]

Some Nage suggested that ebu gogo abducted children in order to eat them. According to an extraordinary idea reported by Julius Poi (which I heard from no one else), the wildmen would steal human infants to raise as their own, but accomplished this by first swallowing and then regurgitating them on returning to their cave. The same informant further stated that children abducted by ebu gogo were always rescued by their parents. It must be stressed, however, that many Nage denied that ebu gogo ever stole children at all. Even fewer supported the idea that they ate human flesh. While some people mentioned the threat to children as a reason the 'Ua people eventually decided to exterminate the wildmen, most claimed this was done simply because they stole from fields. Even people who described ebu gogo as abductors revealed that the creatures were incompetent and thus unsuccessful abductors. Indeed, as shown by the legend of their extermination and the story of Huma Léli's killing of the ebu gogo child, the wildmen were far more often represented as victims than as predators. Although their raiding of cultivated fields threatened local livelihoods, they actually appear to have been quite harmless.

Despite the wildmen's many reputed deficiencies, Nage do credit them with an ability to speak, albeit in a way that was 'unclear', 'impure', or 'jumbled'. Frequently cited passages drawn from standard narratives portray the ebu gogo's language as a simplified form of central Florenese speech. Noteworthy in this respect are the deletion of certain consonants—as in 'a'o' (first person singular, cf. Nage 'nga'o'), 'ako' ('dog', cf. Nage 'lako') and 'ila' ('to flare up, become bright', cf. Nage 'lila')—and dialectal usages (Ngadha 'lawo', referring to a tubular skirt, cf. Nage 'hoba'; 'zede' for 'belly', cf. Nage 'tuka'; and 'ja'o', the first person singular in several central Florenese languages and dialects). Used in Rawe, Ende and parts Ngadha, 'ja'o' occurs in the phrase 'lawo ja'o' ('my clothing'), which in the story of their extermination is uttered by ebu gogo when receiving palm fibre from the 'Ua people. In this context the phrase is always juxtaposed to the nonsensical variant 'liwo ji'o'. The term 'zede' ('belly, stomach'), used by the ebu gogo to indicate the location of eating vessels they had swallowed at an 'Ua feast, is known in Nage, but only in the compound 'ngai zede', meaning roughly 'character trait'. Interestingly, the wildman's pronounciation of 'ako', for 'dog', is also characteristic of the speech of young children.

As passages quoted in the narratives are mostly ones addressed to 'Ua people, ebu gogo speech might suggest an attempt by speakers of a different language to communicate in Nage. Alternatively, the speech could be an imaginative construct, inspired by other dialects and by children's language. Supporting the second interpretation are several inconsistencies: for example, the initial /l/ of Nage words is not invariably omitted; the first person singular is rendered both as 'a'o' and 'ja'o'; and only in one account is /k/ replaced by /t/.[5] Also noteworthy is a Nage story in which monkeys are portrayed as speaking (an ability Nage ordinarily deny them). In a way obviously reminiscent of the language attributed to ebu gogo, the simians are depicted as omitting /k/ and /ng/ from the beginning of pronouns, thus rendering 'kami' and 'kita' (respectively the exclusive and inclusive forms of the first person plural) as 'ami' and 'ita' and, like ebu gogo, pronouncing 'nga'o' (first person singular) as 'a'o'.[6]

Nage narrators also render monkey speech in a gruff voice, but this they do not do when replicating ebu gogo speech. Stories concerning monkeys are a possible source of ideas regarding the wildmen's language. Yet it is equally possible that the converse has occurred and stories of the ebu gogo have provided a model for monkey speech. In addition to speech, Nage regularly characterize ebu gogo as emitting apparently non-linguistic vocalizations. The most frequently mentioned is a deep murmurous sound that may be written as 'gumm' or 'gmmm', but which is sometimes imitated as a low whinny, like the deep, grumbling neigh of a horse. The wildmen were supposed habitually to have uttered this sound at the end of statements; a variant transcribable as 'mm mm mm' forms part of chants they performed while circle dancing at an 'Ua feast.[7]

Nage wildmen in space and time

Nage consistently identify ebu gogo with a single location: the cave named Lia Ula, located on the slopes of the Ebu Lobo volcano about a kilometre above the 'Old 'Ua' ('Ua Olo, also called 'Ua Ola and Ola 'Ua), the now abandoned former principal settlement of the 'Ua people. A short distance to the east of the cave is a stream, the Lowo Lia Ula (Lia Ula river), which forms the boundary of the territory of 'Ua and their eastern neighbours, the people of Wolo Pogo. 'Lia' is the general term for 'hole, cavity, cave'. No one I asked could provide a convincing explanation for 'ula'; but as I later show, it is likely connected with 'ana ula', the name of hominoids similar to the ebu gogo reported from elsewhere in central Flores (see Chapter 3). Lia Ula more specifically denotes the entrance to a cave or series of caves. According to one view, the cave once had as many as seven exits, but most people speak of a single exit, situated near Kusi Badho, in Wolo Pogo territory.

Although I have visited the site of Lia Ula (most recently in 2005), I have never been able to reach the cave entrance. After a continuous ascent over rugged terrain, the closest I got was a spot on a steep cliff edge surrounded by thick forest, which overlooked the reputed entrance on the other side of a steep ravine, about 75 metres to the southwest. Visible on the mountain wall opposite, at a height about equal to my point of observation, are blackened depressions roughly the size and shape of cathedral windows (see Plate 2.4). According to my 'Ua guides and others who have viewed the spot, these are traces of the fire that the 'Ua ancestors set to exterminate the ebu gogo. According to one opinion, to reach the cave entrance from the closest piece of solid, level ground would require a ladder of at least 10 metres; interestingly, the remnant of a bamboo ladder was visible when I visited in 2005. Just below the cave grows a banyan tree and this, I was told, can also be used, with the further assistance of a rope, to gain access to the site. According to 'Ua men, there were formerly narrow 'steps' or ledges that enabled the agile and powerful wildmen to climb up into their cave with ease, but the ledges—which several men said were still present in the 1970s—have since been eroded by seismic activity. Indeed, following the majority opinion, earth tremors in the 1960s and again in 1992, further caused landslides which sealed Lia Ula, so that the entrance has not been visible for some time.

Several older men in their seventies or eighties (among them Julius Poi) related how they had visited the cave in their youth. According to several descriptions, the cave was large, with an entrance the size of a church doorway and, according to one assessment, with a spacious interior as big as a chapel, thus large enough to accommodate a group of fifty or more.[8] By contrast, younger men visiting Lia Ula in the 1970s or more recently claimed they were unable to locate the cave entrance. Among these were two men who said they climbed up to the cave at least twice during the 1980s and 1990s. On one of these occasions, they were accompanied by two (otherwise unidentified) westerners who requested that they enter the cave to obtain 'bones', but they were unable to comply. Another 'Ua man, however, claimed that he and an older companion had been able to find the entrance and, with the aid of ledges still in existence at the time, had entered Lia Ula in 1979.[9] The man was 36 years of age when he spoke to me in 2005 and thus just a boy when he visited Lia Ula in 1979. He further claimed that he and his companion had found humanlike bones inside the cave. They removed two of these, which he described as bones of the forearm—but thinner than an adult human's and no more than a handspan in length. The bones were broken and appeared old; they were of a yellowish colour with black marks—traces, the informant speculated, of the fire that had killed the ebu gogo. In response to my questions, the man further asserted that they had found nothing else in the cave; the palm fibre, he spontaneously remarked, was all consumed in the holocaust, but black marks from the fire were still visible inside as well as outside the cave.

According to this man and others who had heard the story, his companion had taken the two arm bones to the coastal town of Ende. There he had sold them to 'Westerners', for 700,000 rupiah, which at the time would have been equivalent to $700, a very large sum of money by local standards. As he could not (or would not) further identify the purchasers, this part of the account is hard to credit. So too are details provided by his deceased companion, whom I was able to question before his death, in 1988. A person of mystical inclination and a self-confessed 'shaman' ('toa mali') widely known as a teller of tall tales and generally considered unreliable, this man claimed to have entered the cave on several occasions and once to have found there a light-coloured triangular stone adze as well as a long shin bone, which he said had 'turned to stone' and which, he suggested, had belonged to a creature standing 2 metres tall. He was, however, unable to say what had happened to either these objects and his claims were seriously doubted by other local people I questioned.

In 'Ua, there is a general notion that anyone wishing to find Lia Ula and to enter the cave may do so by performing a ritual of offering. Such a rite, addressed to the ancestors, further ensures the safety of prospective venturers, since not only would the climb be difficult, but the cave is now infested by large wasps and snakes. For this reason many people are dissuaded from visiting Lia Ula. Local visitors mostly comprise agile youngsters seeking adventure. Outside of 'Ua, the cave is described as a location where individuals wishing to acquire special powers (including the powers of shamans) might go to conduct a spirit quest (Forth 1998a: 174). Such activity, however, has no direct connection with the history of ebu gogo and the spiritual objects of questers are explictly not the shades of extinct wildmen. 'Ua villagers denied that they themselves have ever used the cave for this purpose, although some suggested that men from distant places, in western and eastern Flores, might do so.[10]

If details of the ebu gogo's reputed abode provide little indication of how the creatures are supposed to have lived, rather more elaborated is the well-known story of how they died, in the fire deliberately set by 'Ua villagers. The holocaust is usually described as following their attendance at a night-long feast. Some versions specify the location of the celebration as the former hamlet of Napu Co'o (now the site of the village of 'Ua Muzi, or 'New 'Ua', founded in 1941); the detail is significant, as this location, described as the furthest ebu gogo would travel downslope, is some 4 kilometres to the north of Lia Ula cave and at a much lower elevation. What is more, most other 'Ua settlements, as well as agricultural fields, would at that time have been located closer to the cave. Once the feast was concluded, either when it was still dark or early the next morning, 'Ua villagers followed the wildmen back to Lia Ula. Sometimes, the tale describes the returning wildmen as groggy, having been plied with copious amounts of palm wine.

It was then that, in an act of treachery, the 'Ua folk gave the ebu gogo large quantities of palm fibre which, proverbially mistaking this for clothing, they gladly received.[11] One version of the story describes 'Ua men as passing the fibre to them on long poles, the hominoids then already being deep inside the cave. According to another version, the ebu gogo were given palm fibre over a longer period and the extermination took place only when the cave was full of the stuff. Yet another account describes the fibre being collected, during some days or weeks, from the neighbouring districts of Wolo Pogo and Légu Deu, as well as from 'Ua (although it is consistently affirmed that only 'Ua men carried out the extermination). Whatever the details, Nage always depict the extermination as a premeditated rather than impulsive act. Once all other exits to the cave were sealed and in some versions once all the wildmen were asleep, the 'Ua villagers tossed in a firebrand, or inserted live coals in the final bale of palm fibre, which immediately burst into flames.

Dry fibre of the arenga palm (*Arenga pinnata*) is highly flammable. While Nage often describe the ebu gogo as simply being burned to death, other details suggest that, with no means of escape, many would have died from suffocation. Consistent with this is the report of numberless maggots emerging from the cave, according to one idea for an entire week and crawling a considerable distance. Also mentioned in myths recounting excessive sacrificing of water buffalo (Forth 1998a: 31–32), this detail could be explained as a simple rhetorical device, emphasizing slaughter on a large scale.[12] Yet the appearance of maggots also accords with a quantity of decomposing bodies, only partially burned if at all and rendered lifeless through smoke rather than flame. In the same vein, Nage say that, after the fire, a tremendous smell emanated from the cave. According to some accounts, moreover, it took a long time for all the wildmen to die and before doing so they made a terrible commotion.

Arguably, this level of detail lends plausibility to the tale. So does the motive. As noted, the primary and perhaps sole motivation for their extirpation was the wildmen's habit of stealing crops. Nage oral history describes the 'Ua people moving into a previously unsettled area and eventually encountering a population of largely vegetarian hominoids. Once the 'Ua folk had planted fields, what would be more likely than that their crops would quickly become subject to the depredations of these relatively large, non-cultivating plant-eaters? And since the ebu gogo were extremely strong and fast, what better way to exterminate the creatures than by trapping them inside a cave, placing flammable plant matter in the entrance and igniting it?

We also need to consider when all of this might have happened. The 'Ua people are among the last of current groups to arrive in Nage. Originally, they lived on 'Ua Hill (Wolo 'Ua), about 15 kilometres to the west of their present settlements. The location now falls within the modern administrative district ('desa') of Solo, near Solo's eastern boundary. 'Ua myth relates how Kea Noi, an 'Ua ancestor wandering far eastward in pursuit of game, once came across an uninhabited forest where he discovered an enormous tuber. Camping in what was to become 'Ua territory, he was able to remain long in the vicinity; for every time he ate from the tuber, it immediately regained its original size. Later, the tuber turned to stone, eventually being inaugurated as a sacrificial pillar ('peo') when the people of Wolo 'Ua decided to move, en masse, to this new and fertile area. Not surprisingly perhaps, their former neighbours in Solo tell a different story. Following conflict over land, they say, their ancestors drove the 'Ua folk out of Wolo 'Ua, thus forcing their eastward move.

Not only its less fantastic character but other details of the history of Solo and also of the neighbouring Rowa region (Forth 1998a: 31), make this second version far more likely. But what is important in the present context is that everyone I questioned in Nage, including present neighbours of the 'Ua people in Wolo Pogo and Nage 'Oga, were quite certain that the present 'Ua territory—a region occupying a higher elevation than neighbouring territories—was uninhabited land. In other words, no Nage group asserts a prior claim to this territory, an affirmation fully supported by the mythical idea that 'Ua's sacrificial instrument (the petrified tuber) grew of its own accord and was not granted by any previous land-owner.[13] These points are important because they confirm that the story of the extermination and the maggots swarming as far as the River Lia Ula do not function as a mythical charter legitimizing the 'Ua people's claim to their territory (even though the stream is in fact recognized as the boundary between 'Ua land and the territory of the Wolo Pogo people). Nor does the legend fulfil any similar aetiological function. As I was consistently told, after ebu gogo were extirpated, everyone simply returned to their villages. The act had no negative consequences. As one man put it, once the wildmen were all destroyed, 'everyone felt more secure and people no longer felt the need to construct fences around their fields'.

The evidence of Nage oral history and genealogy suggests that the 'Ua people began moving to their present territory some time in the eighteenth century and probably in the earlier decades. The most complete 'Ua genealogy I obtained places Kea Noi (the first immigrant, who discovered the giant tuber) eight generations before the informant, a man of about 85. Estimating a generation as between 25 and 30 years (and assuming of course that the genealogy is reasonably accurate), this ancestor would have been born between 1680 and 1720. After arriving in their present territory, 'Ua contact with ebu gogo was probably not immediate. One man stated that a whole generation had resided in the original village of 'Ua before discovering the ebu gogo. Another explained that, as people expanded their occupation of the territory, laying fields and building field huts and palm wine distilleries in the area upslope of 'Ua village, their contact with the wildmen gradually increased. We thus glimpse a scenario which is ecologically plausible. At first, the 'Ua people were unaffected by ebu gogo, or at any rate were able to tolerate their depredations. Early contacts were relatively peaceful, even to the extent that the wildmen would attend 'Ua feasts. But eventually, their stealing from fields—which would have become more regular as these became sited closer and closer to the wildmen's abode—proved intolerable and the cultivators finally decided to take action.

If 'Ua people moved to their present territory early in the eighteenth century but competition with the ebu gogo did not become critical until some time later, then the extermination might well have occurred sometime between 1750 and 1800. However, other details of 'Ua history complicate the issue. The extirpation of ebu gogo is one of two major events in this history. The other was a catastrophic eruption of the volcano Ebu Lobo, which all but destroyed the main 'Ua settlement (Ola 'Ua) and caused a temporary evacuation of the entire territory. This is the only such eruption remembered by the 'Ua and by Nage people in general. By all indications, it was also the first eruption of Ebu Lobo recorded by western vulcanologists, which occurred in 1830. At that time, lava flowed from the volcano's northern breach to a distance of 4 kilometres from the summit—a detail consistent with 'Ua claims that the flow stopped just short of their main village, located just east of the breach.[14]

In view of the importance for 'Ua of both the volcanic eruption and the extermination of the ebu gogo, it is perhaps surprising that people who had definite opinions on the matter were about evenly divided as to which occurred first.[15] Some 'Ua men suggested that the events were not far apart; others insisted that the volcano erupted while ebu gogo were still alive. If the eruption did occur first, this would accord with its conception as a punishment by local spirits angered by the 'Ua people's ignoring agricultural taboos, including dancing and manufacturing indigo dye at planting time. Such ignorance of taboos is arguably consistent with a group that had recently migrated to a new territory and hence suggests a date not long after 'Ua's removal from Wolo 'Ua. Genealogical and historical evidence mentioned above, however, indicates that the move took place rather earlier than 1830, in the early or mid eighteenth century. Also, an eruption preceding the extirpation of ebu gogo leaves unexplained how the wildmen, occupying a cave upslope of 'Ua and thus closer to the volcano's summit, were able to survive the catastrophe.

But the genealogies allow other possibilities. Nine 'Ua men were able to trace descent from lineal ancestors who they claimed were alive at the time of the ebu gogo's extermination. Calculating a generation at 25 and also at 30 years and assuming an average age of 25 for the ancestor at the time of the extermination, this produces a range of dates between 1785 and 1885.[16] Half the dates fall around or before 1830; hence given that genealogies are more often contracted than expanded (especially among Nage, who tend to name people after forebears in the grandparental generation), we might reasonably conclude that the wildmen's extermination—assuming, as ever, that this reflected a real event—occurred a short time before the volcanic eruption.[17] On the other hand, if the extermination did occur after 1830 and if the wildmen did somehow survive the volcanic eruption—perhaps by moving temporarily from Lia Ula—then this would place the ebu gogo's extinction at significantly less than 200 years ago, making an accurate memory of the event and of the creatures' habits and appearance, more likely. Since dates derived from genealogies mostly cluster around 1830, it might even be considered that the wildmen were exterminated in the heat of the eruption and that the legend describing their destruction in a fire set by the 'Ua folk is a mythical gloss on this. All the same, the fact that Nage are equivocal as to whether the extirpation or the eruption came first contradicts any suggestion that the latter was a mystical reprisal for the former. Indeed, Nage I asked consistently rejected this interpretation.[18]

Fantastic elements

Although informants disagreed about certain features, Nage ideas generally present a reasonably coherent image of the creature they call ebu gogo. The attribution of language is inconsistent with the idea that the wildmen lacked technology or material culture. But during the two centuries or more since the creatures disappeared from the Nage region, it is to be expected that some features will have been forgotten or exaggerated, resulting in an image which in certain regards appears quite fantastic. Actually, only two regularly reported features appear truly fantastic: the ability to swallow large (and normally inedible) things whole and the possession by females of breasts so 'long' they could be slung over the shoulders.

Apart from whole fruits and vegetables, things Nage mention ebu gogo swallowing include: dishes and other vessels made of gourd and coconut shell, cooking pots, palm wine containers of bamboo, puppy dogs and piglets and human children. Swallowing vessels made from plants, however, could just conceivably refer to unsophisticated beings mistaking these for food, especially when filled with actual food. Also, the consumption of small animals and children is mentioned infrequently in general descriptions and appears invariably to reflect narratives in which the wildmen's voracious appetites are stressed. The Nage stereotype of ebu gogo as greedy is of course connected with their more credible depiction as persistent garden thieves. Undoubtedly sustaining the wildmen's gluttonous reputation is a standard Nage simile comparing human gluttons and ravenous eaters to ebu gogo. Yet, as I demonstrate in the next chapter, the idea more likely derives from images of malevolent and mostly female spiritual beings and bogeys, which are frequently credited with swallowing human victims by way of a hole in the back and which in several parts of Flores are partly or entirely independent of coexistent representations of hairy homonoids. Whatever the explanation, it is noteworthy how many Nage, including two of my oldest informants, expressed doubt regarding ebu gogo's propensity to swallow large objects, suggesting, in the words of one, that it may be 'something that has been added (to an otherwise credible image)'.

Over-the-shoulder breasts can equally be traced to these same female spiritual images (see Chapter 3). Among primates, long breasts are an exclusively human (or, perhaps, hominin) trait. But while this rules out experience of monkeys or apes as a source of the image, as I discuss in Chapter 10 yet other explanations are possible—provided one treats the idea that ebu gogo could carry their breasts over the shoulders as a gross exaggeration of some empirical reality. In any case, if the Nage wildmen have any historical basis, comparative evidence from elsewhere on Flores indicates that an image of long-breasted female beings would certainly have been familiar to the 'Ua people before they entered their present territory a few hundred years ago and before possibly encountering creatures occupying the cave at Lia Ula.

Although rather less fantastic, several other components of the ebu gogo image nevertheless seem improbable if literally construed. That the creatures were afraid of dogs and fire is not particularly remarkable—although fear of fire (the element which effected their final destruction) may carry some symbolic load and moreover makes literal sense only if the creatures were not human. In themselves, infant abduction and anthropophagy are not at all impossible. But since only a minority of Nage attribute them to the wildmen, they can reasonably be regarded as fanciful accretions. Fear of bamboo combs, a quite specific aversion mentioned with some regularity, is another matter. It could be understood as a hyperbolic expression of the hairy wildmen's unkempt and dishevelled appearance (rather as one might describe a person who does not bathe as being 'afraid of soap'). On the other hand, as I show in Chapter 3, fear of combs and attributions of abduction and anthropophagy, like excessive swallowing and extraordinarily long breasts, probably derive from representations older than the 'Ua people's encounter with the wildmen.

Also bearing on the credibility of the ebu gogo legend is the Nage depiction of relations between the wildmen and local humans. While still extant, ebu gogo were rarely encountered and seem mostly to have been wary of people. Nevertheless, they did occasionally visit 'Ua villages and attend festivals, including the one at Napu Co'o just prior to their extermination. Although the ebu gogo are described as participating in the night-long circle-dancing characteristic of such celebrations, the term Nage usually employ to refer to such attendance, 'moni', meaning 'to watch, observe', suggests a rather less active participation. It is also important that these celebrations and particularly circle-dancing, are performed after dark; thus, implicitly, ebu gogo only visited Nage settlements, in groups, at night. The further idea that they would not dance around a fire may be taken to suggest that the wildmen's dance, evidently an imitative affair, was performed at the edge of settlements. And since fires are lit for circle-dancing in the centre of a village, it may be that the ebu gogo, rather than being fearful of fire, were simply reluctant to fully enter settlements.

It is unclear how the 'Ua people might have induced the wildmen to attend the fatal feast in Napu Co'o. They would presumably have had to tether their dogs; they might also have deliberately placed food at the village margins. Alternatively, the ebu gogo's attendance could have taken the form of scavenging, since edible matter—including such things as coconut shells with remnants of coconut flesh—is especially plentiful around village perimeters during feasts. This suggestion gains credibility when we consider that 'Ua cultivators encroaching on ebu gogo territory would have gradually eroded the wildmen's original food base, resulting in hunger that could have caused them to overcome their fear of humans and of their dogs and thus to raid cultivated fields as well as attend feasts. Participation in 'Ua festivals and concomitant dancing and gluttonous consumption may therefore be elaborations of a rather different kind of activity, consistent with an evolving ecological relationship between the wildmen and Nage humans. It scarcely needs adding that attending feasts could represent an earlier, more peaceful (or at least neutral) phase of human–wildman interaction. It is also in the context of the wildmen venturing more frequently near settlements and cultivated fields and thus field huts, that we may understand Nage apprehension about wildmen abducting human infants and perhaps even consuming them.

Two negative features are relevant to the plausibility of the ebu gogo legend. Consistent with the wildmen's reputed lack of technology, Nage do not report ever having engaged in any sort of exchange relationship with them and they deny that ebu gogo ever inter-married (or interbred) with their own ancestors. It seems, then, that whatever else the creatures might have been, they were not human. On the other hand, one may of course be dealing with a selective memory that has reduced real humans to the level of sub-humans, a possibility I come back to in Chapter 10.

Knowledge and categorization: survival of the image or survival of ebu gogo?

Questions remain as to how Nage are able to provide such detailed and internally consistent descriptions of the ebu gogo, if indeed these were creatures that disappeared some two hundred years ago. If their image does reflect something empirical, we may further ask whether this might be a phenomenon still present in the local environment. Although the wildmen of Lia Ula are supposed to have been the last of their kind in Nage country, comparable traditions from elsewhere in central Flores concern creatures reputedly surviving to recent times (see Chapter 3). What is more, while not consciously linked with ebu gogo, during the twentieth century there have been at least two reputed sightings in the Nage region of creatures which could conceivably reflect the same empirical basis as the ebu gogo representation.

Accounts of the ebu gogo's extermination describe how two individuals, who happened to be absent from the cave, escaped the conflagration. Always a male and a female, they are sometimes specified as siblings and sometimes as mates. Learning of the fate of their fellows, the pair fled as far as Mount Ua (Wolo Ua), also called 'Big Mountain' (Wolo Méze), a rugged territory to the northwest. Usually Nage say they settled in Tana Wolo, a densely forested region on the eastern side of Mount Ua. Less often the wildmen's refuge is specified as Namu, a similar landscape to the north of the mountain, while one 'Ua elder stated that, before proceeding to Mount Ua, the refugees settled for a time in a spot near the village of Libu Nio, in the So'a region (see Map 2).[19]

The idea of an opposite-sex pair of ebu gogo escaping the extermination of course facilitates the possibility that they survived in another place and may therefore be considered simply as an instance of a common mythological or narrative device. At the same time, Nage are generally equivocal about the possibility of ebu gogo's continuing existence. According to one idea, the people of Tana Wolo maintain a special affinity with ebu gogo and hence a particular enmity with the people of 'Ua. Specifically, Nage say, 'Ua people who visit Tana Wolo should not identify themselves as 'Ua but simply as Nage. They further claim that should their identity become known, they could be killed on the spot.

In 2005, I visited Tana Wolo, but no one I questioned had heard of this tradition, nor indeed were they familiar with the name 'ebu gogo'. Nevertheless, some Nage statements suggest a perceived biological continuity between the refugees from Lia Ula and their possible hosts. One 'Ua man remarked how some Tana Wolo people (although not all) had very ugly faces and their women long breasts. Another suggested that the simian appearance of a man he had met from the Mount Ua region could reflect descent from ebu gogo that had interbred with local humans. According to the more common opinion, however, Tana Wolo people do not differ physically from Nage. They are described as darker-complexioned; but this is sometimes qualified as a reference to former times, when marriage with outsiders was less common than at present. Several times Nage men told me how, during a hunting expedition in the 1960s, they had encountered a woman in the Namu region whose lactating breasts were so pendulous she was able to suckle from the hip (with the breast tucked under the arm and a child 'sucking from behind'). Some commentators even described the woman as a 'female ebu gogo', surmising that she may have descended from wildman escapees who had interbred locally. One man further claimed that Namu people possess 'large jaws'; however, there was no indication that the long-breasted woman did so, or that she did not otherwise look like an ordinary human. In all this, Nage state or imply a survival of isolated ebu gogo features by way of miscegenation, rather than the continuing existence in Namu or Tana Wolo of wildmen as a distinct and separate population. The suggestion that some Tana Wolo and Namu people may descend from refugee wildmen amounts to a pernicious rumour and in all likelihood reflects a Nage evaluation (which moreover probably developed only in colonial times) of these populations as rustic, backward and impoverished. Certainly, there is nothing to suggest that earlier Nage experience of them gave birth to the legend of ebu gogo, which in any case is centred not in these regions but in Nage itself.

Quite a different matter is a pair of sightings from the 1960s and 1970s, within the Nage region, of what both informants described as strange human-like figures. Neither informant identified what they claimed to have seen as ebu gogo, but they also did not explicitly construe them as encounters with spiritual beings. One account was related to me by Jon Lulukumi, a civil servant and former head of a high school from the Ngadha ethnic group. His story is all the more remarkable as he was unfamiliar with the 'Ua tradition, or even the name 'ebu gogo', when he first reported the following:

> In September 1975, when he was in his early twenties and just after completing his education, Jon Lulukumi and a friend of the same age decided to hike around the Ebu Lobo volcano. Late one afternoon, they set out from the village of Wudu (in the Wolo Pogo region, immediately east of 'Ua), where they had been lodging with another friend and headed around the mountain towards Flores' south coast. In the light of a full moon and in a region of forest interspersed with gardens some distance

> above Wudu, they suddenly encountered, about 30 or 40 metres away, what 'may have been an old man'. He was naked with long, dishevelled head hair and his belly was somewhat distended (or, as later described, 'sagging'). Appearing from behind a stand of banana trunks, he turned and fled when he saw the youths. Although they too were afraid, they were curious enough to investigate. But when they arrived at the spot they could see no one, nor (as the narrator remarked spontaneously) could they find any footprints.[20]

Mostly in response to my questions, Jon Lulukumi provided further information about the naked man's appearance. Although when spotted behind a banana trunk, the individual may have been crouching or bending over, he nevertheless seemed to be of short stature. (Lulukumi first said 'over a metre', while later, with the hand, he estimated a height somewhere between 1.2 and 1.3 metres.) When I asked whether the man could have been stooped, perhaps with age, the informant thought this was possible; also, when I first spoke with him, in 1999, he had mentioned how the individual's lower arms appeared 'bent' or 'crooked', which he thought might indicate an advanced age. Although the man looked elderly, his head hair was black, not grey. He also appeared quite sturdily built and possibly broad-chested. Explaining that he was able to observe the individual only very briefly and then at a distance, Lulukumi said he did not notice whether the body was particularly hairy, nor whether the man had facial hair. The individual's complexion, however, appeared darker than most local people. All Lulukumi was able to say about the face was that it 'resembled a human's'. He added that, while the individual looked like an ordinary human, he could not be sure whether it really was a human or something else.[21]

Continuing their journey, Lulukumi and his companion had no opportunity to enquire in Wolo Pogo whether what they had seen was known locally, or whether the individual matched any known person, for example, a local eccentric. An enquiry I made in Wudu in 1999 also yielded no result. It was not possible, Lulukumi thought, that the naked man had been bathing, as there was no spring or stream nearby. In fact, there was, according to my informant, no indication at all of what the individual may have been doing in this place and at this hour, about twilight. Nearby, there was no sign of agricultural work in progress, although this was the season for clearing fields. Neither Lulukumi nor his companion greeted or called out to the individual, as they were too afraid. Nor did the individual make any sound.

The second sighting was reported to me by Yohanes Soda Ule, a respected Nage elder and traditionalist whom I have known for over twenty years. His account was in fact offered in response to my mention of Lulukumi's story, several years after I first heard it and without revealing its source. What Lulukumi had probably seen, my informant suggested, was someone 'transforming into a witch', something supposed always to involve a naked individual and to take place outside of settlements. His story is as follows:

> In 1965, when Soda Ule was in his late twenties, he was approaching a stream near a wooded area not far from his village at about 5 o'clock in the afternoon. Suddenly, on the other side of the stream, he saw a being of generally human form apparently fleeing from him towards the forest. Although the individual was initially just ten metres away, Soda Ule only saw its back and therefore could not be certain of the sex; but there was no tail. He was particularly taken by its peculiar proportions. Although the creature was perhaps 1.6 metres and thus hardly short by local standards, the lower legs appeared extremely long and the arms extended to the knees, while the upper legs and torso were accordingly short. As the informant and fellow villagers discovered from a later inspection of the footprints, the feet were also large: the heel was small and pointed while the front of the foot was very wide, with large toes. What struck Soda Ule as much as anything was the way the thing moved. As he demonstrated, it ran bipedally, swinging its arms and legs inwards towards the body rather like a cripple and appeared to move quite slowly. Soda Ule did not attempt to pursue the creature, as its appearance frightened him considerably and made him feel 'strange'. The colour of the naked body, which he compared to a painted object at hand, was a rusty or coppery red, which he also likened to the colour of hot coals. It did not, however, glow or shine and he was unable to say whether this was the colour of the skin or body hair. (He therefore could not confirm whether the being was hairy.) The head hair, however, was 'black', like that of local humans, but not long. Enquiring of fellow villagers, he later discovered that no one else had ever seen anything like this being; he himself never encountered the like again.[22]

A rather different sort of report was offered in 2005 by another longstanding informant, Mikel Goa Na'u. Born about 1925, Goa Na'u is the leader of the prominent clan Mudi centred in Tiba Kisa. His account concerned 'small humans' he claimed to have observed on two occasions as a young man, emerging from a cave near Hobo Bo ('Gewang palm dale'), a place in Mudi territory several kilometres northwest of Tiba Kisa. Goa Na'u identified the creatures as 'ana noa'. Nage generally apply 'noa' to a category of spirits which take the alternate forms of crows and of small, jet black humans with short frizzy hair and by mystical means, cause epidemic disease among large livestock (Forth 1998a). Goa Na'u's sighting, however, concerned something rather different. Hobo Bo is well known as a specific haunt of 'ana noa' and indeed Soda Ule, also, told me he had once seen, at some distance and therefore indistinctly, a group of the creatures heading in the direction of the cave. According to Goa Na'u, the naked hominoids he observed appeared quite human, but stood less than a metre tall. They were not hairy-bodied, but had genital hair like a human's and their skin, not particularly dark by local standards, appeared 'hard' (perhaps meaning 'coarse'). Their head hair was 'black', like local humans', but short and sparse and their faces appeared to be halfway between a human's and a monkey's. They also emitted a 'high-pitched sound'. Partly in this last respect, these ana noa recall creatures reported from Poma and Tana Wolo (see Chapter 3); but they are not entirely distinct from the disease-causing spirits Nage call by the same name ('noa'). And by the same token, despite their troglodytic habitat, somewhat simian appearance and apparent lack of culture, they seem quite different from the ebu gogo.

What is to be made of these reports? Jon Lulukumi's naked man was dark-haired and sturdily built, features apparently at odds with his suggestion that the individual was of an advanced age. The arms of Soda Ule's creature, long and possibly bent, vaguely suggest a comparison with Lulukumi's naked man. Yet I was unable to confirm whether this feature and the awkward gait implied a stooped posture; and other than the generally human appearance, it is clear that the two figures differ considerably. What Lulukumi saw sounds mostly like a small, stocky and dark-skinned human, whereas the human status of Soda Ule's subject is rather less certain, not least because of its odd proportions, peculiar locomotion and reddish colour. Even assuming that it had some perceptual basis, moreover, it is less obviously a continuing source of the Nage representation of ebu gogo than the naked man seen by Jon Lulukumi. But since neither account represents a common experience of people in the Nage region, neither in fact has any definite bearing on the widely known image. While one may be tempted to ask whether Jon Lulukumi's sighting could reflect an extant 'wildman'—that is, a creature or phenomenon that had earlier provided the basis for a general Nage representation of the same sort—far more certain is the character of the witness, an educated, intelligent and consciously 'modern' person with a limited interest in local traditions and, by all indications, not at all given to fantasy.[23]

As for Goa Na'u's report, given this man's generally mystical inclination (he has in fact claimed visual encounters with a variety of spiritual beings), one might well dismiss his claims as fanciful and although possibly based on some experience, as reflecting no more than an idiosyncratic interpretation of a widely known spirit category. Nevertheless, the story does resonate with accounts of similar creatures described in other parts of central Flores. And as I later explain, there may very well be an indirect connection between these and ebu gogo, revealed by the name of the wildmen's cave, Lia Ula.

If orang-utans (*Pongo pygmaeus*) occurred on Flores, one might reasonably infer that the reddish and ungainly creature seen by Soda Ule was a specimen of the red ape, come to drink at a stream in the late afternoon but fleeing on the approach of a human. Orang-utans or other apes are, however, an unlikely source of ideas regarding ebu gogo since, until very recently, Nage have had extremely limited access to audio-visual or print media. As mentioned, Nage knowledge of orang-utans, creatures they sometimes compare to ebu gogo, would seem mainly (if not exclusively) to derive from banknotes issued in 1992.

An equally unlikely source is modern education. Occasionally I heard ebu gogo characterized as 'ancient humans' (Indonesian 'manusia purba'), an interpretation which gained some currency towards the end of the twentieth century among younger and better educated Nage. This, I should emphatically add, refers to the period before the 2004 announcement of the discovery of *Homo floresiensis*. More generally, local paleoanthropological education has always been minimal in the Nage region and no one I questioned could remember seeing pictorial reconstructions of extinct hominins (such as *Homo erectus* or Australopithecines).[24] More to the point, older Nage, the source of most of my information on ebu gogo, had had little schooling and, according to their own testimony, while in school never learnt anything about 'ancient humans' or creatures resembling the local wildmen. As I was often assured, knowledge of the latter has always been passed on locally, 'from mouth to mouth'. Similarly, I found no evidence of informants being exposed to film depictions of prehistoric humans, or archaic hominins. Even if they had been, one would then have to assume that the wildman legend originated in the late twentieth century and everything elderly Nage say about hearing the tradition from their parents and grandparents contradicts this. In addition, prominent physical features of the ebu gogo—long head hair, pendulous breasts and even bipedalism—obviously cannot be accounted for by recent exposure to images of apes. Conversely, modern depictions of prehistoric hominins would surely have revealed that these possessed fire, tools and hunting technology—all things Nage say the ebu gogo lacked.

Names, masks and bogeys

Perceptively, an 'Ua man raised the question of why his ancestors, upon encountering the wildmen of Lia Ula after moving from their old home in Solo, had bestowed the name 'ebu gogo' on them. In fact, circumstantial evidence suggests that 'ebu gogo' once had a different or more inclusive meaning. Only in 'Ua and immediately neighbouring parts of Nage does the name refer to a population of hairy wildmen. Elsewhere, 'ebu gogo' is either unknown or is applied exclusively to an explicitly imaginary bogeyman and to masks that depict this otherwise vaguely represented figure.

'Ebu gogo' comprises two analytically separable components. 'Gogo' can mean 'greedy, gluttonous', a meaning Nage sometimes link with the wildmen's reputed ability to swallow large objects whole. Partly in the same connection, 'gogo' further denotes a mythical figure, a gluttonous monster which Nage parents invoke to threaten disobedient children or scare away bothersome youngsters. Simple mention of 'gogo' alone summons up the idea of something that can be 'called' or 'fetched' and which might seize and swallow children unless they are compliant.[25] The monstrous bogey is more completely designated as 'ebu gogo'. Yet in this context a more common usage is 'gogo meo', where 'meo' is understood as 'cat'.

'Gogo meo' or 'ebu gogo' occur as synonyms for the bogey mostly in 'Ua and other places close to the Nage centre of Bo'a Wae. One sense of 'ebu' is 'ancestor, grandparent' (and also 'grandchild', since the kinship usage is reciprocal); but other compounds suggest instead an honorific applied to singular things which are especially large, unusual, or spiritually powerful, including indeed the volcano Ebu Lobo (Forth 1998a: 240–42). In Nage, 'ngebu', a probable cognate of 'ebu', denotes the crocodile, while in Ngadha and other regions to the west and northwest, the same term designates spiritual beings manifest as snakes and other creatures (Arndt 1930: 825; Molnar 2000: 207). A man in Tana Wolo similarly described something called 'ngembu ngudo' ('visible ngembu') as a frightening being, capable of assuming various forms, including those of odd-looking domestic animals and about which children leaving the village alone in the late afternoon were especially warned. As Nage pointed out, 'ebu' as 'ancestor' makes little sense with reference to the wildmen of Lia Ula, since people definitely do not regard the hairy hominoids as human ancestors (Forth 2006). Of course, the wildmen were not large or in any way spiritually powerful either. But large size and extraordinary power are indeed attributes of the mythical bogey, a circumstance suggesting that 'ebu gogo' primarily refers, or once referred, to this figure (which Nage adults expressly regard as imaginary) rather than to the hominoids which 'Ua folk believe were exterminated by their ancestors.[26] The other more common term for the bogey, 'gogo meo', it is also important to note, is never applied to the wildmen of Lia Ula.[27]

To the west of the Nage, the Ngadha call spiders 'ebu gogo' (Arndt 1961: 208, s.v. 'cebu gogo'). Another Ngadha term for 'spider' is 'toro gogo', a phrase Arndt further translates as 'having large staring eyes' (ibid.: 550; see also Bader 1953: 63) and which people in the So'a region apply to hominoids very similar to the Nage wildmen (see Chapter 3). Yet another Ngadha term for 'spider' is 'fége ré'e' (Arndt 1961: 144), a name employed for a monstrous hominoid image encountered in several parts of central and eastern Flores. In Nage, too, 'ebu gogo' can denote a kind of spider. Similarly, in the Dhawe region (in the far northeastern part of the colonial division designated as Nage), 'gogo meo' refers both to spiders and to a mythical child abductor. All these usages are consistent with a monstrous bogey as the primary referent of 'gogo'. The connection with spiders accords with a general Florenese assessment of arachnids and especially large varieties, as loathsome or frightening creatures.

Inextricably linked with the Nage bogey are masks made of the spathe of the areca palm or of large bamboo, or a kind of light wood ('mési'; see Plate 2.6). The masks are identically named 'gogo', 'gogo meo' or 'ebu gogo'; they are the only masks known to Nage and were formerly more common that at present. Blackened with charcoal and occasionally provided with head and facial hair of palm fibre and with teeth (usually pig teeth), the masks were normally kept well hidden. Instead of simply intimidating disobedient children with the arrival of a notional figure, harried parents could then threaten to bring out the masks, referring to these as 'ebu gogo' or 'gogo meo'. Sometimes, when mere threat proved insufficient, elders would don the masks, at the same time covering the body with a dark cloth to further obscure their identities (see Plate 2.7). They might then make noises like a cat and crouch and move stealthily, usages that illuminate the inclusion of 'meo' ('cat') in 'gogo meo'. Maskers, or unmasked adults trying to frighten or scare away disobedient or bothersome youngsters, might also make deep murmuring sounds ('mmm' or 'gumm') to invoke the image of gogo meo, or ebu gogo—vocalizations which as noted are further attributed to the wildmen of Lia Ula, but which may very well have their ultimate source in deep feline growls or purring.

Apart from the Bo'a Wae region, the story of the 'Ua people's extermination of the wildmen is generally known in Wolo Pogo, Mula Koli and Kéli Mado—all located to the east or southeast of 'Ua—and in the Nage Sapadhi region and Légu Deu, immediately to the west and southwest of Bo'a Wae. By contrast, both the bogey figure and bogey masks occur, or did so until recently, throughout the Nage region.[28] Beyond Bo'a Wae, the bogey and mask are known in the southeastern Nage district of Wolo Wea, in Dhére Isa to the northeast of the Nage centre and also in the western part of the Keo region (see Map 2). In all these places, the masks and bogey are generally called 'gogo meo', although 'ebu gogo' is known as a synonym in Dhére Isa. In Dhawe, 'gogo meo' or 'ebu gogo' is a child-abducting trickster depicted in folktales as a giant man who is eventually out-witted by his intended infant victims. Signalling a radical difference from the Nage wildmen, the character is further portrayed as living in a house and wearing clothes. The name 'ebu gogo' is not generally known in Keo or in Wolo Wea. What is more, neither in these districts nor in Dhére Isa or Dhawe are people familiar with the legend of the wildmen of Lia Ula. Nor do any recognize the existence of cave-dwelling hairy hominoids called by another name. In view of the far more widespread occurrence of the bogey image and associated masks, therefore, it is unlikely that either has its source in the legend of the creatures exterminated by the 'Ua ancestors.

During the twentieth century, bogey masks came to be worn in the Bo'a Wae region in an 'ebu gogo dance' ('jédhe ebu gogo'), a choreographic performance that Nage describe as a recent innovation, originating in the 1970s or 1980s. Initiated by 'Ua converts to Catholicism, the dance has been performed on just two or three occasions, all of them modern celebrations associated with the Church. On what was probably the first occasion, the character of 'ebu gogo' was represented by a single masked figure who uttered murmuring sounds. The exposed parts of his face were blackened with charcoal and he wore a costume of palm fibre simulating body hair (see Plate 2.5). The dance character, however, appears to correspond to the customary bogey at least as much as it does the extinct Nage wildmen. According to an exegesis provided by Andreas Séda, the 'Ua man recognized as the creator of the dance, the ebu gogo figure—one of several characters who together performed what was a sort of pantomime rather than a simple dance—represented the overly strict father of the person whose ordination into the Catholic priesthood the performance celebrated. Nevertheless, since Nage do not describe the bogey figure as hairy-bodied, the palm fibre costume is evidently inspired by the legendary wildmen rather than the disciplinary device.[29]

If the figure of the bogey does not derive from the creatures of 'Ua legend, then either the wildmen are a special, imaginary development from the bogey, or the two have separate origins and one name for the bogey has become secondarily attached to a quite different and rather more prosaically represented category. Apparently supporting the first possibility are obvious similarities between the two images, more particularly the murmuring sounds and an association with palm fibre, which provides facial and head hair for the masks—and body hair for the ebu gogo dancers—and which the wildmen of Lia Ula reputedly mistook for clothing. Similarly suggesting a common symbolic source is the use of charcoal to blacken the bogey masks (and parts of the body of the ebu gogo dancer); charcoal is of course a product of fire, the element used to exterminate the wildmen, who in a sense were encased in the palm fibre that filled their cave. On the other hand, whereas the masks and the clothes of adults who impersonate bogeys are perfectly black, the skin colour of the wildmen is described as no darker than that of Nage themselves.

Evidence against a derivation of the wildmen of Lia Ula from an imaginary bogey or from images manifest in masks arguably outweighs evidence favouring such a development. First, both the masks and the bogey name 'gogo' are far more widespread within the Nage region than is the representation of the extinct wildmen, which is identified with a single location and a single hominoid population. This point gains further support from evidence reviewed in the next chapter, concerning masks and bogeys from other parts of Flores. Second, Nage speak of the bogey they call 'gogo meo' or 'ebu gogo' as a single character (rather like 'Santa Claus'), whereas the wildmen are depicted as a fairly sizeable group. And third, the wildmen are consistently described (by Nage adults at least) as being *extinct*. Related to this, as a reason for the creatures' extinction, is the attribution to the wildmen of child abduction and anthropophagy, traits explicitly associated—indeed, definitive of—the bogey. This might be counted as another similarity and hence as another indication of a common symbolic origin. Yet as noted, most Nage deny that the wildmen were actually abductors and anthropophagers, interpreting this as a fanciful addition to a more basic image of creatures exterminated solely because of their thieving from cultivated fields. It should also not be forgotten that Nage generally depict the wildmen as timid creatures, as wary of Nage as Nage were of them.

The bogey usage therefore hardly precludes a separate and empirical basis for the wildmen of Lia Ula. The world over, bogey figures are constructed from empirical beings—for example, animals (such as cats or geckoes, which Nage also employ to frighten children) or feared or disliked ethnic others. On the other hand, where imaginary figures serve as bogeys, these draw on established and more generally conceived categories, such as malevolent spirits or witches, two sorts of beings that Nage further employ to threaten children.

If the extinct wildmen were a neighbouring group of culturally and technologically distinct humans—or indeed a different species of hominin—their deployment by Nage as bogeys would almost be expected. But it is not even clear that the Nage bogey owes any part of its origin to the wildmen of Lia Ula, whatever their ontological status. What appears more likely is that the *name* of a pre-existing, mythical bogeyman came at some point to be applied to the wildmen as well. By the same token, fantastic features of the wildmen's behaviour, such as swallowing large objects, abducting and consuming children and the murmuring speech (very possibly based on cat vocalizations), can equally be explained as a transfer of traits from the bogey to the denizens of Lia Ula. And as I demonstrate in the next chapter, their extraordinarily pendulous breasts and fear of combs might derive, if not from the bogey, then from an equally imaginary source.

Further supporting the idea that 'ebu gogo' may be a relatively recent name for the wildmen is evidence that Nage may once have called them by another name. The name of their former abode, 'Lia Ula', translates as 'Cave of the Ula' or 'Ula cave'. Most people I asked did not know the significance of the name, but some suggested a connection with 'ule', 'maggot' (after the many maggots that emerged after the extermination) or 'ular', the word for 'snake' in Malay and the Indonesian national language. Linguistically, however, neither interpretation is viable and a more likely explanation is that the cave was once associated with beings which, elsewhere in central Flores, are called 'ana ula' ('ana', 'child, person, member of a group') (see Chapter 3). At present, this name is unknown to the people of 'Ua and most other Nage. However, two men from Tiba Kisa village, just to the west of 'Ua, described 'ana ula' as a former Nage usage. One, Mikel Goa Na'u, whose story is recounted above, actually identified the term with the ebu gogo, while the other man stated that 'ana ula' was a name once applied to 'ana noa', the creatures described in Goa Na'u's story. It would therefore appear that either the wildmen were once called 'ana ula', or their cave was named after another sort of anthropomorphous being—if not by 'Ua people then perhaps by an unidentified group who were familiar with the location and its reputed occupants before the 'Ua folk arrived in their present territory.

Classification: human, animal, spirit (or something else)?

What sort of creature do Nage consider the ebu gogo to have been? Since Nage have terms that correspond to English vernacular senses of 'human' and 'animal', the question of the wildmen's relation to these categories can readily be asked and I often did so. The general consensus was that ebu gogo were not really human beings ('kita ata'); yet most people were reluctant to classify them as animals ('ana wa'). Accordingly, the hominoids are generally enumerated with 'ga'e', the numeral classifier normally reserved for humans (as in 'ga'e wutu', 'four people, persons') rather than with the classifier for animals , 'éko', also the word for 'tail'. (As a couple of people explained, ebu gogo cannot be enumerated wth 'eko' as they had not tails.) By the same token, Nage distinguish male and female wildmen with the terms otherwise reserved for humans ('ana haki' and 'fai ga'e'), rather than distinct terms applied to male and female animals (Forth 2004b).

Besides their physical resemblance, the feature of ebu gogo that most inclines Nage to class them with human beings is their reputed ability to speak. On the other hand, Nage reluctance to identify the wildmen as fully human relates, as much as anything, to their reputed lack of culture, meaning especially tools, fire, houses and clothing. As one 'Ua man put it, they 'looked like humans but did not live like humans'. The same man, incidentally, played down ebu gogo's linguistic ability: while acknowledging that they spoke, he remarked how he had never heard they could utter more than a few phrases (specifically, the ones regularly cited in standard narratives). Nage uncertainty about the zoological status of the hominoids was revealed in a variety of other statements. These included: 'they might be classed as animals as they looked like monkeys, but they could speak, though imperfectly'; 'they were not human, but also not animals'; 'a kind of very dull-witted humanity'; 'possibly a mixture between humans and animals'; and 'humans, but more exactly ancient humans'. The suggestion that ebu gogo were a 'mixture' should not be taken to imply that the wildmen derived from humans mating with some sort of animal. Also, their assessment as 'ancient humans' reflects a recently introduced concept.

Considered as a category of Nage folk zoology, the ebu gogo would appear to occupy a class of their own, intermediate between humans and animals, but on the whole regarded as resembling more closely the former. They could be said to constitute for Nage a 'kind of' humanity, yet one so different from themselves as not to count as fully human beings.[30] By the same token, while they compare ebu gogo to monkeys in regard to certain physical and behavioural features, Nage do not consider the wildmen as a species of monkey, or as ethnotaxonomically more closely related to monkeys than to humans. Nage are thus as equivocal about the human status of ebu gogo as palaeoanthropologists have often been about the human status of pre-sapiens hominins, including members of the genus *Homo* other than *Homo sapiens* (Cartmill 2001). This comparison is intriguing, suggesting a common human response to different, though not dissimilar, images of imperfect humanity. But while perhaps suggestive, this obviously does not entail that ebu gogo were as real even as the reconstructions of human palaeontology.

The regular attribution of several implausible or improbable features to ebu gogo might suggest that the category refers to no more than a variety of 'spiritual' or 'supernatural' beings. Indeed, to the extent that anthropologists have given any attention to such figures, they have uncritically classed them precisely in this way.[31] Nage, however, explicitly deny that the wildmen are—or rather, were—'spirits', a category they generally identify with the term 'nitu' (Forth 1998a). They also articulate a series of features that distinguish the ebu gogo from spirits.

These features are not specific to the Nage category ‘nitu’, but tend to characterize beings anthropologists have described as spirits cross-culturally. For Nage, spirits are normally invisible beings that assume diverse physical guises (including those of animals and human beings) and change shape at will. Ebu gogo could do none of these things.[32] More specifically, people remark how, unlike spirits, the wildmen could be seen by anyone and were once encountered just as one would encounter an animal in the forest. They also state that they possessed no ‘mysterious’ or ‘mystical powers’ (a concept usually expressed with the Indonesian term ‘gaib’); that their history is not a ‘folktale’ (Nage ‘no nange’, cf. Indonesian ‘dongeng’: a story of contestable or uncertain factuality); and that, whereas spirits of all kinds are still encountered by people predisposed to see them, nowadays ebu gogo are no longer seen, having been exterminated generations ago. Indeed, the idea of their extinction strikes a special contrast with spirits which, although sometimes spoken of as suffering individual destruction, are not typically the objects of categorical elimination.

Since tradition confines them to a single location (the cave Lia Ula), the ebu gogo further differ from spirits insofar as spirits occur generally in a variety of mostly uninhabited places, or are associated generically with places of particular kinds—such as springs, large trees, or mountains. Almost by definition, Nage spirits, like spirits everywhere, possess powers superior to those of normal humans, being credited with superior intelligence, ways of knowing and powers of communication. Many are also able to harm people in mystical or unseen ways; and a related and equally widespread attribute is trickery or cunning (even though some spirits can of course be benevolent or especially helpful). In addition, the most encompassing Nage spirit category, ‘nitu’, is most closely identified with beings that take the form of beautiful women and handsome men (Forth 1998a: 67–69). In all these respects, ebu gogo are diametrically opposed to spirits. Not only were they ugly; they were also unintelligent. Whereas spirits are often destructively deceitful, the wildmen were easily gulled by their human adversaries; and far from being able to do serious harm to Nage, they proved vulnerable to human attack—pathetically so in the case of Huma Léli’s seemingly wanton killing of the wildman infant and terminally in respect of their eventual extirpation. In other words, if spirits are absolutely or contextually superior to humans—or are superhuman—the ebu gogo were decidedly inferior, indeed quite literally sub-human, their superior physical strength notwithstanding.

Further indicative of the ebu gogo’s non-spiritual character is the absence of any ritual or religious function attaching to the wildmen. (The ‘ebu gogo dance’ is no exception to this; not only is it a modern and partly Catholic innovation, but as I have explained, in this context ‘ebu gogo’ refers as much to a distinguishable bogey figure as to the extinct hominoids.) Whereas spiritual beings commonly occur in myths and other standard narratives, there are at most three narratives concerning ebu gogo: the one relating their encounter with Huma Léli, the story of the abducted youngsters who eventually escape and the legend of their extermination. Unlike witches and spirits, wildmen do not figure in creation or origin myths. Of course, the fact that ebu gogo are distinct from spirits does not prove they are not fictitious. Nor does it mean that 'spirit' or 'spiritual being', pertaining to a kind of universal representation, is not a useful category for cross-cultural analysis. The distinction does, however, indicate a need to consider the ebu gogo—and, as we shall see, apparent congeners found in a variety of other cultures—as instances of another sort of category whose nature remains undetermined.

Internal comparisons and summary considerations

For Nage, ebu gogo were, ambiguously, either some kind of 'human being' or some kind of 'animal'. Since Nage attribute fantastic or 'spiritual' qualities to a variety of attested species (see e.g. Forth 2004a: 115ff) and indeed to certain humans (shamans, ancestors), the association of such qualities with the wildmen does not contradict this characterization. At the same time, different representations of ebu gogo can be distinguished as 'naturalistic' or 'mythical'. This relative contrast is seen partly in the use of 'ebu gogo' for both the extinct population of Lia Ula and the present-day bogey figure (which is also called by another name and is still living or, perhaps better said, is immortal). Yet the distinction is equally discernible in accounts of the wildmen recorded in 'Ua and neighbouring villages and representations recorded in settlements further away.

To be sure, 'Ua people mention ebu gogo's incredibly pendulous breasts and ability to swallow large objects as often as anyone. Nevertheless, it was people in Nage villages distant from 'Ua (generally, five or more kilometres away—a long distance on rugged Flores island) who made the most fantastic claims about the hominoids. These statements, it should be stressed, were denied by 'Ua and other Nage I questioned subsequently. The following are illustrative:

1 Claiming to have entered Lia Ula as a young boy, an elderly man from Mula Koli, a district to the southeast of 'Ua and Wolo Pogo, described the cave as containing stone 'benches' where the wildmen used to sit and sleep and a special stone 'seat' reserved for their ruler (or 'raja'). He also stated that, in addition to stealing from cultivated fields, ebu gogo would steal people's clothes and, using bamboo spears, would kill and consume their horses and buffalo (an idea possibly drawn from the standard image of 'noa' spirits 'spearing' these animals, which is to say, causing death among large livestock; Forth 1998a: 99). The man also claimed that a human child was burned in the conflagration at Lia Ula; the child's abduction had been the main reason 'Ua villagers had set the fire, but the wildmen refused to surrender their victim. All this was derisively dismissed by people of 'Ua and neighbouring settlements. A similar account of the wildman king and the killing of livestock was offered

by an elder from Kéli Mado, a district even further to the east, who may indeed have heard these details from the Mula Koli man. The latter described ebu gogo as no smaller than local humans. However, the Kéli Mado man was among the minority who described ebu gogo as having been 'large and tall'. Possibly in the same vein, his older brother spoke of the creatures swallowing whole sheep and goats.

2 An elder of clan Deu in Wolo Mako, a village to the west of Bo'a Wae, related a clan history in which he described his ancestor, Lowa Bata and the ancestor's cousin, Koi To (of clan Nila) as taking the initiative in the extirpation of ebu gogo. This they did before going to war against the people of Rawe, in order to protect their children and grandchildren, whom they feared the wildmen might kill in their absence and to save their pots and eating vessels, all of which the creatures might swallow. While this version of the wildmen's extermination was ridiculed even by the narrator's fellow clansmen, a particular criticism concerned the way the story illegitimately attributed the demise of the wildmen (by fire and palm fibre as in other accounts) to Nage ancestors other than the forbears of the 'Ua people. Told by the same elder, another story, evidently a variant of the 'Ua tale according to which the ebu gogo abducted a pair of children, described how the wildmen kidnapped a Nage youngster so that it would teach them how to cook. The plan came to nought when the child, instead of cooking food, heated stones and tricked the ebu gogo into swallowing these, thus killing the wildmen.

3 Further west of Bo'a Wae, another aberrant account of the ebu gogo's destruction was related by a man of clan Dhuge, in the village of Ola 'Ewa. This too depicted the informant's ancestor as the initiator of the wildmen's demise and moreover as the victim of thefts normally identified as the 'Ua ancestor Huma Léli and as the killer of the infant ebu gogo left behind in a field hut. In the usual way, this account attributed ebu gogo's actual extermination to the people of 'Ua, but it included additional, more fantastic elements: it described a marauding wildman not only as swallowing puppy dogs and large containers, but as threatening to devour the Dhuge ancestor himself.

4 A senior man of clan Kisa Ola, resident in Tonga Tei just to the west of Bo'a Wae, stated not only that ebu gogo were exterminated because of a general practice of 'swallowing human infants', but also that escapees from the conflagration at Lia Ula had fled to Indonesian New Guinea (West Papua), possibly by aeroplane, where they had become ancestors of Papuans. Apparently confusing the creatures with a category of spirits ('logo lia'; see Chapter 3), the man further claimed that ebu gogo 'had holes in their backs'. Again, other Nage deemed these ideas absurd.[33]

5 Further afield, an obviously modern story from So'a, an ethnolinguistically distinct region to the northwest of Bo'a Wae, features a female character called 'Ebu Gogo' in whose care are placed the seventy offspring of a 'raja' (an office that did not exist in pre-colonial times). This

> occurs after the children are successively and unsuccessfully, fostered by a series of domestic animals: a pig, a dog and a horse. Fearing that this Ebu Gogo—described by the translators as a 'female witch' and associated in the story with the Nage volcano Ebu Lobo—will devour them, the children conspire to kill her by amassing firewood, dousing it in gasoline and setting it alight after the creature enters a cave (Mommersteeg and Dirkzwager 1999: 194–99). Especially the reference to the Nage volcano indicates that the story reflects So'anese knowledge of the Nage legend, which has been incorporated into a tale featuring not only a colonial raja but children wearing school uniforms and playing modern musical instruments. These modern elements recall stories of the previously mentioned giant trickster-bogey named Ebu Gogo told in the Dhawe region, which is as distant from 'Ua to the northeast as is So'a to the northwest. The Dhawe figure is described as wearing shoes, trousers and a hat, smoking cigarettes, sleeping in a modern bed and visiting a 'town'. Being associated with the Ebu Lobo volcano and being burned inside a cave, however, the So'anese character is far closer to the Nage legend than is the similarly singular Ebu Gogo of Dhawe.

Stories like these So'anese and Dhawe tales, I would emphasize, are not heard in 'Ua or neighbouring Nage villages. Such extra-local representations permit two generalizations. First, the themes of excessive swallowing and voracious anthropophagy are especially prominent. Second, stories linking ebu gogo with specifically named ancestors of other clans, as well as depictions involving modern elements (rajas, aeroplanes, gasoline), detach the legend of the Nage wildmen from its local historical context—the time of 'Ua ancestors perhaps just four or five generations before my oldest informants. Although not particularly prevalent in extra-local accounts, it may also be recalled that the idea that ebu gogo were taller than local humans is proportionally more prevalent among people outside of 'Ua than within.

In regard to seemingly fantastic elements, the difference between local and extra-local accounts of ebu gogo is one of degree rather than kind. But it is discernible nonetheless. The contrast suggests the common process whereby knowledge of a subject is less directly and less regularly communicated the further one moves from its geographical source. By virtue of the greater number of links in the communication chain, over time and distance stories told and retold usually become less accurate. As has long been observed of rumours, moreover, representations, as they spread, typically become condensed or simplified—that is, details are deleted, reduced, 'levelled', or run together, in part resulting in the highlighting of other details (Allport and Postman 1947; DiFonzo and Bordia 2007: 134–42). When compared with stories recorded in 'Ua, the narratives concerning the Deu and Dhuge ancestors clearly illustrate this process. At the same time, such representations can become elaborated or exaggerated, especially in the direction of fantasy—a phenomenon some have seen as particularly characteristic of the formation of 'legends' (Allport and Postman 1947: 154). This process is reflected in all of the extra-local tales of ebu gogo summarized above. In addition, at least two cases (the Deu and So'anese narratives) reveal a process of conflation, possibly an aspect of 'condensation', whereby the ebu gogo legend becomes compounded with or incorporated as part of another tale.

All these developments may explain the relative naturalism of local accounts of ebu gogo. By the same token, they suggest that 'Ua knowledge of the creatures was not only indigenous to 'Ua but possessed some actual historical basis. And to that extent the contrast with more fabulous extra-local representations challenges an understanding of the ebu gogo as entirely imaginary—or the subjects of a 'double fantasy', to recall a point made in Chapter 1. The contrast does not prove beyond all doubt that the original events, or the original referents, were real, or substantially grounded in empirical phenomena. But by the same token, it hardly contradicts the thesis that people residing closest to the wildmen's putative abode are able to describe something more in accord with an empirical being because, over a period of two hundred years or so, they have retained a more comprehensive and coherent collective memory—of something.[34]

The approach of this chapter has been largely heuristic: to determine how far a case can be made for the ebu gogo as a representation grounded in a zoological reality. I do not claim to have fully made such a case. But I would provisionally conclude that the Nage wildman, rather than something completely fictitious, suggests a composite representation focused on an empirical entity which is probably no longer present in the local environment and to which have accreted fantastic elements drawn from other, more obviously imaginary categories. How far this thesis can be sustained can only be determined after reviewing comparable traditions of hairy hominoids encountered in other parts of Indonesia and Southeast Asia. I begin in the following chapter with a review of categories reported from other regions of Flores.

Plate 2.1 Julius Poi, an 'Ua elder.
Source: G. Forth.

Plate 2.2 The volcano Ebu Lobo volcano as seen from the higher reaches of 'Ua territory.
Source: G. Forth.

Plate 2.3 Men of 'Ua and a man of the Nage village of Tolo Pa (on the left).
Source: G. Forth.

Plate 2.4 Lia Ula, the cave reputedly inhabited by the ebu gogo.
Source: G. Forth.

Plate 2.5 A man in an ‘ebu gogo’ costume, as worn in the ebu gogo dance. The costume was specially created for the author.
Source: G. Forth.

Plate 2.6 A wooden bogey mask from Nage, ‘gogo meo’ or ‘ebu gogo’.
Source: G. Forth.

Plate 2.7 Man playing a bogey with an improvised mask of areca spathe.
Source: G. Forth.

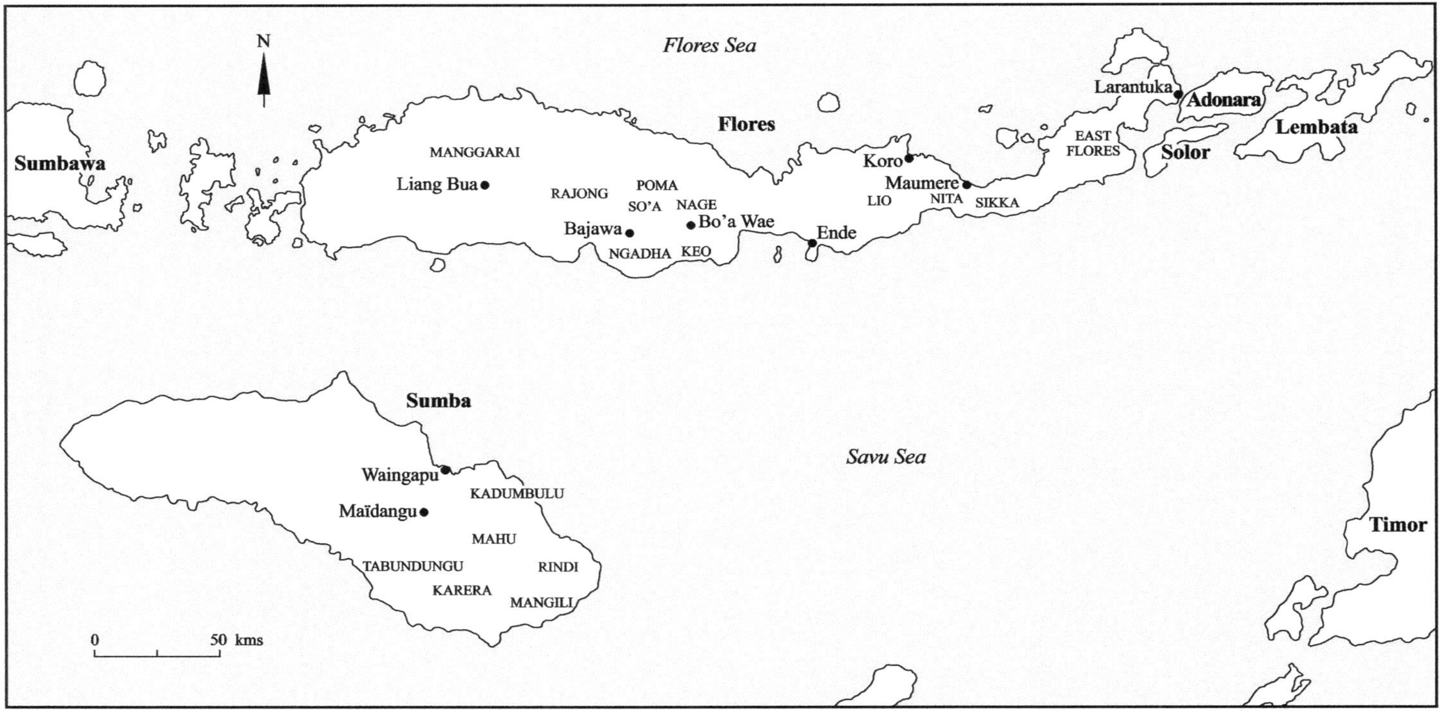

Map 3. Flores and Sumba.

3 Other Florenese hominoids

In this chapter, I review images of small hairy hominoids that are basically similar to the Nage ebu gogo. I also discuss several mostly distinct Florenese representations which, while rather less comparable, nevertheless provide a wider context in which the Nage wildman may be understood. Owing to considerations of length (and to avoid overburdening the reader), it will be necessary to condense much of the material, particularly information deriving from my own ethnography. Throughout, it will be useful to refer to Maps 2 and 3.

The 'ana ula' of Poma and Rawe

Nearly 20 kilometres to the northwest of Nage is the district of Poma, a rugged, partly forested and sparsely populated area. In Poma and in the neighbouring district of Rawe, people speak of small anthropomorphous creatures named 'ana ula'. Like the wildmen of Nage, ana ula are supposed to have been exterminated several generations ago. Yet people also say that some may have survived. A middle-aged Rawe man claimed to have observed ana ula as a boy while hunting porcupines with his father at night in a place called Wolo Piki. Both father and son saw several of the creatures emerging from a cave near a porcupine den, an experience that so frightened them they immediately returned home. An older man similarly described how his long deceased father, peering out of his field hut one day, had seen a group of ana ula; so shaken was he that he would not leave the hut until the creatures had gone, even though he claimed to have observed ana ula before. Yet another story concerned a man from Wangka, to the north of Poma (see Map 2), who in the late 1990s had captured an ana ula in a snare; although the creature suffered injury to its throat, it was nevertheless able to flee to a cave after the man released it. Despite the apparently empirical character of this last specimen, one opinion of such reputed experiences is that they concern spirits of the creatures exterminated long ago and that whereas ana ula were once visible to everyone, nowadays only certain people can see them.

Information from eight informants mostly questioned separately provides a reasonably detailed picture of ana ula and their habits. Estimates of their height ranged from just half a metre to a metre or more, with most estimates falling around a metre. They are also frequently spoken of as being the size of children.[1] Ana ula are consistently described as hairy, with bodies covered in short hair, longer and thicker than a human's but less so than a monkey's. The colour is variously identified with local terms translating as reddish or reddish brown, brown or dark brown and a colour similar to but darker than macaque pelage (usually a mixture of grey and light brown).[2] Ana ula head hair was described as longer than the body hair. Most people further described it as dark (or 'black'), thus darker than the hair on the body. Two informants, however, referred to brownish hair, lighter than local human head hair. Another two described the head hair as kinky or curly and yet another as 'coarse'. Several people further characterized ana ula as possessing genital hair distinct from the hair on the rest of the body and genitalia which were humanlike rather than monkey-like. (By contrast, I never heard Nage mention the genitalia of ebu gogo.)

Apart from visible genitals, ana ula could be distinguished sexually by the breasts of the females. But these were described as small or 'short' and less conspicuous than women's breasts. Just one man referred to 'long breasts', but his account apparently reflected familiarity with the Nage representation of ebu gogo.

Poma and Rawe folk compared the skin colour of ana ula to that of local humans, though two men thought they were darker. Everyone described the creatures as erect, bipedal and tailless. Although they were sometimes explicitly characterized as intermediate between humans and monkeys, simian qualities mentioned most often were their hairiness and small size and only a couple of people suggested ana ula may have been somewhat stooped. By human standards, the arms and legs were in proportion to the body; thus the feet were small, like those of a child. According to one account, ana ula possessed long, sharp fingernails, which they employed when attacking humans.

Ana ula faces were rather more human than monkey-like. Nevertheless, some informants mentioned prognathousness and prominent canine teeth, comparing these to a dog's or a male monkey's. Possibly in the same vein, two informants described some ana ula—but not all of them—as facially 'ugly'. I recorded no reference to receding foreheads, heavy jaws, or the absence of a chin; but rarely did I ask about these specifically. One man spoke of shaggy patches or tufts of hair growing on the faces (apparently the cheeks) of male ana ula; another said that the facial hair was like a monkey's, only finer. Just one man mentioned 'beards' and then only as a feature peculiar to males.

Several informants described the ears of ana ula as proportionally larger and broader than those of local people. Although not often mentioned, the nose was said to differ little from a human's, although a couple of descriptions suggested a smaller, flatter nose. Only one report indicated a difference in the eyes: some male ana ula had red eyes, also specified as 'cruel' eyes. The same man described the eyebrows as coarse and thicker than those of humans.

In accordance with their generally human form, most people I questioned said that ana ula should be enumerated with the classifier for 'humans' ('mori' in Poma) rather than the classifier for animals ('éko', 'tail', as in Nage). Nevertheless, the creatures were occasionally referred to with Indonesian terms for 'animal' ('binatang') or 'male animal' ('jantan'); also, one man argued that, since ana ula were not human beings, they must be animals and should therefore be counted in 'tails'. Applying the human classifier apparently reflects more the generally human form of ana ula than their reputed linguistic ability, which is somewhat less definite than what Nage report of their former local wildmen. A narrative recounting the extermination of the ana ula describes them as speaking but not in a way that humans could understand. Another informant similarly stated that the ana ula had a 'language' which, however, was unintelligible to humans; yet another claimed that they understood Poma speech, but was uncertain whether they could speak themselves. Apparently distinct from speech are other vocalizations attributed to ana ula: high-pitched sounds like the squeaking of rodents but louder, or like shrieks or squeals uttered by Flying foxes (fruit bats) while feeding (cf. Forth 1998a: 102).

Ana ula formerly resided in caves. They are particularly identified with a cave located in a wooded area about two kilometres north of Subi-Wea Wawo, the former principal settlement of the Poma people. Called Lia Ana Ula (Cave of the Ana Ula), the name is virtually identical to that of the former abode of the Nage ebu gogo (Lia Ula). It was in this cave that the ana ula reputedly met their end (see Plate 3.1). According to one opinion, descendants of the creatures that survived the extirpation may still inhabit other caves in Poma or Rawe. People encountering ana ula at present are said typically to do so on rainy days and especially during light rain accompanied by sunshine or when a rainbow appears—both meteorological phenomena identified, in various parts of eastern Indonesia, with dangerous spirits. In a similar vein, a Nage elder familiar with Rawe traditions described ana ula as venturing abroad on cloudy days, when long clouds signal their time and place of journeying.

Before their extermination, ana ula would steal eggplants—their favoured food—and other crops from Poma fields. A stranger object of alleged ana ula thefts was coals from local hearths, but I was unable to confirm that the creatures somehow consumed the coals, as I had earlier heard from Nage (Forth 1998a: 102). Ana ula also ate wild fruits and vegetables and evidently meat as well. A Nage man reported hearing that ana ula used to catch porcupines and wild pigs. However, in accordance with the general representation of the creatures as lacking tools and weapons, he could not say how they caught their prey. Occasionally, ana ula would catch and consume human victims, something they easily accomplished by virtue of their great physical strength (and, according to one account, their possession of long, sharp nails). Lacking fire, ana ula consumed their food raw. One man linked their consumption of raw food to a distinctive body odour, which he was unable to describe. Another likened the odour to that of a monkey and yet another to the smell of a male goat.

Recounting their great strength, local legend describes the ana ula as assisting Poma ancestors in the construction of large stone ramparts, the 'ture ana ula', used partly for defensive purposes in time of war.[3] Only the ana ula were able to lift the large boulders that can still be seen in what remains of these structures (see Plate 3.2). At the same time, the little wildmen were curiously incapable of lifting light objects—a notion revealing an inversion identically expressed in representations of spiritual beings from both central and western Flores (see Forth 1998a: 102, regarding Nage 'noa'; Verheijen 1951: 204, regarding Manggarai 'darat'). In return for their assistance, the ana ula received gifts of food, animal flesh or bones (which according to one account they preferred to meat).[4] However, for reasons no longer remembered, this positive reciprocity came to an end. Not only did the diminutive wildmen then begin stealing from cultivated fields, but they also took to killing and consuming human victims.

There are several versions of the legend describing the extermination of the ana ula. Some specify the cause as the abduction and consumption of an elderly woman named Bupu Tiye, whose kinsmen discovered her fate by following strips of pandanus leaf she had dropped to the mouth of the creatures' cave. Another mentions a mother and child and later an adult man being killed and eaten, while a third refers only to a small child. Yet the response was always the same: villagers from Subi-Wea Wawo built a large fire in the entrance to Lia Ana Ula, the cave in which the hominoids resided, killing many in the smoke and flames. Poma men speared or clubbed others trying to escape the holocaust; accordingly, red marks still visible on large stones outside the cave are considered to be traces of their blood. Details of the fuel employed to fire the cave vary. Of nine accounts, five specified this as wood and the other four as palm fibre. According to a further idea, Poma folk sprinkled the burning wood with chilli peppers, thus producing a noxious, blinding smoke.

Obviously, this story is remarkably similar to the legend of the ebu gogo's destruction at the hands of the 'Ua people; in fact, it is mainly the precipitating events that distinguish the two traditions. Poma people, however, are also less clear than are Nage about when they exterminated their local wildmen. Two informants gave estimates of 80 and 100 years ago. Another placed it in a time before Poma people possessed iron tools, while an elderly man associated the event with an ancestor who lived seven or eight generations ago.

How many ana ula may have escaped the conflagration and the subsequent slaughter is similarly indefinite. But one escapee was a pregnant female. The male child she subsequently bore married a human female and they produced a son named Hapa Dou, whose descendants reputedly survive to the present. One such descendant is the elderly man mentioned just above. When I spoke to him in 2005, he denied that his ancestor—and therefore he himself—had ana ula blood. He further claimed that no surviving ana ula had mated with humans and that all who escaped extermination had immediately fled to the mountainous region of Wolo Méze. Nevertheless, others I questioned—including the man's wife—confirmed the story. According to Poma and Rawe people generally, current descendants of human matings with ana ula can be identified mostly by virtue of shorter stature. Interestingly, hirsuteness was not mentioned in this regard. Although he perhaps stood no taller than about 1.55 metres, the man I met did not strike me as peculiar in any way; and I was never able to locate other putative descendants of ana ula. Not surprisingly, many people I spoke with were reluctant to identify particular individuals, evidently for fear of the embarrassment this might cause. The idea that Rawe people with deformities descend from human–ana ula matings, such as I sometimes heard from Nage informants, appears not to be current in either Rawe or Poma.

Despite striking similarities, ana ula differ from the Nage ebu gogo in several respects: they appear somewhat smaller and less robust than ebu gogo, yet they were evidently more aggressive—and more carnivorous. Informant accounts do not unequivocally credit the ana ula with language. At the same time, suggesting a more human and more naturalistic, representation than the Nage image of ebu gogo are the small breasts of ana ula females and descriptions of their genitalia. Just one version of the legend of the ana ula's extermination refers to their being tricked into assisting in their own destruction, by helping lift large pieces of food to fuel the fire—thus somewhat recalling the ebu gogo's gullible acceptance of palm fibre. This variant, however, was related by a Nage man; and more generally ana ula, unlike ebu gogo, are not described as stupid or unintelligent. In certain respects (for example, their construction of stoneworks combined with their inability to lift light objects) ana ula are somewhat more fantastically conceived than the Nage wildmen. But notwithstanding the idea that current sightings of ana ula reflect spiritual transformations of the creatures, they do not seem significantly more like a category of spirits and in view of their many similarities it is reasonable to consider ana ula and ebu gogo as variants of a single representation—or just conceivably, as reflections of (hypothetical) empirical beings of the same kind. On the other hand, arrestingly similar stories concerning their extermination suggest that these reflect no more than a common narrative tradition, a matter I take up again at the end of this chapter.

Before leaving the ana ula, some attention should be given to Tana Wolo, the district to the west of Poma which Nage consider as the last resort of the ebu gogo. Whether the people of this district recognize hominoids closely comparable to the ana ula is equivocal. They do, however, apply 'noa' (or 'ana noa'), a term used throughout central and eastern Flores for a kind of spirit, to diminuitive beings (just half a metre according to one estimate) possessing a generally human form, great physical strength and possibly—but not definitely—hairy bodies. Other resemblances include: a distinct body odour (like a snake, according to one man); high-pitched vocalizations (once described as like an owl's cry); a habit of stealing eggplants and also live coals (which in this case they are said to eat), a fear of dogs (but only bitches) and a former practice of marrying human females. An ancient marriage between a woman, named Ngina and a male ana noa is still commemorated in Tana Wolo ritual and is further enshrined in the name of a now abandoned village named 'Ngina Noa'. Although not considered extinct, the ana noa are described as less common than in former times. Nevertheless, from several attributes listed above, as well as an attribution to these creatures of inverted feet, the ana noa of Tana Wolo seem rather more fantastically conceived and thus more like spirits, than the Poma hominoids. Indeed, Tana Wolo informants described them precisely in this way. Thus, if Poma were considered the geographical centre of the ana ula legend, just as the village of 'Ua is in regard to the ebu gogo, this difference could count as another instance of a representation becoming less naturalistic the further one moves from its source (see Chapter 2). Much the same case could be made with reference to depictions of ana ula offered by Nage familiar with Poma and Rawe traditions, which were similarly more fantastic than what I recorded in those districts themselves.

'Toro gogo' in So'a

Another story of hairy hominoids burnt to death inside a cave is told in the village of Libu Nio, in the So'a region, about 15 kilometres northwest of the main Nage village of Bo'a Wae. The tale was recounted to me in 2005 by Ruma Séme, a man in his late sixties or early seventies. It concerned his paternal great-great-grandfather and namesake, Ruma Réwo and creatures called 'toro gogo'. The So'anese narrative can be summarized as follows:

> One day Ruma Réwo saw several toro gogo approaching his garden to steal cucumbers; on the way they destroyed a field of ripening rice. Before beginning their foray, they placed an infant creature in Ruma's field hut. Ruma then entered the hut and killed the infant by sticking the teeth of a comb into its ears. When the returning toro gogo discovered what had happened, they set up a tremendous wail. Ruma then released his dogs (which had previously been tied up) and the toro gogo fled with the dead infant directly to their cave. At this time, there also lived a young unmarried woman named Bepe Ruma, whom a grandparent had cruelly beaten (implicitly for some minor infraction). The girl therefore fled from the village. She ran in the direction of the toro gogo's cave, carrying a container she had been plaiting from pandanus leaf. It was nighttime and the moon was full. Detecting her smell (perhaps, specifically, the smell of blood from her wounds), the toro gogo seized the beaten girl and carried her away to their cave. Knowing she was going to die, however, Bepe Ruma took pieces of pandanus leaf, smeared them with her blood and dropped them in front of the cave for her relatives to find. After searching for the girl, her family did indeed discover the bloodied

> pandanus. Realizing what had occurred, they collected a great quantity of firewood and placed it inside the cave entrance. They then piled chillies on top of the wood and set everything alight, fanning the fire and thus creating a great volume of acrid smoke. There was much weeping inside the cave. The smoke and flames killed all of the toro gogo except for a male and female who, being exceptionally strong, were able to escape by burrowing underground. They emerged in the vicinity of Wolo Méze ('Big Mountain'). Thus it is that toro gogo survive to the present in Wolo Méze. Hunters who make camp overnight there should take ritual precautions so that the beings will not disturb their sleep. At certain feasts, people also set aside a portion of rice and meat for the toro gogo, who then descend from Wolo Méze and invisibly carry the food away. When they come, a sound (imitated as 'tuuuu') is heard in the highlands.

There can be little doubt that the So'anese story of toro gogo and the Nage legend of ebu gogo share a common origin.[5] The creatures are also described similarly: toro gogo were somewhat shorter than local humans, standing about 1.4 or 1.5 metres; they were covered in long hair, were very broad and extremely strong and their females had extraordinarily long breasts that they could sling over their shoulders. Ebu gogo's aversion to bamboo combs finds an echo in the use of a comb to kill the toro gogo infant. Indeed, part of the Libu Nio tale reveals an obvious parallel with the Nage story of Huma Léli recounted in the last chapter, all the more striking in regard to the similarity of the names Huma (also locally pronounced as 'Uma') and Ruma.[6] Another similarity is the flight of a male and female hominoid in a northwesterly direction, in the case of the toro gogo to Wolo Méze. In central Flores, the name 'Wolo Méze' ('Big Hill') is applied to at least three different peaks. Although it is not clear that it is this peak which is intended, one is Wolo Ua, the mountain to which Nage mostly say the pair of surviving ebu gogo fled from the conflagration at Lia Ula. (Wolo Méze, it may be recalled, is also the place identified as the refuge of surviving ana ula survivors that escaped the extermination in Poma.)

Perhaps not surprisingly in view of the location of Libu Nio about halfway between Nage and Poma, the toro gogo differ from the ebu gogo to roughly the same extent that they resemble the ana ula. The differences, however, concern not physical features, but mostly details of the extermination legend: the use of wood and chillies (rather than palm fire) to fuel the conflagration and the hominoids' killing of one or more adult humans as motivation. Like the murdered So'anese girl, an elderly woman in one version of the Poma tale dropped strips of pandanus, thus leaving a trail for fellow villagers. In addition, So'anese attributed the curious habit of eating live coals to the toro gogo; and like the ana noa of Tana Wolo, they are described as having been fearful specifically of female dogs. Yet another resemblance may be discerned in the survival of the toro gogo partly as invisible spirits, recognized in ritual food offerings. A fantastic idea not found in other Florenese representations concerns the toro gogo making their escape by burrowing underground. The detail is possibly connected with features of their cave, more accurately described, according to its current appearance, as a crevice in the ground.[7]

As noted previously, some ‘Ua people suggested that the two ebu gogo who escaped the Nage conflagration initially fled to Libu Nio. For a long time, So’a people have been associated with Nage—as political allies, hunting partners and sometimes affines. People I questioned in Libu Nio seemed unaware of the ‘Ua legend; in fact, they described the toro gogo as inhabiting their region before the arrival of humans. Nevertheless, in view of the remarkable similarities between the two traditions—and despite evident influence from representations maintained to the north of So’a (in Poma and Tana Wolo)—it is quite possible that the So’anese tradition reflects the long history of So’a contact with Nage. In fact, resident in the village of Libu Nio for many years has been an ‘Ua man married to a Libu Nio woman, a nephew of the ‘Ua elder who first told me about surviving ebu gogo having fled to Libu Nio. I am therefore inclined to conclude that, if nothing else, the episode concerning Ruma Réwo’s killing of the toro gogo infant derives from the ‘Ua tale of Huma Léli and has in the process been adapted to the genealogy of the Libu Nio narrator. In regard to this linking of the toro gogo with a specific forbear, as well as other, more fantastic features of the So’anese image, one is moreover reminded of the tendency in parts of Nage to an extralocal elaboration which includes connecting the similar figure of ebu gogo with prominent ancestors external to ‘Ua, the generally recognized locus of Nage wildman history. It is not impossible that the ‘Ua tale of Huma Léli has been taken from So’a. Yet in view of the extent to which Huma’s story is known among Nage generally, who invariably link it with ‘Ua, this seems far less likely.

Ngadha variants

To the west of the Nage are numerous communities collectively labelled by the missionary ethnographer Paul Arndt as ‘Ngadha’. In the easternmost part of the region, an area bordering directly on Nage territory, people describe what appear to be two hominoidal images. They also employ two quite different names; yet these names, it is important to stress, do not consistently label the two distinguishable images.

The first designation is ‘ebu ngiu’. A variant, ‘ine ngiu’, incorporates the term employed throughout central Flores for ‘mother’ (‘ine’), thereby suggesting a predominantly female figure. An onomatopoeic name used in various parts of Flores for the Brown hawk-owl (*Ninox scutulata*), ‘ngiu’ refers to a characteristic bird-like sound these creatures are supposed to make.[8] The other name is ‘kedho’.

Information on the referents of ebu ngiu (or ine ngiu) and kedho derives mainly from the eastern Ngadha district of Sara Sedu.[9] The creatures are said to have preceded humans in this region and still to reside on forested mountainsides and in caves. As regards physical appearance, there is nothing to distinguish them from the ebu gogo: they resemble small humans covered in hair, with monkey-like faces, long breasts and long fingernails. Except for the long breasts, ebu ngiu are also generally similar to the ana ula. They are extremely strong, capable of lifting a large man or a water buffalo and are given to capturing humans, including children that they abduct to raise as their own.

As indicated, the foregoing image is associated both with 'ebu ngiu' (or 'ine ngiu') and 'kedho'. Nevertheless, the name 'kedho' appears more often to refer to a more fantastically conceived figure possessing the supernatural ability to change shape. According to what I was told in Sara Sedu, kedho can appear as various animals; they can also take the form of particular humans, especially relatives of intended abductees, who then lead away unsuspecting youngsters to desolate places. Such abductions are temporary, as the children are always recovered; but after their recovery, they usually do not live long. This representation is remarkably similar to what Nage describe as 'being carried off by a witch' ('polo dhoko', Forth 1998a: 61).[10] On the other hand, at least one feature that distinguishes kedho from Nage witches simultaneously identifies them with the Nage ebu gogo. Possibly referring to their original form, a Sara Sedu man described the metamorphosing kedho as giant women with extraordinarily long breasts, an image he connected with a female ancestral figure named Kedho Susu Léwa ('Kedho Long Breasts'; Molnar 2000: 35). In contrast, another source describes this ancestress as both a 'small, hairy, human forest creature' (ibid.: 30, note 12; 39) and the 'founding mother' of a (human) group established in the village of Sedu (ibid.: 39). Further specified as the 'sister' of other human ancestors (ibid.: 35–37), the human affinity of 'Kedho Long Breasts' appears to separate this figure from most central Florenese hominoids and, as we shall see, to link her with hirsute human ancestors later described for Manggarai and eastern Flores.

However fantastic the powers attributed to 'kedho', the name 'ebu ngiu' (or 'ine ngiu') does not consistently refer to an entirely prosaically described being. Creatures thus designated are sometimes credited with inverted feet, a feature attributed to hairy hominoids outside of eastern Indonesia (see Chapters 5 and 7) and universally, like extraordinarily long breasts, to a variety of spiritual beings (see Thompson 1946: 248; Forth 1998a: 102). Also noteworthy is the statement of one man, who described the ebu ngiu as deriving from women who died in childbirth (Molnar, pers. comm., 1996). As we shall later see, this is the origin of a variety of malevolent female spirits (including the Nage 'logo lia') identified throughout Indonesia with the Malay term 'pontianak'.

Arndt (1930: 826–27) used 'pontianak'—a reference to mostly female beings that prey on both adults and children and especially on pregnant women—to characterize a western Ngadha figure named 'ibu ngiu' (or 'cibu ngiu' in that author's peculiar orthography), creatures which resemble eastern Ngadha images in more than name.[11] Residing in caves and dense forests, ibu ngiu assume the form of women with enormous breasts which they are able to toss over their shoulders; also like the Nage ebu gogo and So'anese toro gogo, they are afraid of combs. Using their breasts to bind or suspend victims, ibu ngiu abduct humans and by virtue of their great strength they are able to carry off two or three victims at a time. Since the creatures are described as taking mates, they presumably come in both sexes. Yet in Arndt's account, they are primarily a female representation.

Two Ngadha men from the district of Jére Bu'u whom I interviewed in the 1980s and 1990s provided further details of the ibu ngiu. They described them as short, dark-skinned hominoids about one metre tall with long, unkempt head hair. When seen, ibu ngiu appear dirty and dishevelled and wear old and tattered garments. Despite their small size, the creatures are extremely strong and are able to take enormous strides. Consumed raw, their diet comprises garden produce, which sometimes bears the creatures' teeth marks. One man described ibu ngiu males as bearded. But neither Arndt nor any other Ngadha source characterizes the creatures as hairy-bodied. In several important respects, therefore, not least of which is their wearing clothes, the ibu ngiu obviously differ from other images sketched so far. According to my Jére Bu'u informants, ibu ngiu are mostly significant as bogeys and indeed their use as bogeys appears more pronounced than is the case with other Florenese hominoids. Especially by comparison with the wildmen of Nage, the ibu ngiu are further distinguished by a general malevolence, a quality reflected in their alternative name, 'fége re'e', which Arndt glosses 'the evil Fége' (1930: 827).[12] One of the few seemingly naturalistic features of ibu ngiu is the idea, reported by Arndt, that the beings were formerly more widespread and were no longer heard of by the time he wrote. However, not only does more recent evidence show that Ngadha folk by no means consider ibu ngiu extinct, but as I discuss later on, evidence from other parts of Flores confirms that ibu ngiu—and also some images eastern Ngadha designate as 'kedho' or 'ebu ngiu'—have more in common with a class of malevolent female spiritual beings than they do with the extinct (or virtually extinct) wildmen of Nage or Poma.

Returning to eastern Ngadha figures: I conclude this section with a report of a creature, in this instance identified as a kedho, reputedly captured near the old village of Sara during the first half of the twentieth century (either in the 1920s or 1950s, but according to a version of the story I recorded in 2005, rather earlier and before the arrival of the Dutch). Between about one and 1.3 metres in height and covered in hair except for the face, hands and feet, the creature was bound and taken to the Regency capital in Bajawa (Molnar, pers. comm., 1996). The further detail, that the captive cried 'like a bird or a wailing infant', recalls the vocal reference of 'ngiu' in 'ebu ngiu' and 'ibu ngiu'; it also raises the question of whether the 'ngiu' sound may be associated with the weeping or crying (or perhaps whimpering) which, as will later become clear, is commonly attributed to captured hominoids, not just on Flores but elsewhere in Indonesia and beyond. Who took the kedho to Bajawa, villagers were unable to say. According to one version of the tale (Molnar ibid.), the specimen was a female with long, pendulous breasts. In contrast, the version I recorded in 2005 concerned a male creature which had been abducting local children and which was caught in a pig snare in a forested area near the village. Moreover, rather than being taken to Bajawa (which of course would not have existed as a centre of regional administration if the event were located in pre-colonial times), the creature, I was told, was quickly dispatched and the body buried.

As indicated earlier with regard to the ana ula, tales concerning the capture and killing of small hairy hominoids are also heard in other parts of Flores. Still other instances, from Flores and elsewhere in eastern Indonesia, are reviewed later in this chapter and in the next. For the time being, it may be noted that hominoids in Ngadha (including some explicitly hairy ones) appear, on the whole, to be more fabulously represented than figures recognized in other parts of central Flores and certainly more so than the ebu gogo wildmen of Nage. What is more, available evidence suggests they are subject to a more fantastic representation in western Ngadha than in eastern Ngadha, that part of the region closer to Nage.

Manggarai (western Flores): 'ine weu', 'poti wolo' and hairy ancestors

To the west of Ngadha lies Manggarai, Flores' westernmost region, now famous as the discovery site of the hominin sub-fossil interpreted as *Homo floresiensis*. In view of links several commentators have suggested between *floresiensis* and putative hominoids like ebu gogo (Forth 2005, 2006), it may seem ironic that Manggarai people appear not to recognize a category of small, hairy humanlike creatures closely corresponding to ones encountered in central Flores. If such relatively naturalistic images are discernible in Manggarai representations, they occur as components of two named categories which, on the whole, are more fabulously conceived.

The two categories are named 'poti wolo' and 'ine weu' (or alternatively 'empo weu'), a term essentially identical to Ngadha 'ine ngiu' and its variants 'ebu ngiu' and 'ibu ngiu'. Also a reference to the hawk-owl, 'weu' is synonymous with 'ngiu', while 'ine' is a word for 'mother' in Manggarai as it is in Ngadha and other parts of central Flores. Indeed, beyond the lexical similarity, Manggarai 'ine weu' (hereafter employed as the general term) and 'poti wolo' largely correspond to eastern Ngadha 'ebu ngiu' (or 'ine ngiu') and kedho.

Recalling Arndt's account of the Ngadha ibu ngiu, Verheijen describes Manggarai ine weu, as 'a kind of pontianak' and more particularly as a malevolent long-breasted female being who uses her breasts to capture victims (1950: 64–65; see also Verheijen 1967: 177, where the name is glossed as 'langsuir', a Malay word denoting a female vampire almost identical to pontianak). Confirming their close resemblance to Ngadha counterparts, ine weu can be detected by the sound 'eu', identical to 'weu' as the onomatopoeic name of an owl. At the same time, Verheijen (1951: 213; 1967: 177) has also characterized ine weu as anthropophagous 'forest people' (Indonesian 'manusia hutan', also 'wild people'), a description that accords not only with other Ngadha images, but also with central Florenese wildmen like the ebu gogo and ana ula. Information I more recently recorded in Manggarai depicts ine weu as bipedal hominoids with very hairy bodies that live on stolen eggplants, fruit and chickens. Unlike most central Florenese hominoids, however, these ine weu are not small but of human height, with large bodies and long legs and feet which enable them to move quickly, taking long strides, an attribute linking them with the Ngadha ibu ngiu. They also frighten people by suddenly appearing and disappearing, while according to another report, anyone who sees an ine weu will not live long. A possible further feature—this time connecting them with the eastern Ngadha ebu ngiu—is inverted feet.[13]

Evidently not entirely distinct from ine weu are beings called 'poti wolo', described by Verheijen as 'ape-men' (Dutch 'aapmens') and hairy 'forest devils'; local people, he notes, sometimes compare poti wolo to orang-utans (Verheijen 1967: 515; 1950: 64). One feature the creatures share with hominoids elsewhere is a distinctive odour, in this case resembling the smell of a goat. However, in contrast to hairy and odoriferous ebu gogo and ana ula, poti wolo were generally described by Manggarai people I questioned as 'big and tall'. Glossing the name as 'manusia hutan' ('forest people', a term virtually synonymous with 'orang hutan' and therefore English 'orang-utan'), a man from eastern Manggarai further characterized the creatures as living in trees and, apparently like ine weu, taking long strides when they move. Also like ine weu, poti wolo possess a decidedly spiritual character. Indeed, it would seem that the 'apeman' or 'orang-utan' is but one form which, according to Manggarai, can be assumed by a wider class of malevolent spirits. Noteworthy here is the general sense of 'poti' as 'spirit', including spirits of the dead as well as nature spirits (Verheijen 1967: 514–15). The more specific term 'poti wolo' similarly denotes more than an 'apeman', referring among other things to beings which, in Manggarai origin mythology, consume the first humans whom God places on earth (Verheijen 1951: 150–51).[14]

Further distinguishing poti wolo from more prosaically described hominoids elsewhere is their ability to disappear and change shape, assuming the form of any animal they choose—another trait linking them with the kedho (or ebu ngiu) of eastern Ngadha. Generally characterized as fearful creatures, poti wolo are also conceived as the source of malevolent powers possessed by witches or sorcerers, an idea reminiscent of links between the Ngadha categories and Nage witches. On the other hand, humans are sometimes reputed to 'marry' poti wolo. Verheijen's dictionary (1967: 126) contains a brief reference to an elderly woman from Biting in eastern Manggarai, teknonymously named 'Mother of Neni', who was considered the wife of a poti wolo and who 'caused much fear in central Manggarai some time about 1920'.

Distinct from poti wolo and ine weu are Manggarai images of hirsute ancestors of present-day humans. One is a figure named Empo Paju ('Lord Paju'), or more completely Paju La'e, the ancestor of several clans in the Cibal region of central Manggarai, whose name comprises words denoting the male genitalia.[15] The following summarizes several versions of the story of Paju I recorded in 2005:

> Living alone in forest caves and covered only by his own profuse body hair, Empo Paju was once pursuing a wild pig. After killing the animal, he emerged from the forest and saw a field hut with smoke rising from it. Being ignorant of fire, Paju went to investigate and, entering the hut, encountered two women, a pair of sisters. The women were initially afraid of Paju, as his hairy body suggested he might be a poti wolo. However, when he offered them a share of his kill, they accepted. Unlike Paju, who consumed only raw flesh, the women insisted on roasting the meat. As they were cooking, Paju approached the hearth. The heat of the fire then caused all of Paju's hair to fall off, revealing him to be human. After acquiring all the trappings of human culture, Paju later married the two women and went to reside in their village. From these marriages derive several groups in Cibal.

Empo Paju's hair—about a fistful and described as dark and somewhat longer and thicker than the body hair of modern humans—is preserved in a house in the main village of Cibal (Compang Cibal). It is brought out and inspected only on major ritual occasions.[16]

Indonesian media articles appearing in 2005 identified Empo Paju as the ancestor of Manggarai people who, in the same sources, were tendentiously designated as 'pygmies' and accordingly construed as descendants of the diminutive hominins whose skeletal remains have been interpreted as a new species, *Homo floresiensis* (Forth 2006: 344). In contrast, Manggarai people I spoke with in 2005 described the man Paju not as small or short, but either as large and tall or no different in height from most local humans. Also contradicting any association with *floresiensis*, there can be little doubt that Manggarai regard Empo Paju not as a sub-human—as Nage do the ebu gogo—but, indeed, as a fully human being and an ancestor of many humans living at present. As discussed elsewhere (Forth 2006), traditional Florenese possess no conception of humans having evolved from non-human beings.

Despite his originally hirsute body, Paju is described as in every other way a modern human. According to one of his descendants, his skin was dark, so that many of his descendants are also darker than the local average; but for Manggarai this does not indicate a difference of kind.[17] Empo Paju is not the only Manggarai ancestor to be characterized as dark and hairy. Eventually emerging as the politically dominant group in Manggarai, the Todo people of south central Manggarai tell how their forbears, migrants from Minangkabau in Sumatra, encountered the people of Desu, dark-skinned and hirsute indigenes wearing clothing of tree bark and lacking knowledge of fire. When these Desu folk first saw the Todo ancestors kindling flame, they fled to the mountains—thereby, according to present-day Todo people, relinquishing all rights to the land. Todo also claim that Desu people formerly displayed copious hair on the arms and legs, hair growing from their ears and on their backs and head hair that was sometimes curly. They are further described as 'rather short' and as possessing flat noses and wide mouths and jaws. Occasionally, I was told, these traits are observable at present, but—apparently owing to intermarriage with other Manggaraians—like their original darker complexion they are no longer so evident.[18] Very similar to the Desu and also preceding the Todo ancestors in their present territory (particularly the Pongkor region), were hairy indigenes designated in the older literature as 'Rua' (Van Bekkum 1946: 65–66). Living in caves, eating only raw food and lacking any 'understanding of a social order' (ibid.: 66), these people apparently take their name from 'rua', meaning 'wild' or 'savage' (Verheijen 1967: 572).[19] One ancestor of these hairy folk, named Okong, was entirely covered in hair like a goat. Recalling the myth of the Cibal ancestor Empo Paju, when he first visited the Todo immigrants and encountered fire for the first time, his hair was singed; in the same incident, Okong discovered that cooked food was tastier than raw and, generally impressed by the Todo ancestors' technological superiority, surrendered to them all power over the land (Van Bekkum 1946: 6). Also recalling Empo Paju, who rose to prominence as a military leader of the Cibal people, the Rua folk, described as strong and courageous, provided the Todo immigrants with valuable assistance in wars against other older populations (ibid.).

The name of another Rua ancestor, 'Hairy Dog' (Acu Wulu; 'wulu' is 'body hair'), was identified by my Todo informants as a generic reference to the Desu people. Verheijen (1967: 304) records a similar name, 'Empo Wulu', as that of an ancestor in the northern Manggarai district of Lamba Léda. The notion of hairy ancestors, therefore, appears to be widespread in Manggarai. Moreover, despite their originally hirsute condition and cultural simplicity, neither the Rua nor the Desu, who to some extent still compose an identifiable group, are represented as anything other than human beings. What we encounter in Manggarai, then, are either images of malevolent spiritual beings sometimes, but not invariably, manifest as hairy hominoids (like the ine weu and poti wolo) or representations of hirsute ancestors who were essentially human and who gave rise to human descendants. What one evidently does not find are stories pertaining to sub-human, or ambiguously human, populations like the Nage ebu gogo or Poma ana ula.

An exception to the foregoing may be creatures described in a story from the Rajong region of far eastern Manggarai (see Map 3). This concerns 'niung', a variety of short, hairy 'forest spirit' (Erb 1987: 97) whose name is related to Ngadha 'ebu ngiu', 'ine ngiu' and 'ibu ngiu' and whose females possess long pendulous breasts which they drape over their shoulders. The creatures furthermore 'have a passion for eating eggplants as well as the live coals of banked fires' and produce sounds similar to the cries of a hawk-owl (ibid.: 242). If, by now, this description does not seem particularly remarkable, the story itself is surely more arresting. Translated by Maribeth Erb (1987: 97–99), the Rajong tale runs as follows:

> In the ancient past, a mother left her sleeping child in a garden hut when she went off to pick vegetables. All of a sudden, she heard the screaming of her child and ran back to the garden house. There in the garden hut was a niung . . . The niung had taken the human child and replaced it with her own niung infant. When the human mother entered the garden hut, the niung was crying out in a melancholy voice about the exchange of the two infants, as if she herself had not brought it about . . . 'My child is clean and smooth [the niung exclaimed], but the human being's child is full of fur!' But all the while she was chewing away on the fingers of the human child, who was crying out in pain. The human mother in horror ran to the village of Nanga searching for help from the inhabitants. They ran back to the garden hut. The niung had already run off with the woman's baby, but had left her own child there. The villagers took the niung child and burned it in the fire. The next day they took . . . bark from the enau tree . . . and went to the meadow . . . where niungs were often to be found . . . herding wild buffalo. Near this meadow was a hole that was the opening of their village. The villagers piled up the bark and built a fire there. The niungs started to run out of the hole and were bitten by the villagers' dogs. A great war started where many niung were killed. The spot today is called rate niung, 'the grave of the niung', for many niung were apparently killed there.

Features linking this story with Nage and Poma narratives are too obvious to require mention. (As 'enau' is the Arenga palm, enau 'bark' clearly refers to palm fibre.) The first part of the story is clearly redolent of the Nage and So'anese tales in which particular ancestors (Huma Léli and Ruma Réwo respectively) kill a hominoid infant left in a field hut—except that in this case, it is a human child that is left in a hut and the infant hominoid is killed by burning rather than stabbing. There are also additional elements. One is the theme of changelings: the switching of human and hominoid infants. Another is the child-abducting niung's quite ingenious attempt to pass off the 'smooth' human child as her own, a tactic plainly uncharacteristic of the dull-witted and readily victimized ebu gogo. And yet another is the characterization of niung as herders of wild buffalo, a notion recalling associations of similar creatures elsewhere with wild pigs (see Chapters 5 and 8), but more importantly in the present context, of Nage and Ngadha earth or nature spirits named 'nitu' (Forth 1998a: 70). Whether all the niung were exterminated in the Rajong holocaust perhaps remains a question, though according to Erb (pers. comm.

December 2004), Rajong people believe in the continuing existence of some beings called ‘niung’.

Insofar as the niung clearly differ from images encountered elsewhere in Manggarai and by the same token resemble small hairy hominoids of central Flores, they do indeed appear exceptional in relation to a more general ethnolinguistic pattern. Yet they may be an exception that proves the rule. For Rajong, located in the extreme eastern part of the present administrative region of Manggarai, is in fact more closely related linguistically and culturally to Nage and Ngadha than are more westerly parts of Manggarai. Interestingly, it is also much less densely populated, especially in comparison with central Manggarai.[20]

The ironic absence in Manggarai—or at any rate, central and western Manggarai—of hominoidal representations and traditions closely comparable to the ebu gogo might appear to weigh against any connection between the Nage wildmen and *Homo floresiensis*. On the other hand, geological evidence shows that the hominin population inhabiting Liang Bua, the cave in central Manggarai where the type specimen and other remains were discovered, was rendered extinct by volcanic activity occurring some 12,000 years ago. *Floresiensis* populations living in more easterly parts of Flores may well have survived this volcanic activity, which contrary to what was initially thought, originated not on Flores but well to the west, on Bali (Morwood and Van Oosterzee 2007: 178). Of course, eastern survivors could have later re-colonized western Flores. Yet they may have been prevented from doing so by human (*Homo sapiens*) expansion. Partly because it enjoys higher rainfall and more fertile soils than other parts of the island, Manggarai has long been the most densely populated part of Flores. Accordingly, if any hominoid image hypothetically reflects *Homo floresiensis*, the arguably greater fit with central Florenese exemplars may be better explained with reference to western Flores’ larger human population (cf. ibid.: 178–79). Of course, one cannot ignore the ethnolinguistic associations of different images; but then, as already indicated, ethnolinguistic boundaries largely coincide with geographic and ecological ones.

The ‘lae ho’a’ of Lio

Although situated well to the east of Nage, the Lio region belongs linguistically to central Flores. By the same token, Lionese dialects are quite distinct from eastern Florenese languages spoken in Sikka and Nita immediately east of Lio (see Maps 2 and 3).

Lio people speak of a creature they call ‘lae ho’a’ (dialectal ‘lae o’a’). Designated by this Lio name, the same creature is also known to Nita people who live to the east and northeast of Lio and who in many cases speak Lionese as well as their own language, classifiable as a dialect of Sikka.[21] The sense of ‘lae ho’a’ is uncertain. While ‘lae’ might mean ‘to trick, deceive’, the term ‘ho’a’ recalls various Florenese words for macaque monkeys (e.g. Lio ‘ro’a’,

Nage "o'a', Keo 'yo'a'.). So the name could conceivably mean something like 'false monkey', a sense not inconsistent with local descriptions of lae ho'a. On the other hand, some people thought 'lae' should be understood as an anatomical term, corresponding to Indonesian 'ulu hati' ('pit of the stomach'). In his Lio dictionary, Arndt (1933) gives the inaccurate transcription 'lai hoa' as 'malevolent spirit', a gloss he provides for numerous other categories. I was unable to establish whether there is a Sikkanese equivalent of 'lae ho'a'. A Nita man suggested 'sége re'e', but this term—which, interestingly enough, also derives from Lio—is mostly associated with a rather different sort of being, as I describe below.

I recorded much of my information on the Lio hominoids during a two-month visit in 2005, thus after the announcement of the discovery of *Homo floresiensis* and after a British reporter's brief visit to Lio-speaking villages in October 2004. The reporter explicitly asked people about resemblances between unspecified local images and *floresiensis* and showed pictures of pre-sapiens hominins. It is unlikely, however, that this exposure significantly affected subsequent Lio descriptions of lae ho'a. For one thing, illustrations of *Homo floresiensis*, including the graphic reconstruction by Peter Schouten (see Plate 3.3.), were almost certainly not among the reporter's pictures (Forth 2006: 343–44). For another, there was no appreciable difference between what I heard in 2005 and information I recorded in 2003 (including accounts given by people later approached by the British reporter).[22]

Like certain figures from more westerly parts of Flores (but not of course the Nage ebu gogo), lae ho'a are generally described as a rare but nonetheless extant population of hominoids. Of seventeen people I spoke with, over a dozen claimed to have observed lae ho'a, mostly on a single occasion. Informant descriptions suggest two partly distinguishable images labelled 'lae ho'a'. One resembles a large, ground-dwelling and at least partly bipedal monkey. Three people characterized these as erect but stooped, especially when moving, while another three specified the creatures as moving quadrupedally. Most accounts that conformed to this more simian model described lae ho'a as possessing a tail, though one shorter than that of a Long-tailed macaque (Flores' only known non-human primate) and possibly as short as a hand span (about 18 to 20 centimetres). In one or two instances, the tail and perhaps other features as well, suggested that the person may have observed a Palm civet (*Paradoxurus hermaphroditus*). Yet these animals are as well known in this part of Flores as are the ubiquitous macaques. Several times, I suggested to putative witnesses that they might have seen a monkey, but they consistently denied this.

In accord with a generally more simian depiction of lae ho'a is a view that the creatures should be enumerated with the classifier for animals ('éko', or 'tail', as in Nage) rather than the classifier for humans. I also recorded a reference to a male specimen with the term for non-human animals ('mosa'). (Nage, it will be recalled, use the human classifier for ebu gogo.) At the same time, some Lio were unsure whether lae ho'a should be classified as animals or humans. One man argued that no numeral classifier should be used, since the creatures were, as he said, 'intermediate' between humans and non-humans. He further suggested that that they could be considered intermediate between humans and spirits as well.

An extreme example of accounts of a more monkey-like creature was provided by a 26-year-old man who, during the dry season in 2004, was out hunting. It was about midday when he heard his dog barking ahead of him. Following the sound, he came to a large fig tree, with the dog still barking below. Standing erect on a branch, perhaps 10 metres from the ground, was a creature resembling a macaque, about half a metre tall, but with a very short tail and arms somewhat shorter than a monkey's. The young hunter was about to fire his air-rifle, but, as he was taking aim, the creature vanished. So startled was he by this sudden disappearance—something he mentioned repeatedly—that he became frightened. The creature, he inferred, was no ordinary animal and no ordinary monkey, but a lae ho'a.[23]

The contrasting image of lae ho'a is far more humanlike, pertaining to a rather larger, erect and tailless creature, about one metre in height, with the general form of a small hairy man or a human child and facial features more human than simian. Exemplifying this more human image is a male creature reputedly encountered at close range in 1953 by a retired school teacher, now deceased. The account was related to me by his son, a man about 70 years old and resident in the Lio village of Wa'u Lejo. He had, he said, heard the story from his father many times. Although the father's meeting with the lae ho'a took place at night, he was evidently able to pass on a detailed description. The creature stood erect and had a generally human form. But the face resembled a monkey's: the eyes, ears and nose were small, although the ears may have been positioned as in a human. The forehead was 'sunken' or 'concave', not prominent like a human's; the eyes were also sunken and the brows prominent. The chin, too, resembled a monkey's, although from the cheeks to the chin the face seemed to narrow, as though forming a point. The area above and below the lips was broad, which one might take to indicate a large mouth and prognathous jaw. While the other teeth seemed human, the canines were more like a monkey's. The creature's legs were quite long, as in a human being and the arms were like a small child's. The body was covered in sparse hair, about a centimetre or two in length, but thicker on the chest. The skin was about as dark as local human complexions. Like a monkey, its head hair was not distinct from the body hair and of about the same 'greyish' colour. But it had the genitalia and genital hair of a human male.[24]

It is important to stress that more hominoid and more simian representations of lae ho'a by no means differ in all particulars and, in fact, display considerable overlap. This applies even with respect to height. Drawn from all accounts, most estimates ranged between 80 centimetres and something over a metre, although two eyewitness reports of tailed creatures mentioned heights of just 30 and 50 centimetres. Regardless of other features, lae ho'a faces were invariably characterized as displaying some simian characteristics, including prognathous jaws, low foreheads, receding (or absent) chins, prominent canines (especially upper canines) and facial hair. Several accounts depicting a generally more human image explicitly described the lower part of the head and face as resembling a monkey's (or in one or two accounts, that of a dog), while the part above the eyes appeared more human (a 'round' head like that of a human being, according to three descriptions).

Also generally attributed to lae ho'a was body hair more extensive than a human's, though with regard to thickness and colour details were notably various. While most people spoke of thick hair like a monkey's, others mentioned sparse hair, in some instances describing this as unevenly distributed over the body—for example, thickest on the chest or absent on the legs (with reference to a putative female specimen). Two men specified the body hair as between one and two or three and four centimetres long, but others were vague about length. Evaluations of colour mostly suggested a grey or grey-brown, the colour of macaque pelage. Three people said the hair was dark (or 'black'), while terms corresponding mostly to browns were mentioned twice. Only one man mentioned 'red' (or red-brown) hair. Although comparisons with the colour of macaque hair were common, there seemed to be general agreement that the body hair of lae ho'a is darker than that of monkeys. Nearly everyone who mentioned head hair described it as indistinct from the body hair; only one man (who also spoke of 'black' body hair) claimed that creatures he had seen—a group comprising over ten individuals—possessed long flowing hair, longer than the hair on the body. People able to specify skin colour indicated a 'dark' complexion little different from local humans, though some stated this was obscured by extensive body hair.

Besides the presence or absence of a tail, a contrast of more humanlike and more simian images is revealed in descriptions of the limbs. One man characterized lae ho'a feet as like hands, capable of grasping, although neither he nor anyone else described the creatures as arboreal—the young hunter's putative experience notwithstanding. Nor did anyone explicitly mention long arms, although one informant suggested that arm length was proportionally intermediate between a human and a monkey. Another man described lae ho'a walking bipedally with arms held towards the rear and with the hands bent somewhat backwards. At least four men attributed long fingernails to lae ho'a; but these were not clearly associated either with the more human or the more simian image. No descriptions specified lae ho'a feet as particularly small, but neither did anyone mention relatively large feet or hands. Two informants provided fantastic accounts of lae ho'a limbs and appendages. An elderly man resident near Ma'u Lo'o, in south coastal Lio, described the body of a dead lae ho'a female he claimed to have appropriated over 25 years ago, after frightening away a group of the creatures attempting to bury it about midnight on a sandy beach; he stated that the corpse lacked arms and had just three digits on either foot. The other man spoke of two or five toes pointing 'backwards', thus recalling the 'reversed feet' sometimes attributed to hominoids in Ngadha and Manggarai. But these attributions were exceptional.[25]

One man characterized the lae ho'a as robust, or more particularly as broad-chested. Another described a creature with a broad chest contrasting with a much thinner waist. Yet others spoke of an average build, comparable to a monkey or small human being. At the same time, all Lio informants described lae ho'a as exceptionally strong. As one man stated, human strength is 'meaningless' in comparison, so that one would be ill-advised to strike a lae ho'a. Another informant suggested that parangs or arrows would fail to penetrate the creature's body. In one reputed incident, five adult men had failed to overcome a lae ho'a. As this reveals, there are indeed reports of people attacking the creatures and, as I describe below, even a few cases of people killing them.

Not only do Lio people claim to have seen lae ho'a, they also say they have heard them. The most frequently reported vocalization is a somewhat high-pitched sound reproducible as 'hooo' or 'wooo' (or 'hoo . . . hoo', separated by a one second delay) and similar to a cry men use to call to companions at night or in the forest. Afterwards, another creature might respond. Lae ho'a are also said to mimic similar sounds made by humans. According to another idea, they will mimic words or phrases in a parrot-like fashion.[26] But, as one man pointed out, this ability is reported only in stories relating encounters with lae ho'a. According to his son, the man who encountered a lae ho'a in the 1950s claimed that he, but only he, was able to converse with the creature; its voice was like a small child's, soft and high-pitched. Other informants simply stated that lae ho'a cannot speak human language and have no language of their own intelligible to humans. Vocalizations less commonly attributed to the creatures include a 'whooshing' or 'shushing', a whistle and repetitive sounds that the man who claimed to have discovered a lae ho'a corpse compared to the cry of a raptorial bird, possibly a Sea-eagle. Distressed lae ho'a are reported as crying or weeping. Two men, both from southwestern Lio, described nocturnal sounds of lae ho'a as resembling whimpering or whistling, which appeared attributable to the calls of night birds. The first sound was furthermore reproduced as 'iu, iu' or 'ngiu'—thus recalling the niung, ebu ngiu, ibu ngiu and ine weu of more westerly parts of Flores. (Referring to the 'whistling' sound, one informant attributed this exclusively to the smaller of two varieties of lae ho'a which he alone distinguished in this context.)

Like putative hominoids elsewhere, lae ho'a are mostly associated with highland regions, where they inhabit caves in dense forests. One man stated that lae ho'a do not permanently occupy particular caves, thus seeming to imply a nomadic habit. However, extermination tales suggest an association of some groups of lae ho'a with specific locations. Despite their highland associations, lae ho'a also reputedly occur near coastal caverns and other locations near the sea. One man stated that, whatever their usual abode, all lae ho'a will descend either to the coast or to the banks of rivers in order to bury their dead in sand or soft soil. Monkeys, it should be noted, can be found near the coast as well as in the interior. Both highland and coastal sightings can take place at night or in the daytime, although diurnal encounters seem more often to concern the more simian image. Among Florenese hominoids, burial is attributed only to lae ho'a. Yet in other parts of the island and probably in Lio as well, monkeys also are thought to bury their dead.[27] Reputed encounters with lae ho'a exhibit no clear seasonal pattern. According to several informants, the creatures are seen most often during the rainy season (about November to April), or when people are guarding wet season crops against the depredations of wild pigs. Possibly related to this is the idea that lae ho'a appear when the weather is overcast or there is light rain, a meteorological association also reported for the ana ula of Poma. On the other hand, some stories describe hunters coming across lae ho'a while tracking wild pigs, an activity typical of the dry season. In either case, encounters are described as rare and although a good number of people I spoke to claimed to have seen the creatures, this can be explained by the fact that my attentions were directed to such individuals.

While generally described as inhabiting desolate places, some reports concerned lae ho'a that had supposedly entered settlements and even dwellings to steal cooked food, fowls, or eggs (which they are somehow able to suck dry leaving only very small or invisible punctures). In a similar vein, a couple of stories described lae ho'a invading field huts at night and disturbing lone sleepers. Being omnivorous and apparently opportunistic, lae ho'a are also said to steal maize, tubers, bananas and other produce from cultivated fields. Informants were rather less clear about what wild foods they might consume. Two men described lae ho'a stealing fish, both whole fish and gutted fish left on a beach to dry. Unlike cats, monkeys and other attested animal thieves, which will carry away stolen food before consuming it and will devour an entire fish, lae ho'a, I was further told, eat on the spot and moreover consume only the flesh, leaving the skin and bones behind.

Not all reputed sightings of lae ho'a are of lone individuals. When he was teaching in southern Lio in 1999, a Nita school teacher claimed to have seen a group of seven or eight emerging from a cave, just as he was about to enter to collect swiftlet nests. This was the first occasion he had encountered such creatures and it was only after he related his experience to others that he learned what they were. Another story concerned a nocturnal encounter with a group of more than ten individuals, on a modern road bridge near Watu Néso, also in southern Lio, possibly in the year 2000. These included specimens about a metre tall but also some just 30 centimetres in height, which the informant presumed were infants. Another man described seeing hordes of perhaps twenty or thirty creatures near seaside caves at Ae Wa and Leko Sétan, two uninhabited locations on Flores' north coast.[28] In accordance with the seemingly social nature of lae ho'a, Lio people mentioned how groups of the creatures characteristically travel in straight lines, one following directly behind another and not 'scattered', as was deemed more typical of both humans and monkeys.

Like other central Florenese hominoids, lae ho'a are generally depicted as fearful of both humans and their dogs. A reciprocal human fear of lae ho'a relates largely to a belief that harming or disturbing them can result in mystical retribution in the form of death, illness, or insanity. But Lio also say that meeting such a creature can confer benefits and that favoured individuals can obtain their fingernails, canine teeth, or a quantity of their urine—all things which afford protection, cure illness, or assist in childbirth. Although lae ho'a might give such things freely, I also heard of someone who obtained urine in return for releasing a lae ho'a he had captured by driving a stake into its shadow. The man who claimed to have recovered the corpse of a female lae ho'a from the beach at Ma'u Lo'o showed me a round, apparently metallic object he claimed to have found inside the head, after the body was reduced to a skeleton. He had also given parts of the skull to relatives and friends, in one case to enable a man to become 'invisible' to enemies and, particularly, to avoid detection by officials during an illegal visit the recipient later made to Sabah. Possibly linked with this power to confer invisibility is the oft-mentioned capacity of lae ho'a to disappear and reappear.

Not just this power but, according to a few accounts, the ability to assume the form of other animals apparently distinguishes lae ho'a from more naturalistically conceived hominoids reported from other parts of central Flores. At the same time, several Lio informants denied metamorphosis and even the ability to disappear, at least insofar as this is understood in a supernatural sense. According to another idea, a person must be alone in order to see lae ho'a; but this is similarly contradicted by reports of encounters involving several people (which I describe below). I was also told that people can only observe lae ho'a fortuitously and that if one goes in search of them, they will never be found. Both of these notions are mystically interpreted as reflecting special powers the creatures possess. Yet they could equally reflect chance encounters with a rare and fast-moving animal, whose quick and evasive behaviour could be construed as a mysterious ability to vanish into thin air.

Further revealing a mystical attitude, Lio folk also describe lae ho'a as the objects of occasional ritual practices which should cause them to appear. One man claimed a totemic relationship with lae ho'a, initiated by an ancestor who had received several objects from one such creature. Comprising a container of liquid reputed to be the creature's urine, a metal pendant shaped like a lae ho'a, a small knife blade characterized as its 'weapon' and a detachable wooden hilt also depicting the creature's form, these were the only graphic representations I discovered of lae ho'a (see Plate 3.5). Although the images are hardly decisive, the hilt shows a head with a bridgeless nose descending straight from the forehead. The pendant suggests a generally hominoid creature with a conceivably simian face and four long fingers, possibly representing the frequently mentioned long fingernails. (What may be the thumb is, by contrast, very short.) A pair of protuberances on the chest could be either nipples or very small female breasts. Since lae ho'a are generally described as naked and completely lacking in technology, the notion of a 'weapon' is curious, unless it is a figurative reference to the creatures' powers.[29]

I was able to obtain just one material trace of lae ho'a. This was a bone fragment reputedly taken from the skull of the creature found at Ma'u Lo'o. Provided by my field assistant, Endy Paji, who some years previously had been given it as a protective charm by the discoverer (a close friend of Paji's deceased father), the fragment appears to derive from the foramen (base of the skull) of a small animal, possibly a mammal, but is not otherwise identifiable.[30]

Supernatural powers aside, descriptions of the physical form of lae ho'a are largely naturalistic. Indeed, some putative observations might reasonably be explained by experience of monkeys. Particularly noteworthy in this connection is the absence from local reports of long breasts of the sort characteristic of hairy hominoids in more westerly parts of Flores. Although some Lio accounts mentioned small breasts, including breasts of an apparently non-sapiens form, many suggested that the sexes are not readily distinguishable. Another difference between lae ho'a and other Florenese hominoids concerns odour: no one I asked had ever heard that lae ho'a possess a distinctive smell. Nor is abduction a prominent feature of the Lio representation, although as will presently become clear, the theme does figure in certain stories. Capture of infant or adult humans is generally attributed to beings other than lae ho'a and the creatures, I was told, are not usually employed as bogeys. Only once did I hear of lae ho'a mating with humans. A Nita man related how his grandfather's second wife, a childless woman, was descended from lae ho'a. Although he provided few details, the woman's derivation was evidently not manifest in any physical difference, as she only revealed it to her husband shortly before her death. Supposedly attesting to her peculiar heritage was the further detail that, before she could be buried, her corpse mysteriously disappeared.

Possibly the most significant connection between lae ho'a and other central Florenese hominoids are stories recounting how local humans once attacked and in some instances extinguished, a group of lae ho'a by setting fire to their cave. In the Sikka Regency, tales of this sort are known in both Lio and Nita-speaking communities. One account was provided by a Lio man in the south coastal settlement of Wolo Wiro, who had heard it from his mother, a woman from the Nita-speaking village of Jalo. Concerning events reckoned to have occurred around 1935, the story can be summarized as follows:

> Somewhere near Jalo, parents once left their young child in a field hut before setting out to weed their garden. Lae ho'a from a nearby rocky area came and took the child, leaving one of their own infants in its place. Later, Jalo villagers found the baby in a cave, being suckled by a female lae ho'a. After recovering the child, they sealed the cave exit (which was situated in a more elevated spot) and placed a quantity of palm fibre in

> the entrance. They then ignited the fibre. Trapped inside the cave, the creatures made a terrible noise; all were killed and none escaped. (What became of the lae ho'a infant left in place of the human child was not mentioned.)

A similar extermination is supposed to have taken place somewhere west of the Nita village of Wukak, located about 2 kilometres due west of the centre of Nita. Since Wukak is not far from Jalo and since the only local estimate I recorded placed the event over sixty years ago, the Wukak story may in fact refer to the same location as the Jalo tale. Told by a man from the village of Ri'it, some 3 or 4 kilometres from Wukak, it differs from the Jalo narrative only insofar as lae ho'a are not actually described as abducting a child; hence the changeling theme also does not occur. Wukak people are described as taking action simply from fear of possible abduction, after spotting one of the creatures entering a field. Again, palm fibre was the material used to set the fire, although firewood was also mentioned.

A slightly different tale forms part of the account by the now deceased Lio man who, in the 1950s, encountered a lae ho'a outside his field hut in the middle of the night. Not only does this recount the putative experience of a particular, historical individual; it also concerns a single lae ho'a. After the man and fellow villagers he had called to assist endeavoured unsuccessfully to capture the creature, they searched throughout the night and discovered it the following morning hiding in a cave. After collecting bundles of palm fibre, the men set fire to the cave; but since one of the cave exits was left unsealed, the lae ho'a managed to escape.

Also relating to the early part of the twentieth century, a somewhat similar story told by an elder from Detu Soko in Southwestern Lio described how a prominent man once drove a group of lae ho'a from a cave situated near a spring that fed his rice-field. This, however, he accomplished not with fire but with an explosive device locally known as a 'bamboo cannon', which is normally used to announce a death. This did not kill the creatures but merely caused them to flee.

Another story involving burning of lae ho'a concerns an incident that reputedly occurred as recently as 2001 or 2002:

> A number of men from the Lio region of Feo Ndari were hunting wild pigs in a wooded area, when they came across a lae ho'a fleeing in the direction of a cave. After chasing the creature down and with great effort (since despite its small size it proved exceedingly strong), the hunters were finally able to overcome it, bind it with rope and cast it into the cavity. They then collected a quantity of dry brushwood, tossed it into the cave and set it alight. Subsequently, the men discovered that there had been a second creature inside, apparently the mate of the first, that had already taken cover upon hearing the hunters' dogs. Both creatures 'screamed and wept' as they were consumed by the fire.

I investigated this story in 2003 and again in 2005. On both occasions, my principal source was a Nita man residing in the village of Koro, on the north coast of Lio. He had heard about these events not long after they had reputedly occurred, from several Feo Ndari men who claimed to have been eyewitnesses. These were men who, for a share of the crop, annually assist the Koro cultivator in harvesting rice. Before my departure from Flores in 2005, however, I was unable to locate any of the individuals involved; nor could I find anyone in Feo Ndari who would confess to hearing the story. Its grounding in any empirical occurrence therefore remains questionable.[31] With regard to the detail of killing lae ho'a by confining the creatures inside a cave and setting this afire, the tale obviously bears some relation to the other stories. Equally obviously, all these tales of cave-burning bear a striking resemblance to stories from more westerly parts of Flores.

Another kind of story, similarly widespread on Flores, concerns the inadvertent capture of hominoids in animal traps. In 2003 and again in 2005, I heard of an incident that reputedly took place about ten years previously in Kota Baru, a new settlement of Lio and Nita people not far from Koro, on Flores' north coast. The tale runs as follows:

> Kota Baru people thought wild pigs had been destroying their ripening maize; so they set a bamboo spring trap. One morning they discovered a creature caught in the trap, hanging upside down by the leg. Crying and weeping like a small child, it was not a pig but something resembling a human being. The sight of it caused some people to flee. Most were not familiar with this sort of creature. Some thought it might be a kind of monkey; but older people present said it was something 'more like a human' and was called 'lae ho'a'. They then performed a ritual and released the lae ho'a, which fled into a nearby forest.

One man explained that, had the rite not been performed, some sort of 'catastrophe' could have been visited on people for trapping such a creature. Similarly, the narrator of the tale concerning the Feo Ndari hunters expressed considerable shock at their killing of a lae ho'a, describing this as a senseless act that could well have had negative consequences.

In another recent instance where a lae ho'a was reputedly caught in a trap, a man from southern Lio told me how he had been called in to determine its identity. He declared it to be nothing other than a large male monkey. Like the young hunter's tale related earlier, this would suggest that encounters with macaques lie behind at least some reputed sightings of lae ho'a. Yet it is unlikely that the category as a whole simply reflects experience of these very familiar primates, or even of large, lone specimens that appear somehow odd or more hominoid than conspecifics. For one thing, people elsewhere in Flores seem rarely if ever to mistake monkeys for instances of local hominoid categories, just as (unlike some Lio observers) they do not describe the latter as possessing tails. For another, Lio extermination stories appear mostly to concern not the more simian images of lae ho'a, but indeed a more human image, corresponding more closely to images described elsewhere on Flores.

A note on East Flores

Categories of hairy hominoids comparable to lae ho'a do not clearly occur in the far eastern administrative region of East Flores (Flores Timur; see Map 3). Arndt (1951: 31) mentions beings named 'wae belatung', which live both in forests and near the coast and subsist on snails and worms. Although implicitly plural, the creatures seem only to be imagined as long-breasted females ('wae' apparently means 'girl, young woman') and they are not specified as hairy. An adjacent reference to 'nitu belatung' (ibid.: 31–32; 'nitu' is a term used throughout Flores for 'spirits'), as well as the creatures' magical powers of healing and ability to cause rain and windstorms, moreover suggest that wae belatung are more accurately interpreted as a category of spirits. On the other hand, Arndt records a story describing the capture of one such female being by a man who takes it to his village. Subsequently, the village is affected by adverse weather, an outcome reminiscent of the 'catastrophe' that potentially threatened the Lio villagers of Kota Baru who inadvertently caught a lae ho'a. Also like lae ho'a, the wae belatung are represented as a source of medicines.

If hairy hominoids elsewhere ultimately reflect experience of some sort of empirical creature, then their absence in East Flores might be explained by the region's drier conditions and lack of forest cover relative to Lio territory and other parts of the island. At the same time, representations of hirsute *human* ancestors occur in eastern Flores, as they do at the opposite end of Flores, in Manggarai. Central to the origin myth of the ruling house of Larantuka (see Map 3) is a large and powerful female figure, covered in hair and employing her long fingernails to hunt game. After shaving off her hair, the myth's male hero discovers the creature is a woman, marries her and thus gives rise to the princely line (Dietrich 1995: 120–21). Similarly, to the east of Flores on the island of Lembata, a Kédang hero is described as marrying a hairy female named Bota Ili (Barnes 1974: 40–41). This he does after setting his dogs on her, getting her drunk on palm wine and shaving off her body hair. Dogs and intoxication recall stories of central Florenese hominoids, including the Nage legend concerning the extermination of the ebu gogo. However, in both the Larantuka and Kédang traditions, the hairy woman is a single individual, not an instance of a category or member of a population and is represented as a human being rather than a sub-human creature like the Lio and Nage hominoids.

Wildmen, bogeys and 'pontianak'

Employing the Nage ebu gogo as a reference point, similarities and differences among Florenese hominoids are summarized in Table 3.1. Particularly

Table 3.1 Florenese-named categories including populations of small hairy hominoids.

Name	*Location*	*Physical Features*	*Status*
Ebu gogo	Lia Ula in Nage, central Flores	Short, robust, hairy Extremely long, pendulous breasts Propensity to swallow large objects Fear of bamboo combs Stupid, unintelligent	Extinct; reputedly exterminated by fire some 200 years ago The name also refers to a separate, or originally separate, bogey figure (also called 'gogo meo')
Ana ula	Poma and Rawe in central Flores	Small (1 metre), hairy Smaller, less robust and more aggressive than ebu gogo Not associated with pendulous breasts, excessive swallowing or fear of combs	Mostly considered extinct May survive as spiritual beings
Toro gogo	So'a in central Flores	As ebu gogo	Legendary, extinct creatures, exterminated by fire Survive as spiritual beings
Ebu ngiu *or* Ine ngiu (ine, 'mother')	Eastern Ngadha in central Flores	Small, hairy Long-breasted Inverted feet	Still extant Spoken of as individuals occasionally encountered rather than as a population occurring in groups Partly equated with kedho
Ibu ngiu	Western Ngadha in central Flores	Female, long-breasted *Not* hairy or naked	Mostly or entirely a female representation
Kedho	Eastern Ngadha in central Flores	Small, hairy Long-breasted Inverted feet	Partly equated with ebu ngiu or ine ngiu The name also refers to a fantastic shape-shifter than can assume various animal and human guises, in order to abduct people.

Ine Weu *or* Empo Weu	Manggarai in Western Flores	Hairy No smaller than humans Pendulous breasts Inverted feet (sometimes) Fantastic powers (sometimes)	Not entirely distinguished from poti wolo (see below)
Poti Wolo	Manggarai in Western Flores	Hairy 'ape man' (Verheijen) Larger than humans	Similar to kedho, poti wolo is a shape-shifter which can take the form of a dangerous anthropomorphous creature Also a general term for malevolent spirits
Niung	Rajong (administratively part of Eastern Manggarai, bordering Ngadha regency)	Small, hairy Long-breasted Similar to ebu gogo and toro gogo; in some respects resemble more the ana ula and ebu ngiu	One group was extinguished by fire but not the entire population
Lae ho'a	Lio in east central Flores	Small, hairy Smaller than ebu gogo Does *not* have pendulous breasts Some reports mention small tails and quadrupedalism Some reports mention fantastic powers (transformation, healing power of body parts)	Still extant, but described in stories as being killed by humans who set fire to their caves

Note: The categories are presented in the order in which they are discussed in the text. Note that not in all cases do the terms exclusively denote hair-covered bipedal hominoids displaying simian morphological features.

in regard to physical images and spiritual associations (or rather, the relative lack thereof), I take these several categories to be fundamentally similar. A question may be raised in regard to the lae ho'a of Lio: while in several respects among the most naturalistically conceived of Florenese hominoids and subject to the most detailed descriptions and reports of recent sightings, these beings, paradoxically, are accorded supernatural powers not associated, for example, with the ebu gogo and ana ula. This difference, however, is illuminated by the fact that, whereas the latter are generally spoken of as extinct, the lae ho'a are claimed still to exist. Creatures no longer existing (physically or, in the case of ebu gogo at least, as surviving spirits) are obviously unlikely to be represented as a source of spiritual danger or benefit. We might also remember that fully attested animals can be the object of spiritual practices and beliefs at least as fantastic as those associated with the hominoids of Lio.

In a broader comparative frame, Florenese categories can be seen to comprise two distinct yet in some way related kinds (or, perhaps better, 'congeries') of anthropomorphous beings. Analytically, these can be called 'ideal types', since probably nowhere are they found in a pure form. The first includes small, hairy, non-human or not fully human hominoids reputedly existing—or having formerly existed—in groups. It is for the most part these that are included in Table 3.1. Although all are credited with some fantastic qualities, local people mostly describe such creatures in a prosaic way. The best examples are the ana ula of Poma, the Lionese lae ho'a and the ebu gogo of Nage. To these may be added the toro gogo of So'a and the niung of Rajong and even some—but evidently not all—of the eastern Ngadha images labelled 'ebu ngiu' or 'kedho'. (More doubtful still are the ine weu, or empo weu and poti wolo of Manggarai.) Clearly, this more naturalistic sort of image is most prevalent in central Flores, a region definable on linguistic grounds as comprising speakers of Nage and Ngadha (including So'a, Poma and Rawe) and Lio.

Exemplifying the second kind of image are more fantastically conceived beings, often spoken of as single figures rather than populations or collectivities and moreover as specifically female. Instances include the ibu ngiu of Ngadha, other Ngadha and Manggarai images and several figures so divergent from categories of hairy wildmen that I have yet to introduce them. Although some are represented as hairy-bodied, most are not, or are not consistently so. Even hirsute exemplars, moreover, are not unequivocally credited with simian features. And in further contrast to the smaller hominoids, either the size of these beings is not mentioned or, where it is, they are depicted as large. Creatures assignable to this analytical category are particularly significant as malevolent abductors, of adults as well as children and especially with regard to the threat they pose to youngsters, they often function as bogeys. With considerable justification, Arndt, Verheijen and other earlier ethnographers interpreted these beings as varieties of the Malay 'pontianak'—mainly female spirits found in many parts of Malayo-Polynesian-speaking Southeast Asia that are commonly described as deriving from women who died in childbirth or from stillborn children. Like the pontianak of western Indonesia, the Florenese creatures are distinguished by their possession of a hole in the back or of exceptionally long breasts, or occasionally both (see Arndt 1932: 148–51). In some instances, the ventral cavity is specified as the place where they 'store' human victims, or even the means by which they 'consume' or 'swallow' them (ibid.; Skeat 1900: 326).[32]

Yet another link with pontianak elsewhere is the association of several Florenese images with birds, especially night birds. As noted, 'weu' and 'ngiu', included in the names of certain Ngadha and Manggarai images, denote a kind of owl. Ventrally deficient female figures from East Flores and Sikka can similarly assume the form and cries of birds and can fly (Arndt 1932: 150; 1951: 32), while flight and alighting in trees are attributed to the back-hole spirits Nage call 'logo lia' (Forth 1998a: 87–88). According to Peninsular Malays, both the pontianak and closely related 'langsuir' spirits take the form of owls (Skeat 1900). Among Malays of eastern and central Sumatra, also, pontianak— here as elsewhere distinguished by a hole in the back—can change into large birds (Moszkowski 1909: 995).[33]

Consistent with their essentially spiritual nature, such malevolent and mostly female beings are often able to change shape and to disappear. Especially in regard to these attributes, the Manggarai poti wolo and ine weu (or empo weu) can be included in this class. (Characterized by Verheijen as a 'pontianak', the ine weu, it will be recalled, is further credited with inverted feet.) So for the same reason can certain images the Ngadha identify, regularly if not consistently, with the names 'kedho' and 'ebu ngiu' ('ibu ngiu', 'ine ngiu'). Other instances include: Lionese categories named 'ine léke' ('mother léke'; 'léke', 'large vine, liana'), 'susu deba' ('susu' is 'breast') and 'longgo mbenga' ('back hole'); the Sikkanese 'duä toé robong' ('woman with a hole in the back'); and the 'kurung sanak' of East Flores (Arndt 1932, 1951). To these should be added the largely female malevolent beings known in Lio, Sikka, Ngadha and So'a as 'sége re'e' (an alternative name for the Sikkanese 'duä toé robong') or 'fége re'e' (an alternative name for the Ngadha 'ibu ngiu'). The malevolent figure So'anese call 'sége re'e' or 'fége re'e', for example, is described not only as a sort of 'pontianak', but also a 'female witch' (Mommersteeg and Dirkzwager 1999: 133–42).[34]

It is worth emphasizing how the several Lio categories listed above are all named and conceived quite separately from the small hairy lae ho'a. Back-hole spirits and other malevolent female images, however, are recognized throughout Flores (including Nage, where they are called 'logo lia'; Forth 1998a: 87–88). And if they appear more prevalent in the far eastern and far western parts of the island, this may partly be due to the absence in these regions of distinct representations of more naturalistically conceived hominoids. At the same time, it is equally clear that categories recognized in certain parts of central Flores and also in Manggarai, in varying degrees combine features of the two general kinds of images. One only has to think of the hairy but otherwise quite supernatural poti wolo and ine weu of Manggarai, the pendulous breasts and extraordinary swallowing capacity of the Nage wildmen, or the fear of combs attributed both to these and the far more fantastically conceived ibu ngiu of Ngadha. Similarly, although the features were mentioned infrequently and only by a minority of informants, the Lionese lae ho'a are sometimes credited with bird-like vocalizations, metamorphosis and other mysterious powers—all qualities more closely associated with 'pontianak'. Finally, while the subjects of Florenese tales of capture are most often small hairy hominoids like lae ho'a and ana ula (a generalization further demonstrated in Chapter 4), not only does the wae belatung of East Flores—an instance, apparently, of a less malevolent variety of long-breasted female spirit—also figure in such stories, but so does the birdlike, but equally long-breasted, Sikkanese pontianak named 'duä toé robong'. A story recorded by Arndt (1932: 149) recounts the capture of one such creature in a net, which occurs when it enters a cave with an abducted infant stuck in the hole in its back.[35] After it is netted, the duä toé robong offers 'medicine' to its captor, thus further recalling the East Florenese wae belatung and also the Lionese lae ho'a. Conversely, small hairy hominoids are, of course, often represented as abductors just as are Florenese pontianak.

How, then, is this overlap to be explained? In the previous chapter I showed how, as a reference to the Nage hominoids exterminated in the cave at Lia Ula, the name 'ebu gogo' could have derived from a separately conceived bogey figure. This implicitly imaginary figure is not explicitly a malevolent female spirit; nevertheless, like such spirits from other parts of Flores, it is imagined as a child-threatening monster. The Nage bogey is moreover embodied in a mask. And, throughout Flores, it appears to be only bogey figures and not small hairy hominoids like ana ula and lae ho'a, which are identified with such masks. Senses of Sikkanese 'séngge re'e' (a variant of 'sége re'e') thus include 'images that frighten children' and 'making an ugly face in order to scare small children' (Pareira and Lewis 1998: 170, 182), while 'séngge re'e ala ilok' ('sége re'e with a polished head'; ibid.: 182) denotes a 'wood carving which frightens small children'. In Manggarai, the bogey figure and masks which represent it are both called 'empo gorak' (or simply 'gorak'; Verheijen 1967: 125), an image, quite distinct from either short, hairy hominoids or malevolent female beings, which is more comparable to the legendary but otherwise fully human head-stealers known in Nage as 'mo gele' (Forth 1991; Erb 1991).

Especially revealing in this context is the category 'gogo benga' or 'logo benga' found in the western part of the Keo region, immediately south of Nage territory. Small hairy hominoids like the wildmen of Lia Ula are not recognized in this region—an area of relatively high population density, like many parts of Manggarai. The Keo terms denote a monstrous child-stealing female with a hole ('benga') in her back, in which she inserts abducted infants. The image is further realized in a mask, also called 'gogo meo', or just 'gogo', which like the identically named Nage devices, Keo adults employ to threaten recalcitrant children. As well as the mask, the equivalence of 'gogo' and 'logo' ('anatomical back') implicit in the names of the Keo image point to connections with both the Nage bogey and the Nage back-hole spirits called 'logo lia' (a virtual synonym of Keo 'logo benga'; see also Lio 'longgo mbenga'). In addition, a link with both the back-hole spirits and the Nage wildmen (called 'ebu gogo', as often is the bogey) is suggested by the Keo idea that the gogo benga employ their ventral cavities to 'swallow' kidnapped infants, a notion obviously paralleling the voracious habits of the former denizens of Lia Ula. Excessive swallowing is also attributed to the So'anese 'pontianak' called 'sége re'e' or 'fége re'e', a female figure whose swallowing of eating and drinking vessels, pots, rice mortars and pestles is even more reminiscent of the ebu gogo. Described in a So'anese narrative, the objects all emerge from her belly after she is stabbed with a bamboo blade, an object recalling the bamboo combs of which the Nage wildmen were reputedly averse.

Although the Nage bogey is portrayed as seizing and swallowing children, how far this figure is related to the female 'back-holes' and similar malevolent beings found elsewhere on Flores is not entirely clear. Nevertheless, in this context, the idiosyncratic remark of one Nage man noted in Chapter 2, that ebu gogo had holes in their backs, arguably acquires a special significance—and one equal to the combination of bogey and back-hole in the Keo figure named 'gogo benga'. In regard to the Nage wildmen, the comparative evidence moreover points to a conflation, involving a transference to the otherwise more prosaically described hominoids, not just of the name 'ebu gogo' but of several fantastic attributes—notably, the excessive swallowing, pendulous breasts and fear of combs. If these traits were not original to the homonymous bogey, they have a more obvious derivation in the malevolent female 'pontianak', which in other parts of Flores are denominated and represented quite separately from small hairy wildmen.[36] The comb phobia, specifically, is comparable to the aversion, displayed both by Malay pontianak and their Florenese congeners (Skeat 1900: 325, 327; Arndt 1951: 32; Forth 1998a: 87–88), to sharp or pointed objects. Although these especially comprise items of metal (such as needles, nails and knives), Ngadha women travelling at night will carry a bamboo comb to ward off ibu ngiu (Arndt 1930: 827). And Nage women will take along a comb of buffalo horn to scare off 'back-hole' spirits. It is useful to recall that this fear, as well as excessive swallowing and long pendulous breasts, are *not* attributed to the ana ula and lae ho'a. In these respects, then, these other central Florenese hominoids appear more naturalistic than the ebu gogo wildmen of Nage. With regard to their smaller size and such specific features as the humanlike genitalia shared by lae ho'a and ana ula males, they may also be seen as corresponding more closely with the palaeoanthropological interpretation of *Homo floresiensis*.

Certain traits, notably the hairiness of certain more spiritually conceived malevolent female figures (for example, the Manggarai ine weu), could represent a transfer in the opposite direction. At the same time, pontianak are 'hairy' insofar as their ventral holes are described as covered by their long head hair (Skeat 1990: 326, Forth 1998a: 87)—a condition that could be reinterpreted as a more general hirsutism. Other than a conflation of the two general sorts of anthropomorphous beings, one might alternatively interpret all images discussed in this and the previous chapter, the more naturalistic as well as the more fabulous, as variants or derivatives of a single image, or proto-image. If this were a generally fantastic, basically spiritual and specifically female image—indeed, something like the pan-Indonesian pontianak—then some derivative images would necessarily have evolved in a far less spiritual, or more naturalistic, direction. And they would have done so even while continuing to exist alongside presumably more conservative, spiritual variants. On the other hand, one might consider certain derivatives—the more fantastic and usually glabrous figures—descending from a naturalistically conceived wildman proto-image and then coexisting with more prosaic descendants of the same. Either way, this sort of explanation is far less likely than one that assumes two originally (and perhaps ontologically) distinct images that have become partly conflated. As we shall see in Chapter 6, the hypothesis that the two kinds of image were once separate gains further support from nominal and conceptual distinctions between malevolent female spirits and hairy hominoids recognized by western Indonesians.

Yet another factor distinguishing small Florenese wildmen from partly comparable malevolent spirits—and indeed from human enemies, attested animals and other spiritual beings—is their exclusive association with legends of extermination in a cave effected by a holocaust usually fuelled by palm fibre. With the apparent exception of a Sri Lankan tradition I discuss in Chapter 7, this particular theme moreover appears exclusive to Flores and then specifically to the island's central region. Among the several variants, the Nage story is evidently the most detailed or elaborated, comprising an initial episode in which the voracious wildmen attend a feast, engage in circle dancing and drink to excess; the reference to their peculiar speech; and their gullible mistaking of palm fibre for clothing (linked with their oft-mentioned lack of intelligence). On the other hand, the changeling motif, a narrative device also associated with hominoid representations on the island of Sumba and in Mainland Asia, is confined to Lio and Rajong variants. And only in a single Lio tale does a hominoid female suckle the kidnapped child, a practice ascribed to wildmen of Central Asia. Nevertheless, infant or adult abduction—of which the changeling motif is of course a variant—does occur in the So'anese and Poma legends and in a qualified way even in the Nage tradition. Also, the hairy infant left by a hominoid mother in a field hut in Nage and So'anese stories suggests a further variant, if not a remnant, of the changeling idea.[37] For the most part, therefore, other central Florenese variants of the hominoid holocaust can be understood as simplifications or condensations of the Nage legend—much in the same way as can versions of the latter related by people living at some distance from 'Ua, the centre of the Nage tradition. Further suggesting condensation is the Rajong combination of the holocaust theme with the incident in the field hut, in contrast to their separation in Nage.

There are, of course, other differences between the several stories. The element of trickery, whereby victims are duped into participating in their own destruction, occurs only in the Nage story and in one version of the Poma legend of the ana ula. Moreover, perhaps only in the Nage variant is local extinction the unequivocal outcome. Yet there can be no question that the various tales reflect a single narrative tradition. Although in Southeast Asia the specific theme of the extermination of small hairy wildmen appears exclusive to central Flores, doubt is further cast on the story's status as an accurate depiction of historical events by elements that recur in tales told by other Malayo-Polynesian (or Austronesian) speakers, in which monkeys, spirits, or giants are deceived and subsequently burned to death.[38] In northern Sumatra, for example, an Angkola tale describes how a 'dwarf' tricks an anthropophagous 'forest spirit' into gathering palm fibre; after binding the spirit with rattan, the deceiver then sets the fibre alight, thus burning the spirit to death (DeVries 1928: 97–99). In some ways more similar to the Nage legend, a Ngaju tale from southern Borneo relates how a cultivator, abducted by a group of bothersome, crop-stealing monkeys, tricks the animals into releasing him (Klokke *et al.* 1988: 41–45). After inviting them to climb a rambutan tree and partake of the fruit and persuading the animals to stack wood around the trunk to facilitate their ascent, the man then ignites the wood, thus destroying all but one of the monkeys, a pregnant female.[39] Also relevant are stories from the Philippines, Oceania and Taiwan, regions culturally and linguistically related to Indonesia, which concern the extermination of 'dwarfs' or the killing of giants or other hairy hominoids, if not by fire then in other circumstances comparable to what is described in Florenese legends (see Chapter 9).

Yet however 'mythicized' Florenese stories of hominoid holocausts may be, it is still possible that they have some basis in actual events occurring somewhere on the island and that the story has spread from this location, perhaps becoming attached to creatures with an equally empirical basis elsewhere. That the location could have been Nage country and specifically the cave named Lia Ula, finds support in the more elaborate character of the Nage story, as well as its apparent historical groundedness, its connection with local genealogies and migration stories outlined in Chapter 2. As noted earlier, the So'anese story of the toro gogo, though also bound up with local genealogy, could well have originated in Nage. So too could Lionese variants. Nineteenth-century Nage leaders regularly employed Lio mercenaries in local wars and returning Lionese could then have spread the story to neighbouring villages of Nita-speakers. The occurrence of essentially the same tale in Rajong and the appearance of the additional theme of the changeling both there and in Lio, are less easily explained. Also, by way of the sort of contact just described, the Nage variant could equally have been derived from Lio. This, however, would leave the question of why the story, once introduced to Nage, should have become associated specifically with the 'Ua folk—a group marginal to the centres of political power ever since their arrival in the Nage region—or the cave named Lia Ula.

Clearly, a good deal of comparative evidence suggests that Florenese stories of wildman holocausts are largely fantastical and at best loosely connected with actual historical events. Whatever their basis, it is equally possible that the Nage variant is not original to Nage. But even if the stories are quite fictional, one cannot infer that the creatures to which they refer are any more imaginary than are, for example, the monkeys in the Bornean narrative. In other words, it is not impossible that the tales serve as a dramatic way of accounting for the disappearance, from one or more parts of Flores, of creatures that died out or moved elsewhere, possibly because of less dramatic human action or less direct competition from humans leading to more gradual population decline. To understand further the nature of these images and particularly their ontological status, we need to expand our comparative scope. This task is begun in the next chapter, which considers comparable figures from others islands of eastern Indonesia.

Plate 3.1 The ana ula cave in Poma.
Source: G. Forth.

Plate 3.2 Stoneworks in Poma reputedly built by 'ana ula'.
Source: G. Forth.

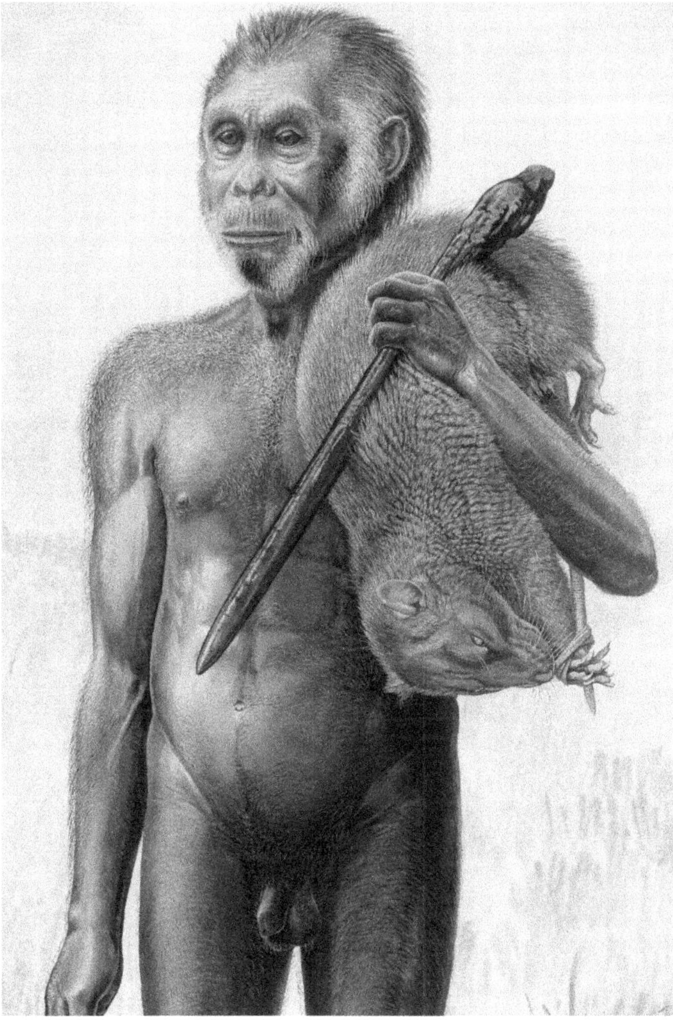

Plate 3.3 Peter Schouten's reconstruction of *Homo floresiensis*.
Source: With permission of the University of Wollongong.

Plate 3.4 Lio man and boy reckoned to be about the same height as a lae ho'a.
Source: G. Forth.

Plate 3.5 Pendant and knife hilt depicting a lae ho'a and a bottle reputedly containing the creature's urine.
Source: G. Forth.

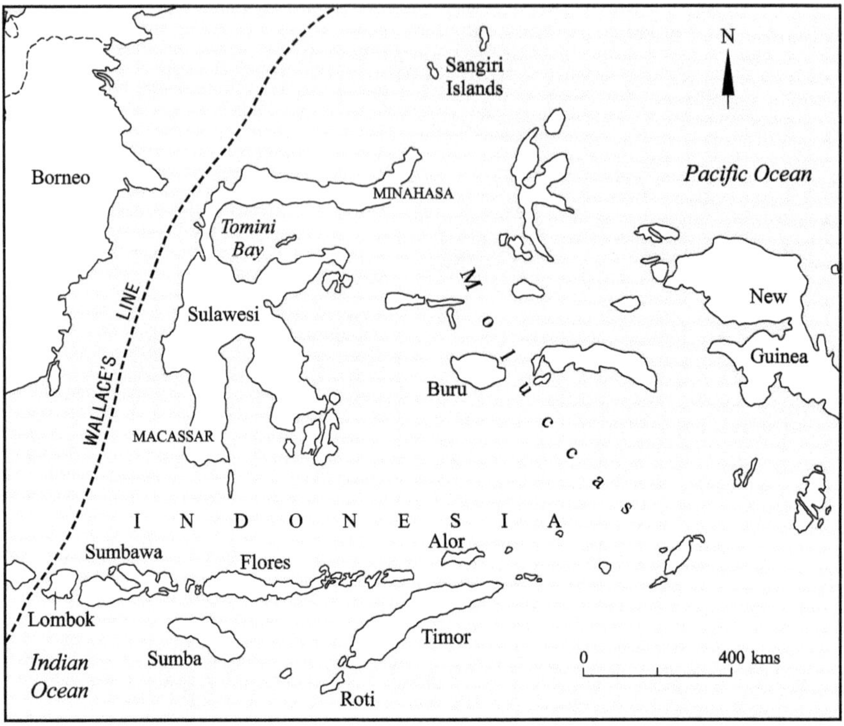

Map 4. Eastern Indonesia.

4 Other eastern islands

In this chapter I review hominoid images from the islands of Sumba, Sumbawa, Timor and the Moluccas and Sulawesi (see Map 4). The focus, I should stress, is on accounts of small, hirsute and group-living creatures whose descriptions largely correspond to Florenese categories described in the previous chapter. By the same token, I am not concerned with mysterious, usually glabrous and typically more fantastically conceived beings assuming a human guise and possessing supernatural powers, which appear to be represented everywhere in the archipelago. In recent years, such diminutive 'little people' may actually have become more common in the eastern islands, owing to the spread of an originally Javanese but now pan-Indonesian image named 'kurcaci', a sort of elf or brownie sometimes observed inside towns and villages and only partially corresponding to indigenous spirit images (such as the 'nitu' of Flores). To judge from local reports, while these kurcaci have proliferated in recent times, hairy hominoids, by contrast, have become decidedly scarcer.

Sumba and the 'mili mongga'

Particularly in eastern regions of Sumba island, people tell of hairy hominoids largely comparable to the Florenese ebu gogo, except for one detail: the Sumbanese wildman is described as larger than a human being. The creature is known by a variety of names, including 'mili mongga', 'makatoba' ('what is mad, deranged', 'madman') and 'meu rumba' ('wild cat', a name also applied to feral specimens of the domestic cat). Although its meaning is less transparent, the name 'mili mongga' is widely known in eastern Sumba; hence it is the one I employ throughout.[1]

Sumba and especially the eastern region, is a dry and barren island; yet it retains areas of extensive forest, the depths of which inhabitants of this sparsely populated island rarely enter. Mili mongga are thought to reside in several of these large forests, particularly in caves and rock cavities (see 'tau omangu', 'forest people' and 'makatoba omangu', 'forest madmen', two further names for the creatures). Rarely if ever do mili mongga venture near the coast, although one report had them living, in the distant past, in caves near the estuary of the Kaliongga River in Mangili. According to various accounts I recorded during a two-year residence in the Rindi district in 1975–76 and again during a visit in 2005, mili mongga, although currently in decline, may still be found in large forests in the interior districts of Mahu, Karera and Tabundungu (see Map 3). At the same time, the wildmen occur regularly in folktales and myth, where they are subject to a considerably more fabulous representation than the more naturalistically conceived creatures of Sumbanese mundane discourse.

Separate from myths and other standard narratives are stories of encounters that reputedly occurred during the last one hundred years. The following summaries should give a flavour of these.

> Around 1930, a man of Lambanapu, in the Kambera region, was burning charcoal in the late afternoon when he was approached by a female mili mongga wanting to warm herself by his fire. Addressing the man as 'grandchild' ('umbuku'), the creature said she was cold. The man naturally fled in terror.
>
> Sometime about mid-century, another Kambera man was out walking after dark when he saw what appeared to be a naked man, about two metres tall and covered in red hair, emerging from a forest. Fearful, he picked up a large stone and threw it, striking and killing the creature. As everyone was afraid to approach it, the corpse was left to rot at the forest edge.
>
> A Rindi man I knew in the 1970s claimed that, as a child, he was befriended and assisted in his chores by a female mili mongga involved in a clandestine relationship with his maternal uncle (Forth 1981: 112). The creature, which addressed him as 'grandchild', wore what appeared to be a shiny skirt.
>
> In 1999, a group of Rindi men were camping overnight in a forest, where they were cutting wood. They slaughtered a calf to provide themselves with meals. In the night, the men heard the sound of something gnawing on large bones they had discarded—in order, they assumed, to extract the marrow. Although nothing was seen and evidently no one was bold enough to get up to investigate, in the morning they found what they took to be mili mongga footprints.
>
> Around 2002, a group of men were hunting in Maïdangu (a region north of Tabundungu). Their dogs caught wind of three mili mongga, a male, female and a child. The trio fled towards a cave in a steep cliff face. They entered the cave and were thus able to escape. Much smaller than the other two, the 'child' wept at the approach of the dogs.

These stories are clearly quite various. By the same token, some appear less credible than others—to Sumbanese as well as (presumably) to westerners.

In 2005, long after the informant's decease, the story of the child befriended by a female mili mongga was dismissed by his relatives as fantasy inspired by well-known mythical narratives. Indeed, references to clothing and sexual involvement with humans are atypical of stories of reputedly historical individual encounters with mili mongga. The Kambera tale in which the charcoal burner was addressed as 'grandchild' is similarly indicative of the mythical genre. Nevertheless, other details suggest a certain naturalism worthy of closer attention.

Although mili mongga are commonly described as very tall, a number of people questioned specifically about stature said they were no taller than local humans and one claimed they could even be shorter. In 2005, I recorded over a dozen detailed accounts of the creatures' physical appearance. With surprising consistency, six educated informants—all government officials or school teachers residing in the main coastal town of Waingapu—characterized mili mongga as extremely tall, with four suggesting heights in the range of three to four metres. In contrast, rural villagers gave rather lower estimates, the highest being 'two metres or more'. In the interior district of Mahu, one of the most forested regions of eastern Sumba and one most closely associated with the hominoids, residents, who included people claiming to have seen mili mongga, stated that the creatures (there usually called 'meu rumba') did not exceed humans in height. At the same time, Mahu folk further described them as much broader, in fact 'twice as wide' as a human being. What is most remarkable about this contrast is that the apparently more naturalistic image was articulated by villagers who mostly related their descriptions to reputed experiences (if not their own, then of kinsmen or neighbours). On the other hand, the fantastic representation of a huge creature comes from educated town-dwelling Sumbanese who apparently and in some instances explicitly, drew on a representation of mili mongga contained in myth. As two men also remarked, the idea that mili mongga are extraordinarily large may stem from their use as bogeys (about which more below).[2]

While representations of mili mongga size suggest a bimodal pattern comparable to what one finds in other parts of Asia (see Chapters 6 and 7), the creatures nevertheless appear larger than hominoids reported from Flores. In other physical respects, the Sumbanese wildmen are not obviously different, or at least are well within the range of variation encountered on Flores. In fact, in everything but height, they are strikingly similar to the Nage ebu gogo.

By contrast to most Indonesian wildman figures, mili mongga are represented graphically. Visual depictions include textile motifs and a single stone statue found in the Rindi district (Forth 1981). One textile motif (see Plate 4.1) reveals a triangular shape on the figure's forehead and, above the eyes, a row of angled lines. I was unable to discover what the triangle might signify, but a woman thought the lines might be eyebrows or brow ridges. A man, on the other hand, said they depicted the dishevelled head hair characteristic of mili mongga. If this is correct, then either the hair hangs over or grows from the forehead, or the forehead recedes considerably, as it does in the statue depicting the creature.

A less stylized image of a mili mongga, the statue is a stone carving dating from around 1900 (thus before the Dutch took full colonial control of Sumba) and forms part of a large tombstone found in the main Rindi village of Parai Yawangu (Forth 1981, Plate 4a; see Plate 4.2). Standing just under 70 centimetres, but presumably intended to represent a larger creature, the stone wildman is kneeling and hunched forward, suggesting a definite stoop. Other features include an absent neck, a very low forehead, thick brow ridges, a large jaw, sunken eyes and large ears. Small round protuberances either side of the chest suggest breasts or nipples.[3]

Verbal accounts usually do not describe mili mongga as imperfectly erect; in fact, the creatures are reputedly capable of travelling bipedally at great speed. On the other hand, except for the humanlike chin and mouth, the accounts mostly affirm the facial features shown on the statue. The statue does not show evidence of body or head hair, nor of large teeth, another feature typical of informant descriptions. But the absent hair, at least, may reflect no more than artistic limitation—as may the chin, arguably the stone figure's most human feature. Lest it be thought that Sumbanese generally associate mili mongga with the dead—a connection that might also be inferred from a story recorded by Wielenga (1913: 214) where a being named 'Meo Rumba' manifests someone's deceased grandparent—the mortuary statue is actually a visual pun. The first occupant of the tomb was a Rindi nobleman considered mentally unstable who therefore bore the nickname 'makatoba' ('madman'), a common Rindi name for the creature otherwise known as mili mongga. The front of the grave displays a large and amorphous animal head, identified as that of a 'wild cat' (meu rumba), another term for the wildman. Hence the entire tomb is named Reti Meu Rumba, or 'wild cat grave'.

Sumbanese generally describe mili mongga as possessing a hominoid, even human form. Some accounts suggest a more ape-like appearance; thus one man said mili mongga faces were exactly like monkeys', while another identified the creatures with an orang-utan he had seen in a Javanese zoo (Forth 1981: 446). Yet another informant described wide and prominent cheek bones. A village woman said the feet were monkey-like, while a man referred to long simian toes. Nevertheless, most people described the creature as having a face and form rather more human than simian and accounts often gave the impression of a large hairy man. Accordingly, Sumbanese regularly apply the Indonesian term 'manusia' ('human') to mili mongga and they enumerate the creatures with the local classifier otherwise reserved for human beings ('tau'). One man even identified them, speculatively, as the first 'people' to inhabit the world. Another suggested mili mongga may derive from human beings who had gone 'wild' ('matàmba'). Conversely, a myth recorded in Rindi concerns a male mili mongga who, after overcoming his fear of fire, is transformed into a 'normal' human being (Forth 1981: 112–13).

Although the last story recalls human ancestors in Manggarai mythology losing their hair after coming into contact with fire, it is not obvious that Sumbanese regard mili mongga as essentially human. Nor do they subscribe to a belief in transformation as a general capacity of the wildmen. More notable points of comparison between Sumbanese and Florenese hominoids concern physical attributes. Probably the most significant is the attribution to female mili mongga of pendulous breasts, so long they can be tossed over the shoulders. Mili mongga are also covered in hair, usually described as straight and long. One man specified the body hair as noticeable only on the chest, back, shoulders, lower arms and lower legs. The face, he said, is quite hairless, but another man spoke of a beard. The head hair is generally described as much longer; according to one account, the hair on the top of the head is thinner and shorter than the longer, unruly hair at the back and sides. Employing colour terms otherwise reserved for horses, Sumbanese mostly classify the head and body hair as 'red' or 'reddish brown' ('mbau' or 'katungu'). Other specifications included 'black' or 'dark' (mentioned in two cases), 'monkey-coloured' (another two cases) and 'off-white' or 'cream-coloured' ('kanuhu'). A large male reputedly observed with a red-haired female and infant possessed a light greyish coat. I obtained no definite information on skin colour.

Suggestive of an erect and bipedal creature, the limbs and trunk of mili mongga are described as humanlike. The creatures are, however, much broader and more robust and are accordingly characterized as extremely strong. Sometimes compared to a horse's, the teeth are large and are often specified as white, thus contrasting to the blackened teeth of traditional betel- and areca-chewing Sumbanese. Several stories describe ancestors encountering and killing mili mongga and retaining their teeth (incisors, molars, or canines). Other material remains include large bones, reputedly buried in Oka Watu, an ancient village site in the Mangili district. Connected with this burial is the belief that mili mongga assisted in the construction of large stone walls that surround the village, as they are also said to have done in Parai Karopangu, a now abandoned settlement in the Rindi district. As shown, similar ideas form part of the representation of the much smaller Flores hominoids named ana ula (Chapter 3). According to a Mahu tradition, another mili mongga grave is located near a settlement, which is itself called 'Meu Rumba' (an alternative name for mili mongga) in commemoration of assistance the ancestors received from one such creature during an enemy attack.

Sumbanese describe mili mongga as capable of speech and as speaking the local language, albeit imperfectly. More particularly, they insert 'ha' between words, in a way somewhat comparable to the insertion of 'mmm' in statements attributed to the Nage ebu gogo. The creatures are also said to repeat words and stutter. According to another idea, mili mongga speech was only heard by, or was only intelligible to, people of former generations. Especially in standard narratives, mili mongga address humans as 'grandchild' and are conversely addressed as 'grandmother' or 'grandfather'.[4] Apart from speech, mili mongga produce a roaring or bellowing sound, variously likened to the lowing of cattle or the roar of a tiger.[5]

Another auditory manifestation of mili mongga is the sound of striking or rapping on wood heard deep in the forest, or of dry branches breaking underfoot. Recalling the Nage ebu gogo and the Manggarai poti wolo, an additional, but equally alarming, indication of their presence is an odour like a goat. But if people are scared of mili mongga, the wildmen are equally wary of humans. Although in myths mili mongga enter settlements and freely associate with people, they are otherwise described as fleeing at the slightest sign of human presence. And like virtually all hominoids sketched so far, mili mongga display a pronounced fear of dogs, animals ubiquitous in Sumbanese villages. Connected with the infrequency of mili mongga sightings is a notion that the creatures are bound by an ancient oath to keep to desolate places and not to show themselves to humans—an oath that seems to have followed a breach of formerly more positive relations with the wildmen (Forth 1981: 111). Should a person encounter a mili mongga, therefore, it may legitimately be killed, as several Sumbanese ancestors are reputed to have done. At the same time, there is the idea that the creatures were once more common than at present—their current rarity and possible extinction, sometimes being ascribed to a recent depletion of forests and the building of motor roads.

Informant accounts suggest that abducting humans is not a pronounced attribute of mili mongga. The theme does occur in myth where (as in stories Nage tell about the ebu gogo) victims usually manage to escape their hirsute, dull-witted abductors. Traditional narratives also describe mili mongga attempting to abduct adults of the opposite sex, to take them as marriage partners—another endeavour in which they typically fail. In addition to episodes revealed in a genre of 'orphan myths' ('ana lalu'), abduction of children is implicit in reputedly historical accounts of changelings left in place of human infants. A comparable mythical theme has mili mongga abducting adults (in one instance a bride) and attempting, ultimately unsuccessfully, to take their place within a community (Wielenga 1913). Further linked with notional child abduction, mili mongga, like some Florenese hominoids, are employed as bogeys. Adults tell disobedient children that a mili mongga is coming to eat them and to add drama to the threat, they might also hide at the back of the house and make a roaring noise. Nevertheless, as with abduction, most people I questioned either denied man-eating as a habit of mili mongga or claimed never to have heard of it. Another narrative theme I never encountered in everyday discourse about mili mongga has a creature (always a single individual) entering a human settlement and partaking of a meal, but consuming not only food but also containers in which it is cooked and served (Forth 1981: 446; cf. Wielenga 1910: 311; 1913: 264; Onvlee 1925: 39, 41, 45). This of course is a hallmark of two Florenese figures: the Nage ebu gogo and the witchlike female being So'anese call 'sége re'e' (see Chapter 3).

If anthropophagy and the swallowing of dishes and pots are fantastic or exaggerated attributes of mili mongga, they nevertheless coexist with more naturalistic ideas about their diet. Mili mongga are mostly described as subsisting on wild plant foods (crop stealing, interestingly enough, seems not to figure in the representation). They are also said to consume slugs and snails. According to a Mangili man, the creatures are therefore easiest to observe during light rain or drizzle, when they come out to collect large quantities of the gastropods. In part, the idea parallels Florenese accounts, according to which the ana ula of Poma and the Lionese lae ho'a are similarly seen most often during light rain or cloudy weather. A further report, that mili mongga feed on 'rotting wood', might be construed as an indication of the consumption of invertebrates such as grubs and termites.

As should by now be clear, in mundane discourse mili mongga are represented in a largely prosaic way, quite different from their portrayal in standard mythical genres. By the same token, the greater role they play in myth and folktale accounts for their generally more fantastic character in comparison to many of the Florenese hominoids described in the preceding chapter. Like the Manggarai figure of Empo Paju, mili mongga are regarded as ancestors by a few Sumbanese clans (Forth 1981: 446). But unlike Sumbanese ancestral deities ('marapu'), for example, they are not classified as spiritual beings. Rather, as one man spontaneously expressed it, mili mongga 'belong to nature' (Indonesian 'alam punya').[6]

Nevertheless, some attributes of mili mongga indicative of a fantastic representation do not obviously pertain to standard narrative contexts. Referring to the construction of traditional houses, an elder in the interior district of Tabundungu once told me that a mili mongga will always come in the still of the night to inspect the building; the creature is never seen, but its footprints may be found the following morning. If a mistake has been made in the construction, the nocturnal visitor will displace the erroneous part and leave. At such times, village dogs should be tied up to allow what is evidently considered a necessary inspection. This idea plainly contradicts the general representation of mili mongga as cultureless cave-dwellers lacking technology, who certainly do not construct houses of their own. In a similar vein, a coastal Sumbanese stated that mili mongga first taught humans to fish with nets. How general these beliefs might be I cannot say. According to evidently more widespread ideas, only some individuals can see mili mongga and then only while alone, as the creatures 'disappear' when other people came along. This claim, however, was contradicted in the case of the family group in Maïdangu, which was reputedly observed by a number of individuals. According to another common belief, not everyone is spiritually suited to observe mili mongga. Thus, if someone should see one, he or she should place salt on the tongue; if the salt can be tasted, the person will survive, but if it is tasteless, death will surely follow (Forth 1981: 112). Further suggesting mystical powers, mili mongga can provide individuals with 'medicine' or invulnerability. That these benefits can be secured by obtaining the creatures' teeth recalls Lio notions concerning similar benefits derived from body parts of the creatures they call lae ho'a. Yet ideas of this sort are attached the world over to a variety of attested animals (bezoar stones obtained from animal carcasses provide one obvious example), so, ontologically, their connection with putative hominoids cannot be definitive of the categories in question.[7]

From a comparative perspective, what is most remarkable about the Sumbanese mili mongga is their resemblance to the Nage ebu gogo. Features common to both representations include imperfect use of human language, long breasts, an animal-like odour and a habit of consuming eating and drinking vessels. Also noteworthy are indications of sub-human intelligence and a concomitant susceptibility to human deceit—and even an association of both with cats (although in the Nage case this pertains specifically to a bogey figure named 'gogo meo', largely represented separately from the partly homonymous Nage wildmen). Interestingly, these latter characteristics are less evident in other Florenese hominoid images. In just one major respect do mili mongga and ebu gogo differ: the larger size of the Sumbanese figure. Possibly connected with this difference is a circumstantial contrast. Whereas mili mongga occur regularly in Sumbanese myth and folktale, the only standard narratives about ebu gogo concern their clash with local humans and subsequent extermination at a particular point in Nage history. Speech and the swallowing of vessels are also mentioned in the Nage legend. Yet in the Sumbanese case, especially the last feature could, by way of elaboration within an evolving mythological tradition, very well have resulted in an expansion in the physical size of the mili mongga. As noted, this seems to have occurred in the interpretations of a minority of Nage, who described the pot-swallowing ebu gogo as 'big and tall'. The question may therefore be not so much why the Sumbanese creatures are represented as large, as why their Nage counterparts are not.

The putative swallowing of large objects is arguably the most striking similarity between the Nage and Sumbanese hominoids. Although otherwise quite different Florenese images suggest excessive consumption (in some instances, of human victims by way of a hole in the back), consuming vessels and other material objects is an attribute only of the mythical female creatures the So'a people of central Flores call 'sége re'e' (see Chapter 3).[8] What is more, nowhere outside of Sumba and Flores (and specifically central Flores) does this voracious habit appear to be attributed to wildmen. Especially in view of this identical attribution, how then might resemblances between the mili mongga and ebu gogo be explained? The simplest explanation is diffusion. For a long time, central Florenese people have had contact with Sumba, particularly in recent centuries by way of a slave trade prosecuted by slavers from the settlement of Ende on Flores' south central coast (see Maps 2 and 3). Besides exporting slaves from Sumba, Endenese, who by the nineteenth century had established settlements on Sumba's northern coast, sold Florenese slaves to Sumbanese rulers (Kapita 1976: 18; Forth 1981: 427–28, citing De Roo 1890: 582). Among these were slaves from

Manggarai, in western Flores, the region of the long-breasted ine weu or empo weu. Even more pertinent is the employment on Sumba of central Flores mercenaries, designated in Sumbanese as 'Tau Nggundungu' ('mountain people', from Malay 'gunung', 'mountain') or 'Kawaü Nggundungu'. Kawaü is the Sumbanese name for Ende and by extension the whole of Flores, while according to the Sumbanese historian and linguist Oemboe Hina Kapita, 'nggundungu' refers specifically to the regions of Nage and Keo (1982: 207).

It is quite likely, therefore, that in a previous century images of hairy hominoids were brought to Sumba by central and western Florenese and perhaps even by Nage mercenaries carrying tales of the ebu gogo. Consistent with the distance of its hypothetical source, in time as well as space, is the generally more fantastic character of the mili mongga and what may be called the Sumbanese figure's far greater mythologization. The latter process could equally account for the mili mongga's most distinctive feature, its much larger size. That is, in mythologizing their wildman to a greater extent than the Nage have theirs, Sumbanese may have considerably exaggerated its size and especially its height. This process may even continue to the present, as suggested by variation in estimates of size between interior villagers and often more educated coastal Sumbanese.

An objection to this interpretation might be that, if Florenese images of hairy hominoids had been introduced to Sumba in the nineteenth century, they would not likely have become so quickly incorporated into Sumbanese narrative traditions. However, it should be considered that such introduced hominoid images could have been grafted onto an existing mythical 'giant' that lacked simian features, a hairy body and long breasts. Relevant here is the fact that neither Wielenga nor Onvlee mentions long breasts; nor do these appear in any of the Sumbanese texts they record, which in fact depict the mili mongga (called 'meu rumba') only as large, strong, having big teeth and consuming pots and other large objects. In just one place, Wielenga (1910: 310) says they are 'hairy all over', but this is nowhere alluded to in the narratives.

The idea that the creatures have survived to the present or until very recent times also does not contradict the mili mongga's possibly extraneous origin. If the category had no local empirical source, then once people had come to accept mili mongga as entities existing in nature as well as in narrative, there would be no reason to doubt their continuing survival—except perhaps with reference to the radical environmental transformation that has characterized Sumba's development in the last few decades. Still, for imported hominoids to become a putative part of local fauna would seem to require an experiential catalyst. Recalling 'tau omangu' ('forest people') as a reference to the mili mongga, one possibility is humans whom Sumbanese describe as fleeing to forests during the frequent slave-raiding and incessant warfare of the nineteenth century. Although it relates to a being rather smaller than the mili mongga, a particular story may be illuminating.

The tale concerns not a category but an individual, a female who came to be known as Apu Kalita ('Old Woman Kalita'). Reputedly caught in a trap in the late nineteenth or early twentieth century, she died perhaps in the 1930s or 1940s.[9] Apu Kalita was a young woman at the time of her capture and when found was naked, mute and according to some accounts, covered in hair. Drawn from several accounts recorded in 2005, her story may be summarized as follows:

> In Lai Taku, in the interior region of Karera (other accounts mentioned Mahu), a cultivator whose garden bounded on forest found that something had been stealing his newly-planted maize. So he set a trap. One morning people found a young naked woman with her hand caught in the trap. Trying to release herself, she had bitten into her hand rather than gnawing at the snare (thus attesting to her stupidity, according to one commentator). Known thereafter as Apu Kalita, the woman was given garments, which she discarded and it was a long time before she would wear clothes. She also refused cooked food. Periodically, the woman would return to the forest to eat the white grubs ('kawatu') that infest rotting wood, or the wood itself; these were her natural foods. Sometime later, Apu Kalita was sold as a slave to the raja of Kambera, who then sent her with other slaves to work his lands in Kadumbulu (the eastern coastal region where her story is mostly known). In Kadumbulu, Apu Kalita became pregnant. She gave birth to a girl, but hid the baby in the forest, thus causing its death. Also begotten illegitimately, another daughter was later born to the strange woman. Named Harabi Loda, the child was quickly taken from her mother and raised by others. Harabi Loda grew up, was married and lived to an advanced age, dying childless in 2002 or 2003.

Five people, four of them from Kadumbulu (including three elderly men who claimed to remember the woman), were able to give details of Apu Kalita's appearance. All said she was short, but estimates of her height—apparently ranging from 1.3 to over 1.5 metres—were vague. Her daughter, Harabi Loda, was also short, perhaps shorter even than her mother. According to three accounts, Apu Kalita had a hairy body. Two specified 'long body hair'. The one man who mentioned hair colour described this as reddish ('mbau', the colour commonly attributed to mili mongga hair) or greyish-brown. He also said her head hair was the same colour as her body hair; but another man claimed her head hair was 'black'. Even people who described Apu Kalita as hirsute said she was considered 'attractive' ('manandangu'), owing largely, it seems, to a light complexion—a major criterion of female beauty for Sumbanese. Her head hair, while long, was shorter than the hair of local women; according to one report, it was wavy, in contrast to the straight hair of most eastern Sumbanese. Informants were divided over whether Apu Kalita ever learnt to speak. One man who claimed she did remarked how she had then told people that she had 'many companions in the forest', evidently referring to others of her kind.

What this 'kind' might have been is anything but clear. Possibly, Apu Kalita was nothing more than a deranged or mentally retarded girl and her body hair, if not an individual peculiarity, was simply an invention or exaggeration. Something of the sort is suggested by a discrepant account from an elderly Kadumbulu woman who claimed Apu Kalita had been born locally, became deranged as a result of sexual abuse and in this condition had later become caught in the snare. Nevertheless, the reference to reddish hair is reminiscent of the Sumbanese wildman, as is her initial nakedness, natural diet (particularly her feeding on invertebrates and 'rotting wood') and habit of withdrawing to the forest. The title 'Apu' ('grandmother, old woman') also recalls the mili mongga and particularly the figure of Apu Kami, the female mili mongga often featured in orphan myths (Forth 1981: 113). In regard to indications of Apu Kalita's mental abnormality, so too does the use of 'makatoba' ('madman, insane person') as an alternative name for the mili mongga.

However, no one I questioned identified Apu Kalita as a mili mongga, owing mostly, it would seem, to the large size of the creatures so named. It is also uncertain how her singular story could relate to general representations of small, hirsute and cultureless hominoids found in other parts of Indonesia. At the same time, the story of Apu Kalita bears comparison with a series of other capture tales reported from the region, as will presently become clear. In several respects, the legend also reveals an extraordinary resemblance to the story of 'Zana', a large hairy female captured in the Caucasus in the mid-nineteenth century (see Chapter 7). Obviously, then, relevant comparative contexts for this sort of image extend well beyond Southeast Asia.

Stories from Sumbawa

A large island located immediately to the west of Flores, Sumbawa is divided by a major linguistic boundary. Spoken in the western half is Sumbawan, a language belonging to the Western-Malayo-Polynesian group and therefore related to Balinese and Javanese. In contrast, inhabitants of Sumbawa's eastern region speak Bimanese and other languages classified with the Central-Malayo-Polynesian languages spoken in Sumba and Flores. The ethnography of Sumbawa includes virtually no information on indigenous representations of wildman figures from any part of the island; hence it is impossible to know how such images—if they exist as definite cultural constructs—might coincide with ethno-linguistic groupings. In recent years, however, Sumbawa has been the site of two non-indigenous reports of mysterious primates and in relation to hominoid images recognized elsewhere in eastern Indonesia, these are suggestive.

Referring to eastern Sumbawa, the first concerns an experience recorded by anthropologist Michael Hitchcock, during fieldwork in the Wawo highlands in 1982.[10] On a visit to Tarlawi village, Hitchcock was told of a large decorated stone located on top of a hill, in a thickly forested area about a kilometre away. One morning he set out, alone, to find the stone, but was unsuccessful. Ascending, he found himself on a path surrounded by low-hanging vegetation, so low he had to crouch and eventually crawl. In dim light (although this was mid-morning, the vegetation obscured his view), Hitchcock was confronted by a noisy group of what his field notes describe as 'angry monkeys'. Numbering at least three (others may well have been out of view), the creatures were evidently not going to allow him to pass. Hitchcock therefore slowly backed off and returned to the village.

Professor Hitchcock's experience might simply be attributed to monkeys, more specifically Long-tailed macaques—except for several details. These, he has confided, have always troubled him and by the same token have kept the incident fresh in his memory. For one thing, the creatures, which he observed from a distance of just a few metres, were all standing erect. Although in the dim light these were little more than silhouettes and although the creatures were facing him, he recalls being unable to see any tails. Also, while the others looked smaller, the tallest stood nearly a metre, the height of the overgrown trail. Even for an elderly male macaque, this is very tall indeed. Also noteworthy is how, according to Hitchcock, all of the creatures appeared to be simultaneously erect, although they were 'bobbing up and down' (pers. comm., December 2005); while macques will occasionally stand for a short time on two legs, it would be difficult to imagine some not maintaining or resuming a quadrupedal posture. Finally, the noise the creatures made, an angry shrieking, sounded lower-pitched than what is usual for a monkey (pers. comm., November 2004).

I should stress that Hitchcock makes no particular claim regarding what he observed; he has, however, allowed me to cite his opinion that the creatures did not look like macaques. There was, he says, 'something odd about them' and although he has 'seen monkeys in all manner of contexts . . . this experience was different'. They were 'definitely primates' but although his field notes describe them as monkeys, 'there was something unfamiliar'. At the same time, Hitchcock notes, quite sensibly, 'how difficult it is to identify primates in dim light and how easy it would be to make mistakes' (pers. comm., November 2004).[11]

The Wawo people (Dou Wawo) apparently have no category of small hairy hominoids corresponding to Hitchcock's experience of mystery primates. Also, carved stones such as the one he was looking for near Tarlawi are not linked with any category of non-humans, hirsute or otherwise. Interestingly, though, the majority Bimanese people regard the Wawo hills as the haunt of small anthropomorphous beings with very short tails. They furthermore appear not to distinguish these entirely from the Wawo folk, whom they consider culturally quite distinct, less than fully human and also as possessing small tails (although not, it seems, other simian or animalistic features; Hitchcock, pers. comm., October 2005). For the present enquiry, Hitchcock's experience of relatively large, bipedal and apparently tailless primates—or what could be taken as such—is of particular interest in relation, for example, to Lionese reports of encounters with creatures they call lae ho'a. If nothing else, it indicates that such experiences are available not only to eastern Indonesian villagers but to western anthropologists as well. This is not to argue that the western experience confirms their empirical reality, only that there may indeed be something in these islands that can generate such experiences. (What else might be required is a question I reserve for later chapters.)

Reported in *Kompas*, a high quality Jakarta newspaper on 16 September, 1996 (p. 3), the second story of apparently erect and bipedal primates derives from western Sumbawa, more particularly from a forested area in the district of Taliwang. The following is my translation of the Indonesian original.

> ***Wild animal that walks like a human will be investigated***
> The accuracy of a report regarding an encounter with an animal that walks like a human in a forested area in the Taliwang district of the Sumbawa Regency will soon be investigated by the Provincial Museum of West Nusa Tenggara.
>
> A resident in a forested area of Taliwang recently encountered four of the strange animals. The animals in question walk with an erect posture like a human being.
>
> The director of the Provincial Museum, Drs. V.J. Herman, explained how local people reported the encounter to his office. Citing a resident named Ratu Singaraja, Herman said the four animals included an adult male and female and their two infants. Their entire bodies were covered in hair.
>
> However, the person who came across the animals was able to capture only the two infants and the male and the mother animal disappeared.
>
> "Unfortunately, it was not long before the two young animals died, because they would not eat the food offered to them. They were given human food, cooked rice and side dishes," said Herman. The skeletons of the two young animals have thus far been kept by local residents.
>
> Asked whether the animals might possibly be a kind of gorilla or orang-utan, the museum director said he could not yet say. In order to confirm what species the strange animals belong to, experts will have to investigate.
>
> At the end of this month Herman plans to leave for the Taliwang district with Ratu Singaraja in order to look at the skeletons of the two infant animals, so that information regarding the find can later be passed on to the central museum in Jakarta, which it may be hoped will send a team of experts to the field.
>
> "If according to the findings of the team of experts, the skeletons of these animals are 'hominidae' or a kind of primate which is also within the biological family of Homo or human creatures, then this will be an extraordinary discovery," said Herman.

The only physical details mentioned in this report are hairy bodies and bipedal locomotion. Yet it is noteworthy how the writer consistently describes the creatures in a completely naturalistic, matter-of-fact way and refers to them only with Indonesian terms denoting wild animals ('satwa', 'binatang'). There is also the suggestion that they may be closely related to known apes, while the museum director seems to hold out the possibility that they are 'hominids'. As the only primate known on Sumbawa is the Long-tailed macaque, the monkey also found on Sumba and Flores, the discovery of an anthropoid ape would of course have been spectacular enough. However, like many reports of this kind, the story seemed simply to fade away. The only follow-up was a brief piece published on 20 September, 1996, also in *Kompas*, which simply stated that the provincial Office of Resource Management (Kantor Balai Konservasi Sumber Daya) was also investigating the report.[12]

That the story was basically a rumour is especially suggested by the fact that the local person who encountered the hairy creatures and captured the two infants is not named. In fact, all information about the creatures seems to derive from Ratu Singaraja, apparently a locally prominent man; but other than as a resident of Taliwang, his status is nowhere specified. Whatever its source, the story is comparable in all fundamentals to stories that in recent years have circulated in other parts of eastern Indonesia. Conceivably it could also reflect a representation similar to one found in the vicinity of the Tambora volcano, in north central Sumbawa (a Bimanese-speaking region). Climbing Tambora in 1941, the Swedish explorer Rolf Blomberg was warned by his Sumbawan guide of mean and ugly 'forest people', no larger than children, possibly hairy and living in trees. One of Blomberg's native bearers even claimed to have encountered these 'forest people', after getting lost in jungle for a night and a day and being found in a confused state by his companions (Blomberg 1948: 88).[13]

Timor and the Moluccas: a bogey from Buru

Images of short, strong and hairy hominoids like those encountered on Flores are not apparent in ethnographic accounts from Timor and the Moluccas.[14] Nor for that matter are larger forms corresponding to the Sumbanese mili mongga. A possible exception are the 'little people', no taller than a 'cubit' and residing in 'caves underground' on an island named 'Aruchete' (apparently near Alor), who are mentioned in the sixteenth century report of a Moluccan pilot, cited by Pigafetta. Other reputed features of these folk include ears as large as their bodies, shrill and thin voices and bodies 'shaven and quite naked'; they also run swiftly and eat fish and something 'which grows between the bark of trees' (Pigafetta 1975: 147). The curious reference to their shaven bodies implies a naturally hirsute condition. But the enormous ears and diminuitive size suggest something lacking any basis in local nature.

A similarly questionable case is the 'gibar bohot', a category reported from the Moluccan island of Buru. According to a late nineteenth-century native administrator, the term denoted 'very hairy' humanlike beings inhabiting caves in the Tomahoe mountains of western Buru. The creatures vocalized, but whether they could actually speak was uncertain. Occasionally, they stole produce from gardens, but none had ever been captured (Kopstein 1930: 46).

Related by a Burunese administrator, the native Regent of Fogi, to a European identified only as 'Dr. W.' (possibly the Dutch ethnologist G.A. Wilken), this description was eventually relayed to a member of a scientific expedition to the Moluccas identified by the zoologist Felix Kopstein only as 'Dr. D.' (ibid.: 45–47). By all indications, Dr. D. was the German geologist Karl Deninger who, according to Kopstein, began in 1906 to investigate the 'gibar bohot' in the convinction that this might be a living 'missing link', a surviving specimen of '*Pithecanthropus erectus*', the ape-man of Java discovered by Eugene Dubois in 1891. In mocking tones, Kopstein—whose sole source was a letter from an anonymous former missionary on Buru—goes on to relate how Deninger eventually realized that the native administrator 'had taken him for a fool'. 'Gibar bohot', it turned out, simply means 'bad person' and in this case referred specifically to fugitives from other islands who had taken refuge in forests and caves. They lived by stealing from fields and during these forays had sometimes abducted a woman or a child. In retaliation, the Buru people killed off the fugitives and in fact the anonymous missionary was able to direct Deninger to a spot where the remains of two of the victims were to be found. Examining these, the geologist concluded that they belonged to *Homo sapiens* and his search for an ape-man thus came to an abrupt end.[15]

What would account for the Fogi Regent's specification of 'gibar bohot' as hirsute and evidently primitive creatures is uncertain, although in view of the date it could conceivably have been influenced by news of Dubois' discovery on Java. However that may be, the interpretation of the term as referring instead to fully human fugitives is borne out by a study of the Masarete language of Buru published in 1897 (thus nine years before Deninger's expedition). Transcribed as 'gèba bòho', the phrase is there defined as 'miscreant, especially wrongdoers who from fear of consequences eschew the company of others and live exclusively in rock cavities and caves' (Hendriks 1897: 31, s.v. 'bòho', 'bad, mischievous, nasty'; 'gèba' means 'man, human'). Yet the explanation of 'gibar bohot' offered by Kopstein may be only marginally more credible than their identification as a kind of undiscovered hominin. For their depiction as desperadoes who abduct women and children recalls a probably imaginary category of kidnappers and head-stealers encountered throughout Southeast Asia who, particularly in the twentieth century, were reputed to take victims as 'construction sacrifices' in connection with modern, colonially initiated building projects (see Drake 1989; Forth 1991). At the same time, the fact that the human remains discovered by Deninger were headless (Kopstein 1930: 47) suggests that these objects of local retaliation may have been victims of an actual and more traditional, practice.[16]

Whatever else is to be made of this curious tale, it illustrates how a western scientist, a geologist apparently influenced by Dubois' virtually contemporary discovery, could interpret the gibar bohot (or one local representation thereof) as a primitive hominin comparable to the palaeoanthropological image labelled 'ape-man' (*Pithecanthropus*). Insofar as Deninger hypothetically identified the gibar bohot with *Pithecanthropus*, he was by the same token evidently open to the possibility that this was not solely a creature of the distant past—as Dubois insisted—but something yet surviving. In this respect, one might see a parallel with recent suggestions regarding the possible survival of *Homo floresiensis* and the competing interpretation of the hominin as a physically, rather than socially, pathological human being.

Sulawesi: historical reports and local legends

Located directly north of Flores, the island of Sulawesi (formerly 'Celebes') is the site of what may be the first published reference to a tailless, bipedal primate anywhere in eastern Indonesia. In his seventeenth-century account of the kingdom of Macassar, in the island's southwestern peninsula, the French priest Nicolas Gervaise described the region as plagued by many aggressive 'monkeys and baboons' (1701: 25). Some among these were without tails; and while some were quadrupedal, others walked 'upright like men' and never made 'use of any but their own hinder feet for that purpose' (ibid.). Some of the monkeys were also 'white' and these included specimens 'as big . . . as an English mastiff' (the large dog). Unfortunately, Gervaise's description does not make clear whether bipedalism correlates with taillessness, or either trait with a light colour or greater size, although the curious canine comparison might suggest quadrupedalism as an attribute of the larger variety. Since true apes are not known to occur east of Wallace's Line (see Map 4), a zoological basis for a tailless primate in Sulawesi could only be the partly misnamed Celebes black ape (*Macaca nigra*), a dark-haired monkey with a very small, almost invisible tail. The species, however, is found only in northeastern Sulawesi; moreover, consistent bipedalism would suggest not an ape or monkey but a hominin. Curiously, Dobson (1953–54: 3), apparently citing Gervaise, implicitly takes his report as an early European reference to the orang-utan.

Gervaise (1701: 25–26) goes on to describe the pestilential primates of Makassar as collectively raping and murdering lone human females. Insofar as this almost certainly reflects what locals told him rather than anything the author himself witnessed, I am reminded of a central Flores tale in which Long-tailed macaques have designs on a woman whose husband they think has died (see Chapter 10).

If not rape, then something approximating abduction of human females forms part of a hominoid image from northeastern Sulawesi. Known to Tontemboan-speakers of the Minahasa region, the creature is called 'lolok'—glossed as 'little forest man' or 'forest spirit'—and resembles a child-sized human with long, trailing head hair (Schwarz 1907: 114–15). Recalling the kedho of eastern Ngadha, lolok are able to disorient people, so they lose their way in the forest and fall under the loloks' power (ibid.; see also Schwartz 1908: 248, s.v. lolok, itjalolok, 'to become like a lolok, to lose one's way in the jungle, to disappear in the forest'). Schwarz mentions two cases—one concerning a girl who became lost for seven days and the other a widow who wandered in the woods for a month and recovered from a deranged state only after her eyes were rubbed with chillies. In a more positive vein, Minahasa stories describe individuals gaining magical powers from lolok, thereby becoming invulnerable or able to harm others, or capturing lolok and in return for their release obtaining continuous success in hunting or agriculture. Unlike the belief regarding disorientation, however, these ideas appear to be taken from particular myths recorded by Schwarz, so there is a question of how they might pertain to a more general representation. In one such narrative, a lolok wears a green headcloth resembling moss, and, by taking possession of this item, a man, whose palm wine the creature regularly steals, gains power over its owner. Tontemboan myths further depict lolok as inhabiting villages located inside rock crevices, or residing underwater in springs, a detail especially reminiscent of Florenese spirits called 'nitu'.

Drawing on these ideas, Schwarz portrays lolok as a variety of spiritual being, remarking particularly on their similarity to 'spirits of the dead' in regard to their consumption, mentioned in one Tontemboan narrative, of only the steam of cooked rice (ibid.: 115). By contrast, in an unpublished report dealing mainly with the Sumatran 'orang pendek' (see Chapter 5), the colonial administrator Coomans de Ruiter (1929) provides a more prosaic account, describing lolok as 'long-haired', bipedal hominoids about 1.5 metres tall and usually occurring in male–female pairs. The creatures, he adds, leave small, humanlike tracks; are very shy and flee at the approach of humans; and feed on a kind of moss—a detail recalling the moss-like headcloth worn by the captured lolok in one of Schwarz's narratives.[17] Unfortunately, Coomans fails to clarify whether the lolok's long hair grows on the body or only on the head (as Schwarz seems to specify).

While Schwarz speaks of people disappearing in the forest as a consequence of disorientation caused by lolok (or 'possession' by lolok, as he interprets it), Coomans, in accordance with his naturalistic approach, describes a particular disappearance as a straightforward physical abduction. This concerned a four-year-old girl reputedly kidnapped by a pair of lolok from her home in Amurang, a town in South Minahasa. Named Martina Rau (or Rau van Wijk), the girl returned the next day apparently unharmed. When asked where she had been, she replied in Malay that she had gone off with her 'grandparents'. Years later, Coomans endeavoured to question an elderly Martina Rau about her experience, but she was too ill to receive him.

Coomans' brief report obviously leaves several questions unanswered. One concerns the ethnicity of Martina Rau: was she Dutch, a local person, or someone of mixed descent? We also do not know the evidence for her 'kidnapping'. She might simply have wandered off and later manufactured the story of abduction by lolok. Or since she was just four years old at the time, the interpretation could have been provided by others. Indeed, the Amurang child could be one and the same as the disoriented girl referred to by Schwarz, were it not for the description of the latter as suddenly wandering into the jungle while at work cutting 'woka leaves', a circumstance that suggests she was considerably older than the four-year-old Martina. Another intriguing detail concerns the child's reference to her abductors as 'grandparents'. This recalls Florenese naming of both wildmen and bogey figures with grandparental terms, as well as the use of such terms in addressing the Sumbanese mili mongga. As many of the Florenese and Sumbanese figures are similarly depicted as abducting children, these comparisons provide further support for Martina Rau's story being at least partly rooted in conventional imagery.

Recalling features of both the lolok representation and the aggressive primates described by Gervaise, another report from Sulawesi concerns an 'ape' that carried off a three-year-old child in Parepare, in southwestern Sulawesi about 100 kilometres north of Macassar (Tchernine 1971: 120). Although later recovered, the child thereafter behaved like an ape. The brief report gives no indication of the source, the date of the supposed occurrence, or the child's sex. Nevertheless, despite the 800 kilometres separating Parepare and Amurang, the site of Martina Rau's alleged abduction, the child's age suggests that the Parepare story, if not a reference to the same incident, belongs to the same genre. As noted, the only 'ape' on Sulawesi, the Celebes black ape, is actually a virtually tailless monkey. Ironically, the monkey does not occur in the vicinity of Parepare, but only in the extreme northeastern corner of Sulawesi, thus the region inhabited by the lolok, the abductors of Martina Rau.

Quite different from Gervaise's bipedal simians and the quasi-mystical lolok are named populations of seemingly more human beings mentioned in the traditions of present ethnic groups. These are recorded in the writings of several early Dutch ethnographers, some of whom construed the representations as possible reflections of aboriginal peoples. In northwestern Sulawesi, inhabitants of the mountainous Dirijo district describe a 'wild mountain tribe' called To Ipono ('To' is 'human, person'). Although still extant, these people are extremely shy, fleeing whenever a stranger approaches. Their bodies are entirely covered in hair and they wear only breech-clouts. No mention is made of their size (Adriani and Kruyt 1898: 488). Also in northern Sulawesi, the To Mini folk claim their ancestors once shared their territory with a population named To Uta (apparently 'people of the forest'). Residing in caves and subsisting on 'inner parts' of the 'seho' tree (probably the sago palm), these beings stood just one metre tall. According to To Mini, some To Uta, although now 'invisible', should still survive in forests and on Una-Una, a volcanic island in Tomini Bay (Riedel 1871: 301–02; see Map 4).[18]

In central Sulawesi, To Lage and other Bare'e-speaking groups describe a similarly small and possibly cave-dwelling people with whom their forbears once waged a protracted war. They were called To Kaneke, meaning 'very small people'. So tiny were these beings they could jump up on the sword sheaths of To Lage, strike them with their own swords and later hide themselves with ease, thus proving invincible. Evidently more aggressive and dangerous than most of the categories described so far, To Kaneke would also kill unsuspecting To Lage in the darkness of night. To Lage therefore determined to drive away the To Kaneke by placing derris root and excrement in their drinking water, thus causing them to flee (Kruyt, letter of 18 February 1903, cited in F. Sarasin 1906: 47–48). The story recalls attempts to exterminate or otherwise counter local populations of small hairy hominoids on Flores and as we shall later see, also Sherpa tales of killing yeti by poisoning beer (Chapter 7). Possessing weapons and also houses reputedly built on 'stone pillars', To Kaneke were evidently cultured beings. However, if the tradition has any empirical basis, certain details have obviously been elaborated; noting how the stone house pillars, still in evidence, actually comprise natural pillars of chalk-stone, Sarasin plausibly interprets this part of the image as a reference to residence in caves (ibid.: 47–48).

Perhaps the most significant detail of local understandings of the To Kaneke derives from Kruyt's further questioning of Bare'e-speakers. Referring to local spirits that assumed the form of dwarfs, Kruyt asked whether the To Kaneke were not spiritual beings like these. Informants consistently denied that the To Kaneke were spirits, claiming they were humans like themselves, only smaller (Sarasin 1906: 48, citing Kruyt's letter of 5th May 1903). Once again, therefore, we encounter a clear distinction in local thought between 'spirits' and putative populations of hominoids physically smaller or otherwise distinct from local people.

Also once inhabiting To Lage territory was another group of small humans, less well known than To Kaneke but of a similar size and evidently more friendly. Named To Ligowi, these folk would regularly attend To Lage feasts. On one occasion, however, misunderstanding over the appointed time caused the tiny people to arrive late. They therefore held their own feast at a spot on the Lage savannah, slaughtering the pigs they had intended to contribute to the To Lage celebration and performing their own ritual dance. Afterwards they left the region for good and what became of them is no longer known (ibid.). Former participation in feasts initiated by present-day peoples is reminiscent of the Nage legend of the ebu gogo and especially the wildmen's habit of attending feasts held by 'Ua villagers and participating in circle dancing. At the same time, other details of the two traditions are obviously different and comparison is further restricted by the brevity of these early ethnographic descriptions. In regard to physical appearance, none moreover describes hominoids that are simultaneously small and hairy; only the To Iponi of Dirijo are specified as hirsute and they are the only group not explicitly characterized as small or short.

Citing the same early Dutch accounts, the Swiss anthropologists F. and P. Sarasin interpreted local traditions of small cave-dwelling folk from northern and central Sulawesi as possible traces of To Ala (Toala)—or more correctly To Ale' ('forest people')—a small, formerly cave-dwelling group of food-collectors resident in the mountainous Lomontjong region of southwestern Sulawesi who may have maintained a stone technology within the last one thousand years (F. Sarasin 1906; see also Sarasin and Sarasin 1903). Insofar as evidence supports To Ala existence as a separately developed ethnic and economic entity surviving until relatively recent times—an interpretation contrary to a current anthropological orthodoxy inclined to view such groups as recent derivatives of dominant agricultural populations (see e.g. Pelras 1996)—the Sarasins' hypothesis would seem perfectly reasonable. How far groups like To Ala might be a source of representations like the lolok, or indeed the tailless, bipedal monkeys described by Gervaise, is, however, quite another matter. Nor is there very much linking the latter with unfortunately brief ethnographic reports of tiny humans or hominoids like the To Uta, To Kaneke and To Ligowi.

Tales of capture, some provisional conclusions and a Flores retrospect

Apart from general physical similarities, material reviewed in this chapter reveals another connection between hominoid images encountered on Flores and elsewhere in eastern Indonesia. In all regions surveyed so far, one hears stories wildmen either abducting people or being captured by humans and sometimes both. Putative historical accounts of hominoid capture most often concern accidental entrapment in a snare set for other animals. In instances from Flores, the captive has sometimes turned out to be a monkey. Yet the occurrence of Long-tailed macaques throughout the region (and a virtually tailless macaque in one part of Sulawesi) is hardly sufficient to explain stories of hominoid capture and in the absence of apes and bears, they lack other obvious zoological referents. Of course, human beings accidentally or deliberately caught in traps could provide a basis for some such tales, especially if they were children. Also noteworthy is the fact that, like many parts of Southeast Asia, all of the regions where the stories occur have a recent history of people regularly being captured by local enemies or slavers (recall that the accidentally captured Apu Kalita was sold as a slave), a practice in reference to which either captives or captors might, conceivably, be imagined as hairy wildmen.

To carry these comparisons further, we need to go back to Flores. As demonstrated in Chapter 3, during the twentieth century, captured hominoids have been reported from several parts of the island's central region. In late 2004, another story was circulating among Nage, concerning a recent incident which most versions located in the vicinity of Namu (see Map 2). A man was out hunting porcupines at night and, after spotting signs of wild pigs, had set a trap. However, instead of a pig, he snared a hairy infant hominoid.

Accounts I recorded included few details of the creature's physical appearance. Closest to the source was a woman resident near the Nage centre of Bo'a Wae, whose sister was married to the captor's elder brother. She described the captured creature as possessing 'the body of a monkey and the head of a human' with long head hair; but like other people relating the story, she could say nothing definite about its size. As in similar tales from Lio and Ngadha, the Namu hunter was alerted to something caught in the snare by the victim's weeping. Shortly after releasing the young creature, the man was approached by the mother animal, which spoke, saying that she and her infant would thereafter have to follow him. Despite his protests, they did so—as far as his village. Once many villagers had become aware of the hominoids' presence, it was decided that they should be taken to the Regency capital in Bajawa.

Although some accounts mentioned people seeing the creatures in Bajawa, specifically in a police cell, authorities I was able to question denied all knowledge of any such thing. And there, much like the 1996 report from western Sumbawa, the trail goes cold. As I had to leave Flores in July 2005, I was unable to follow the story to its reputed source. I was also unable to determine to what local category, if any, Namu people might have assigned the hominoids. Several versions of the story, however, described Namu villagers as being unfamiliar with creatures of this kind—as evidently were the Lionese who reputedly captured a lae ho'a in Kota Baru (Chapter 3) and the Sumbanese captors of Apu Kalita.

Revealing all the hallmarks of western 'urban legends' (Brunvand 2001), an obvious feature of this recent tale is the extent to which it replicates the similarly inconclusive story of the kedho caught in a pig trap in eastern Ngadha and also taken to Bajawa, sometime early in the twentieth century. Suggesting another local precedent is the story from Wangka, a settlement not far from Namu, which concerned the accidental capture and release of an 'ana ula' in the 1990s (Chapter 3). If these stories point to likely sources of the Namu tale, probably another is the Nage legend of ebu gogo. Namu, it will be recalled, is one place identified as the refuge of two wildmen who escaped the conflagration in the cave at Lia Ula. Others include Tana Wolo and So'a, two further locations which, interestingly enough, were mentioned as alternatives to Namu in the 2004 rumour. In fact, several Nage referred to the captive as an 'ebu gogo', or an 'infant ebu gogo'. But this need not imply belief in the continuing existence in these regions of the Nage wildman, if only because Nage who related the tale all appeared to doubt its veracity. Moreover, the creature was alternatively represented as a feral human child (perhaps not unlike the Sumbanese Apu Kalita) and, recalling one version of a Lio tale, a large male monkey.

Further suggesting influence from the legend of ebu gogo, two versions placed the 2004 incident within the Nage region itself. One man reported hearing that a female creature and infant had been captured that year fleeing from a brush fire somewhere in Nage 'Oga—the administrative unit which now includes the 'Ua people, whose ancestors are supposed to have exterminated the ebu gogo by firing their cave some two centuries ago. Referring to the same area, a man from Dhére Isa, a district to the northeast of Bo'a Wae, similarly related how people had captured an 'orang hutan' (Indonesian for 'wild man' and nowadays possibly a reference to the orang-utan) on the slopes of the volcano Ebu Lobo and had conveyed it to Bajawa, the Regency capital. A third version, situating the event near the main Nage village of Bo'a Wae, found its way into an Australian newspaper (*Sydney Morning Herald*, 6 December, 2004). Drawing on a story related to an Australian visitor by a Nage elder in Bo'a Wae, the newspaper article not only identified the creature as an 'ebu gogo', but also linked the reputed capture to the discovery of *Homo floresiensis*, reported in the media just a few weeks previously. According to the Australian report, the creature was an attractive female with long pendulous breasts. When I questioned him six months later, the Nage elder denied ever mentioning breasts. He had, however, relayed a description of the Namu captive (or perhaps its mother) as 'attractive', even though he himself claimed not to find this credible. In this detail, then, one glimpses yet another intriguing parallel with the Sumbanese story of Apu Kalita, the small hairy female captive, similarly described as attractive.

What can be concluded from all this? The appearance of the 2004 rumour was most likely catalyzed by the discovery of *Homo floresiensis* in the western Flores region of Manggarai and by members of the palaeonthropological team visiting Bo'a Wae in the same year and enquiring about possible links between the newly discovered hominin and the figure of ebu gogo (Forth 2005: 14; 2006: 345–46). Yet it is equally certain that stories of this sort have been circulating in eastern Indonesia for some considerable time and, judging by the legend of Apu Kalita, for at least one hundred years. Where such stories refer to established categories, they may be assumed to reflect and sustain pre-existing cultural representations. But in several cases (as in the Namu tale and the story of Apu Kalita), the subjects are represented as previously unknown; hence an equally plausible explanation may be some actual, albeit unusual, occurrence. It is also noteworthy that, even where a category (such as 'ebu gogo') is available, reputedly captured creatures are not invariably accepted as instances.

Florenese capture tales do not only concern hairy hominoids. We might recall, for example, the East Flores story of the glabrous, long-breasted and supernaturally powerful female creature named 'wae belatung' (Chapter 3). Several details of this story—including the refusal of human food, continual weeping and a diet of snails and worms—obviously recall seemingly historical tales and not least the Sumbanese story of Apu Kalita. Thus some common origin is by no means ruled out. Recent rumours of hominoid capture might further be viewed as part of a wider genre also encompassing older stories of killing or extermination. Extermination tales themselves include an element of capture inasmuch as victims are confined, usually in a cave, before they are killed. However, not only are hairy hominoids the most usual victims of capture, they also appear to be the sole victims of cave-burning. And, as noted previously, stories of hominoid extermination by fire appear peculiar to Flores, at least within Southeast Asia. These circumstances, then, emphasize both the distinctiveness of wildman images in relation to more fantastically conceived spiritual beings and of eastern Indonesian instances in relation to hominoids recognized elsewhere.

Although 'rumour' and 'legend' admit no ultimate distinction, the first rubric is certainly appropriate to the 2004 reports of hominoid capture, especially as Nage I spoke with were either sceptical or incredulous of their veracity (cf. Kapferer 1990: 11–12). The story also had a relatively short life: after emerging sometime in 2004, by the time I returned to Flores in May 2005, it was well on the wane. Nevertheless, it is perfectly possible that variants will resurface in the future and, by the same token, that similar stories from other parts of Flores, perhaps the 1996 report from Sumbawa and perhaps even the Sumbanese story of Apu Kalita, all represent earlier instances of a single 'diving rumour' (Allport and Postman 1947: 171)—one that periodically appears, disappears and reappears. Rumours, however, contain another possibility. Over time, they can transform into legends, just as legends may inform rumours (as has apparently occurred in the case of ebu gogo). Apart from the rumour recently circulating among Nage, Florenese legends regarding the extermination of hairy wildmen in all probability also illuminate the recent report of Lio hunters casting a captured hominoid into a cave and setting it on fire.

In spite of suggestive similarities with hirsute or diminutive beings reported from Sulawesi, Sumbawa and elsewhere, the most important finding of the present chapter is that, in eastern Indonesia, detailed, established and consistently named representations of hominoids comparable to ebu gogo and other Florenese exemplars are definitely attested only in Sumba. Moreover, comparative and historical evidence suggests that the Sumbanese image may well have derived from Flores. This is not to say that similarly detailed representations are unknown to inhabitants of other eastern Indonesian islands, only that their presence is not revealed in a generally extensive ethnographic literature. As I show in the next two chapters, physical images comparable to eastern Indonesian hominoids are encountered in more westerly parts of Indonesia and on the Southeast Asian mainland. Yet even the best known of these similarly leaves hardly a trace in ethnographic writing. This is the 'orang pendek' of Sumatra, a creature whose earliest description, coincidentally enough, includes reference to the capture of a male specimen and whose origins, as I show in Chapter 5, may partly lie in human forest dwellers previously taken as slaves.

Plate 4.1 The 'mili mongga' (wildman) motif on a Sumbanese textile.
Source: G. Forth.

Plate 4.2 The 'wild cat' grave (Reti Meu Rumba) in Rindi, eastern Sumba.
Source: G. Forth.

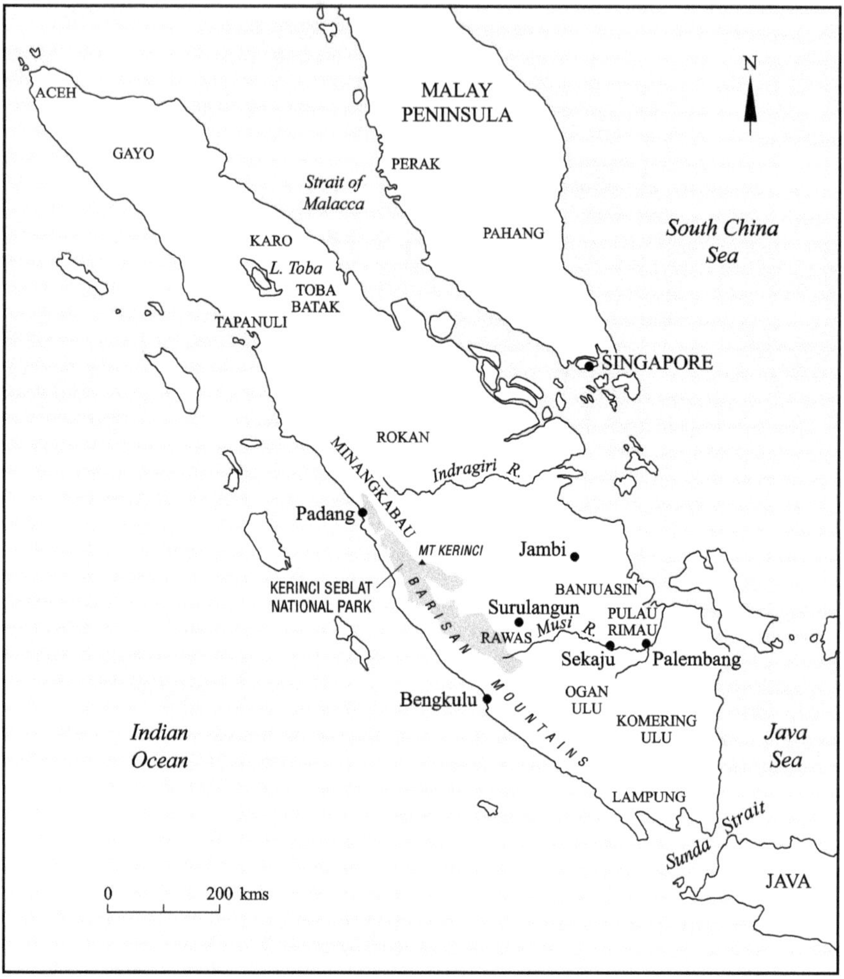

Map 5. Sumatra and the Malay Peninsula.

5 The 'short man' (orang pendek) of Sumatra

Of short stature, covered in hair and with a generally hominoid form and bipedal locomotion, the orang pendek of southern Sumatra bears a *prima facie* resemblance to several eastern Indonesian figures described in previous chapters. The comparative importance of this representation lies in evidence for its probable grounding in an empirical creature, either a variety of orang-utan or another primate. Insofar as local zoological experience can be shown to be crucial to the category, therefore, it cannot be construed as a purely imaginary or completely mythical creature, even though—as with all figures of this kind and like many zoologically documented animals as well—the orang pendek is partly subject to a fantastical representation. If *Homo floresiensis*, as a one time contemporary of modern humans, was a real historical 'wildman', the orang pendek might by the same token be taken for an equally empirically grounded, modern 'wildman'. But this of course is not to say that the two creatures are closely related; nor indeed that orang pendek are in any way connected with legendary Flores hominoids like ebu gogo.

Thought still to exist by local people, the Sumatran creature is known by several names, most of which display a regional distribution. As a general designation, I employ the Malay term 'orang pendek' (literally 'short person', 'short human'), a name first recorded by Schlegel and Müller (1839–44) and popularized in the early twentieth century.[1] Despite its physical resemblance to hominoids described in previous chapters, the orang pendek differs from these in at least two circumstantial respects. First, the category has been known to Europeans at least since the eighteenth century and some westerners have claimed to have seen creatures corresponding to the native representation. Second, the Sumatran creature has recently been subject to investigation by amateur and professional primatologists (Martyr *et al.* n.d.). Reflecting this attention, the orang pendek has received qualified mention as a possible unrecognized primate species in works on Sumatran zoology (Whitten *et al.* 2000; see also Rijksen and Meijaard 1999). According to one primatological interpretation, orang pendek may represent a remnant or recently extinct population of orang-utans (*Pongo pygmeus*) in southern Sumatra. According to another, it reflects a large, mostly bipedal and hitherto unrecognized ape more closely allied to the gibbons. In either case, the significance for eastern Indonesian categories is clear: since there are no apes of any sort east of Wallace's Line, neither of these primatological interpretations can apply to figures like ebu gogo. On the other hand, if orang pendek could be shown to designate a hominoidal image essentially identical to wildmen of Flores and Sumba, its link with Sumatran apes, known or unknown, would be called into question.

Local and colonial representations

In what follows, I first review local Sumatran descriptions of orang pendek, drawn mostly from Dutch colonial reports from the first half of the twentieth century. I then discuss claimed sightings by Europeans of creatures apparently corresponding to the native category and thereafter summarize descriptions of orang pendek derived from local sighting reports recorded by recent western investigators and from the investigators' own reported field observations. An obvious aim is to determine how far these different sources coincide. I separate European from Sumatran testimony not because I consider one more reliable than the other, but simply because of their obviously different cultural and historical contexts. As will be seen, only Sumatrans credit orang pendek with seemingly fantastic attributes like inverted feet. On the other hand, during the early twentieth century, European understandings of orang pendek were bound up with a popular view of human evolution largely influenced by Dubois' then recent discovery of a fossil 'ape-man' (*Pithecanthropus*) on Java and the concomitant notion of a 'missing link'. It is relevant as well that motivation for falsification, including false reporting and the creation of hoaxes, arguably differs in the two cases. Whereas for Europeans these can be quite readily identified in terms of gaining fame (or notoriety), making money, discrediting science and so on (see Daegling 2004), why Sumatrans, particularly in the nineteenth and earlier twentieth centuries, might have falsified evidence is rather less clear—although in one famous incident, one or more Rokan natives certainly did so, driven possibly by the prospect of a monetary reward.

Probably the first published reference to orang pendek is by William Marsden (1783: 35), an English official of the East India Company stationed during the 1770s in the Bengkulu (Bencoolen) region in southern Sumatra (see Map 5). Referring to them with the Bengkulu name 'gugu' (or 'orang gugu'; the author's original transcription is 'googoo'), Marsden describes the creatures as one of two separate forest-dwelling 'species of people' described by Malay inhabitants of the Labun district.[2] The other is the 'Kubu' (or 'orang Kubu'), now known to ethnology as a fully human population of hunter–gatherers inhabiting lowland forests mostly between Palembang and Jambi. Although Marsden characterizes both 'species' as 'dispersed in the woods and as avoiding all communication with the other inhabitants', he describes the gugu as much rarer than the 'pretty numerous' Kubu. They are also covered in 'long hair' (a feature he appears not to subscribe to the Kubu) and differ 'in little but the use of speech from the orangutan of Borneo'. Marsden says nothing about the size of the creatures. As though making good this omission, another early description of 'orang pandak' refers to a 'small wild man' covered in hair which natives of Indrapura (on the coast, west of Mount Kerinci) said had appeared some years before. Contrary to Marsden, however, this creature was unable to speak (Schlegel and Müller 1839–44: 12).

Marsden's characterization of the gugu as a 'species' of human may well reflect no more than the inclusion of 'orang' (Malay for 'person') in one version of the name. Despite its brevity, his account comprises several noteworthy details. Apart from the creatures' reputed scarcity (an idea replicated in much later accounts) and their ability to speak (an idea that is *not* so reflected), Marsden relates how the Labun folk, who had reputedly encountered gugu in just two or three instances, had claimed once to have captured a specimen, a male that had subsequently mated and had offspring by a Labun women. While hairier than Labun people, these offspring subsequently married members of the local population, from whom they were otherwise physically indistinguishable. Referring especially to this episode, Marsden concludes his account by suggesting that 'it probably has some foundation in truth, but is exaggerated in the circumstances'.

Like Marsden, many subsequent writers provide only generalized descriptions of native images of orang pendek and relatively few cite accounts by particular informants. One exception is De Santy (1925), a colonial officer who recorded the accounts of five named villagers from the Banjuasin region of southeastern Sumatra. Among these was a fisherman who claimed to have encountered a dead female orang pendek, designated with the local name 'sedapak'. Although the size of a child of about ten years (a fellow-villager said twelve), the creature was evidently mature, for she had humanlike breasts. The body was covered in hair about a 'span' (20 centimetres) in length, while the head hair, roughly the length of a forearm (or one 'cubit'), was much longer. The hands, feet and nails were like a human's, but the middle finger extended well beyond the others; also the heel of the small foot was much more pointed. The creature had 'long eyebrows' and lacked a cleft in the upper lip. In the previous night, before discovering the body, the informant had heard a sound like a human weeping. From its condition, particularly an abnormally swollen belly, the man inferred that the creature had died during a failed attempt to give birth.[3]

Including De Santy's article, native images of the orang pendek are described in over thirty accounts found in numerous publications appearing in the twentieth century and mostly between 1917 and 1938. Nearly all sources are in Dutch; the majority were written by men associated with the Dutch colonial administration, while others are newspaper articles. Both more generalized accounts and reports of specific native sightings concern four areas: the mountainous region around Mount Kerinci and Mount Kabah, Bengkulu and points south (Komering Ulu, Ogan Ulu), the swampy lowlands between Palembang and Jambi (including the eastern delta region named Pulau Rimau, or 'Tiger Island') and the Rokan district of central Sumatra, near the northern extremity of the Barisan mountains (see Map 5). In more westerly areas, the majority of sightings were reported from cultivated fields or by Malays visiting highland jungle to collect forest products.

Most early sources refer to the creatures as 'orang pendek' (or variants, e.g. 'orang pandak', 'uhang pandak'), a term used in Kerinci and other places in the central Barisan range. Others employ different regional names for what is apparently the same representation. Among these, the most common are 'sedapa', 'letjo' and 'gugu'. These and other names, together with details of their occurrence and possible derivation, are listed in Table 5.1. The Bengkulu name 'gugu' appears to be onomatopoeic; however, Martyr (pers. comm., June 2004) has suggested a connection with Malay/Indonesian 'guguh' ('to knock on'), referring to the creature's reputed habit of striking tree trunks. The similarity of 'gugu' to Florenese 'gogo' (in 'ebu gogo') is almost certainly coincidental. Although several meanings have been suggested for 'sedapa', none is particularly convincing. Analysis of the names thus hardly illuminates local Sumatran classification of orang pendek, in relation to humans or other animals. 'Orang' of course means 'human, person', but as it is further applied to non-human beings (see e.g. 'orang halus', referring to spirits), the designation does not definitely attest to a local conception of the creatures as human.

Since almost all early twentieth-century accounts portray orang pendek as small, hairy bipeds, it is appropriate to begin a survey of the creatures' physical form with stature and details of the hair.

Table 5.1 Other names for the 'orang pendek'

Name	*Where Used*
Sedapa, or **sedapak**	Sumatra's southeastern lowlands
Letjo (cf. modern transcription 'leco')	Rokan district. Sometimes dubiously glossed as 'chattering'. (Moussay, 1995, gives Minangkabau 'leco' as 'a kind of dwarf; cf. Malay/Indonesian 'léco', 'dwarf, midget')
Gugu, or **segugu**, **senggugu**	Bengkulu and elsewhere in the most southerly part of Sumatra
Atu rimbu, or **atu rimbo**, 'jungle dweller'	Rawas district
Sebaba	South Bengkulu and conceivably related to Minangkabau 'baba', meaning 'stupid, slow' (Moussay 1995; for other names see Rijksen and Meijaard 1999; Martyr 1994:75–76)

Height

Most descriptions (28 of 35) refer to the creature's height. None describes a creature larger than a human, although some estimates (for example 1.5 metres) extend into the height range of local people. Beginning with the shortest, estimates include 0.8–1.5 metres (Maier 1923), 1.2 metres, one to 1.5 metres (Dammerman 1924; Anon. 1924b) and 1.5–1.6 m (Van Herwaarden 1924, reporting on information from Kubu hunter–gatherers). None of these distinguishes sex, although Dammerman (1924) mentions a female specimen measuring one to 1.2 metres. Other descriptions are less precise but nevertheless accord with numerical estimates: for example, 'intermediate between a human and a siamang' (*Hylobates syndactylus*, the largest of the gibbons, with a male height of 90 centimetres; Volz 1922: 55), 'as tall as a ten or twelve or thirteen year old child' (De Santy 1925, Rookmaker 1932), or simply 'small', 'of small stature', or 'like a small human' (e.g. Coomans de Ruiter 1929).[4]

Height estimates thus mostly range from about one metre to 1.5 metres. Assuming the reference is an empirical creature, the range could reflect, apart from variation in informant estimates, the age of creatures observed, sexual dimorphism, or the broad, stocky build which, as I later describe, is attributed to the orang pendek. It is also interesting how the upper and lower limits correspond closely to the heights of two Sumatran apes: a male siamang and a male orang-utan (with a standing height of up to 1.5 metres).

Body hair

If estimates of height seem reasonably consistent, descriptions of body hair are more variable. While Marsden's gugu was long-haired and De Santy records a hair length of 20 centimetres, some twentieth-century accounts indicate a body covered in short hair. But what is meant by 'long' and 'short' is obviously unclear, especially as very few reports specify an actual length. Some sources suggest thick body hair, but others describe 'thin', 'light', or 'sparse' hair (e.g. Ridderhof 1929). Referring to native accounts from the Rokan, one report even describes the creature (in this case called 'letjo') as 'hairless' or 'covered in very thin hair, like a human being', or displaying hair only in the genital region (Rookmaker 1932). This is also the only source to mention distinct genital hair. The colour of body hair is variously given as black, dark brown, dark brown to black, dark, brown, red-brown and black or dark grey. 'Black'—the most common descriptor—probably translates Malay 'hitam', which can refer to darker shades of several colours. Nevertheless, there seems to be general agreement that the creature is covered in dark hair.

Head hair

Although few sources mention head hair specifically, many which do characterize it as long, or at any rate longer than the body hair. Three also describe hair falling over the forehead and eyes (De Santy; 1925; Coomans 1929; Westenenk 1932). The impression is mostly given of relatively straight hair; thus there are references to 'long black hair hanging at the back like a mane' (Maier 1923), hair hanging to the shoulders (Anon. 1926a), or simply 'long' or 'fairly long' hair (Rookmaker 1932). On the other hand, the last source records one informant's mention of curly hair which does not fall to the shoulders (ibid.), while Dammerman (1932), apparently drawing on the same information, speaks of 'slightly curly' as well as 'brownish' head hair. Brown hair, it should be noted, would appear to exclude Sumatran populations as referents of orang pendek, as local human hair would normally be classified, by Europeans and natives alike, as 'black'. No report indicates a pronounced colour difference between head and body hair, although one describes head hair as 'black' and body hair as 'dark brown to black' (Maier 1923). The foregoing may suggest a creature with head hair quite distinct, in length or texture, from the body hair. Yet since the majority of accounts are silent on this point, such a generalization is hardly warranted.

Complexion

Only one source specifically mentions skin colour. This moreover reports local descriptions recorded in the Rokan, where the skin of the creature called 'letjo' is characterized as 'pinkish brown' or 'brownish pink' (Rookmaker 1932). Apparently summarizing the same information as Rookmaker, Dammerman similarly speaks of 'light-coloured skin with, here and there, a reddish lustre' (1932). Unfortunately, this description has found its way into several secondary sources (see Napier 1972: 25; Heuvelmans 1995: 124) and as a general characteristic of orang pendek, or the image thus labelled, it is probably misleading.

General body form and build

Besides details of height and hair, other physical features of orang pendek are largely absent from native reports recorded in early twentieth-century writings. Also, features they do record are quite various. Nevertheless, taken together, the accounts suggest a generally hominoid form and an erect, bipedal creature often more humanlike than ape-like. Several accounts describe a creature that is heavily built and stocky (Westenenk; 1932; Pasteur 1960), or thick set (De Santy 1925). Other specifications consistent with a robust and powerful build include a short, square neck (Rookmaker 1932), a broad chest (Dammerman 1924) and broad shoulders (Anon. 1924a; Westenenk 1932). Accordingly, the orang pendek is commonly described as a creature possessing tremendous strength. Drawing on Rokan native accounts of the 'orang letjo', only Rookmaker (1932) describes the creature as 'not robust' and a 'small, slender being'.

How male orang pendek might be distinguished from females is quite unclear. Only one writer refers to male genitalia resembling a human's, adding somewhat dubiously that monkey genitals are equally manlike (Anon. 1924b). Just two sources (ibid.; de Santy 1925) mention female breasts, but in neither instance are these described as particularly large, or long and pendulous like those of some Florenese and Sumbanese hominoids. As the absence of a tail is arguably implicit in the depiction of the orang pendek as generally hominoid, it is perhaps not surprising that just one author reports the absence explicitly (Maier 1923). On the other hand, another source, referring to the region south of the Indragiri River, mentions a 'rudimentary tail'—but then further describes this as a feature not uncommon among the human population of Sumatra (Anon. 1924a).

Limbs, hands and feet

While two accounts describe orang pendek arms as proportionally similar to a human's (Anon. 1924b; Anon. 1927a), a larger number specify them as longer (Maier 1923; Anon. 1926b; Vogelpoel 1930; Rookmaker 1932; see also Martyr 1990). Maier's statement that the arms are 'not so long as a monkey's or ape's' similarly implies they are nevertheless longer than a human's. When mentioned, the hands and nails are mainly described as hominoid, although as noted, De Santy's specimen had a long middle finger (1925).

Scarcely any reference is made to the length of the lower limbs, though one account refers to short legs (Vogelpoel 1930). By contrast, much attention is given to the feet, generally characterized as short and wide and resembling those of a human with a rounded or sometimes sharp, pointed, or narrow heel (a feature which De Santy claims has given rise to a belief that the 'sedapak' has no heels). Apparently the first recorded European observation of a print corresponding to this shape is described by Hagen (1908: 33). A district officer in Sekaju named Saijers informed Hagen of a footprint resembling a small child's which he had found in deep jungle, far from human settlements. Such prints, according to Saijers, belonged to 'orang guguk' (cf. Marsden's 'gugu'), creatures as small as young children. The Dutch administrator added that Malays are very afraid of these creatures, which he claimed, apparently incorrectly, occur only in the district of Sindang, bordering on the Rawas country. Although much of the early twentieth-century evidence for orang pendek comprised prints corresponding to the ones described to Hagen (see Jacobson 1917; Volz 1922; Luitjes 1925; Sody 1925), it is now generally agreed that all or most of these were made by Malayan sun-bears (*Helarctos malayanus*), as Dammerman (1924) argued early on. By the same token, it is now well established that local Sumatrans regularly interpret bear tracks, particularly ones not showing claw marks and therefore especially humanlike, as prints of orang pendek (see Plate 5.1). Nevertheless, Rokan natives describe another foot shape, distinguished by a large toe that 'sticks straight out' (Rookmaker 1932) and thus suggesting something quite different from a human or a bear's foot. Similarly, a Dutch police captain named Bor is reported to have come across a print with a very prominent big toe (ibid.), possibly in Pulau Rimau, while another European who observed tracks attributed to orang pendek explicitly described these as not like a bear's, but like the prints of a 'manlike animal' or a 'large monkey or ape' (Dutch 'aap', Vogelpoel 1930). The oft-mentioned local idea that the feet of orang pendek are inverted, or turned the wrong way, is discussed below.

Head and face

A few accounts indicate a face resembling a human's (e.g. Anon. 1924b; Dammerman 1932). Also reported, however, are morphologically primitive features, such as heavy 'eyebrows' (Rookmaker 1932; cf. de Santy's 'long eyebrows'), the absence or near absence of a chin, a flat nose (Rookmaker 1932), a cleftless upper lip (De Santy 1925) and large or long canine teeth (Van Herwaarden 1924, describing Kubu accounts). Further specified is a forehead which 'reclines sharply' or 'slopes strongly to the back' (Rookmaker 1932). Another account describes a pointed cranium with a sloping forehead (Vogelpoel 1930). Although facial hair rarely receives specific mention, one source suggests it may be shorter than the body hair (Anon. 1924b). On the other hand, an orang pendek reputedly shot dead by a native was later described by the killer's son as possessing a beard ten centimetres long (Vogelpoel 1930).

Vocalization

Native accounts described in colonial sources attribute several vocalizations to the orang pendek: a hissing sound characteristic during flight (Maier 1923), a nocturnal sound transcribed as 'brr brr' (Volz 1922), a two-toned 'u-u' (Anon. 1924a) and a possibly identical call reproduced as 'hu-hu-hu' (Coomans 1929: 19–20).[5] As several writers have subsequently remarked, some of these are reminiscent of calls made by Sumatran apes, including siamangs, gibbons and orang-utans (see Bakels 2000: 124, regarding the siamang's 'rising hu-hu-hu'). At the same time, De Santy's informants (1925) described the 'sedapak' as crying or weeping like a human, as well as uttering a 'loud sound'.

Contrary to Marsden (1783), orang pendek are generally not credited with speech (Westenenk 1932). On the other hand, a native account from the Rokan refers to creatures possessing a language or 'at any rate communicating with sounds' resembling the rapid vocalization of monkeys (Rookmaker 1932). Possibly relevant here is the interpretation of 'orang letjo', a Rokan name for orang pendek, as 'jabbering man'. A reputed native eyewitness from the same region described a mother 'letjo' addressing her infant with a sound like 'tjanggoa' and the infant replying with 'tjena'. This, however, is contained in a report of the killing of the infant and the production of a specimen which later turned out to be a hoax (Dammerman 1932) and so is suspect on that ground alone.

Locomotion

In accordance with their erect posture, orang pendek are generally described as moving bipedally, sometimes at considerable speed. Nevertheless, one Sumatran villager described a specimen fleeing 'on hands and feet' (De Santy 1925). Another source mentions the arms being used to push aside tall sugar cane by an animal running away on two legs (Coomans 1929: 28–29). Yet another describes a creature similarly grasping onto branches of nipah palms while moving bipedally in marshy lowlands (Anon. 1926a). Suggesting the same habit is a report that muddy hand prints always appear, at a height of 1.5 metres, on trees growing beside paths traversed by orang pendek (Coomans 1929: 8).

Habitat and sociability

Reputed native sightings reported during the colonial era occurred in highland forest (where the creatures were mostly observed by Malay collectors of forest products), in or near cultivated fields and near isolated huts, including lowland fishermen's huts from which the creatures are described as stealing fish (De Santy 1925). Orang pendek are said usually to avoid settlements and accounts generally confirm Marsden's description of the creatures as rarely encountered and of sporadic occurrence (e.g. Anon. 1924a). Both reputed sightings and other information indicate a creature active both in the daytime and at night (Volz 1922; Dammerman 1924; Rookmaker 1932). Throughout its notional range, the orang pendek is described as ground-dwelling and is said not to inhabit or move in trees (Maier 1923; Anon. 1924a). Accordingly, 'letjo mats', constructed of vegetal material and reputedly used by the creatures for sitting and sleeping, seem always to be found on the ground (Rookmaker 1932; Dammerman 1932).

A few accounts describe orang pendek as inhabiting caves (Coomans 1929; Rookmaker 1932; Dammerman 1932). A story from Bukit Nanti—a peak near the boundary of Ogan Ulu and Komering Ulu (see Map 5)—describes how a group of 'segugu' inhabiting a nearby cave engaged in a 'bloody battle' with six Kubu foragers who accidentally stumbled across their abode sometime in the early 1930s (Coomans 1929: 4). All except one of the Kubu were killed.

The Bukit Nanti story is one of several suggesting that orang pendek form either groups, in one case including four to six individuals, or pairs (Maier 1923; Anon. 1924b; Rookmaker 1932; Pasteur 1960). A 1927 sighting from southern Sumatra concerned two animals, one a pregnant female (Coomans 1929: 12), while in an account of a native shooting, the victim was a male encountered with a female in a mango tree (Vogelpoel 1930). Coomans also describes an incident in which several people, lost in the forest and obliged to spend the night in a tree, observed some eight creatures emerging from a nearby cave at Batu Dua'ara, apparently not far from Bukit Nanti (1929:

13–15). Nevertheless, most native sightings refer to lone individuals (see e.g. De Santy 1925).

Diet and feeding

Reputed native observations of orang pendek frequently concerned creatures in search of food or feeding. The attributed diet appears largely vegetarian, consisting of bamboo shoots and, in the lowlands, possibly nipah fruit (Anon. 1926a). Other dietary sources include produce from native fields, notably sugar cane, bananas and durians, but also maize, tubers (Maier 1923; Anon. 1924b) and sago meal (De Santy 1925). Orang pendek are further described as catching fish (e.g. Vogelpoel 1930) and shellfish in creeks and small streams and moreover as feeding on larvae, termites, worms, snakes, slugs and snails. The latter they mostly acquire by turning over fallen logs and rotten tree trunks, a practice facilitated by their great strength (Jacobson 1917). In addition to fresh fish and stolen dried fish (De Santy 1925), orang pendek diet reputedly includes meat, in the form of chickens stolen from villages (Coomans 1929, referring to the south Sumatran 'segugu') and even the flesh of rhinoceros killed in native pitfalls (Jacobson 1918). While most food is consumed raw—orang pendek not being credited with use of fire or knowledge of cooking—a few accounts describe the creature entering forest collectors' huts and stealing cooked rice (Rookmaker 1932; Anon. 1933).

Temperament

While several accounts characterize orang pendek as shy and departing quickly when approached by humans (e.g. Maier 1923), another describes a creature making off 'in no great haste' (Jacobson 1917). Besides the 'battle' at Bukit Nanti, a capacity for aggression is indicated by reports from Komering Ulu and the Rokan of orang pendek throwing stones (Anon. 1924b; Vogelpoel 1930; Rookmaker 1932) and a 'surprise skirmish' with Malay labourers, resulting in one man being injured (Pasteur 1960). Aggression is further implicit in the local idea that male 'segugu' rape human females (Coomans 1929: 5).[6]

Overall assessment and classification

Several sources provide information on native assessments of the creature's humanity and its possible place in local ethnozoological classification. Some indicate that Sumatrans regard orang pendek as 'human' or 'a sort of human' (Anon. 1924b; De Santy 1925; Anon. 1926b; Coomans 1929). Yet others suggest something more apelike, or intermediate between ape and human: a sort of ape that walks on two legs (Anon. 1926a), 'a manlike being but different from a human' and 'not a monkey or a mawas [orang-utan]' (Westenenk 1932). Further suggesting an intermediate status is a story told by the Rejang people of Bengkulu; this describes orang pendek as deriving from a monkey or ape (Dutch 'aap') and a young woman placed in a pit with the simian, as punishment for wrongdoing (Ridderhof 1929). The story recalls the seventeenth-century report by Bontius, that Javanese consider the 'ourang-outang' (often taken as a reference to the ape *Pongo pygmaeus*) as originating in intercourse between 'Indian women' and monkeys and apes (1931: 285).

None of the early publications mentions the Malay use of numeral classifiers that distinguish 'humans' from 'animals', or how orang pendek might be categorized in this system. Deborah Martyr, however, has stated that Sumatran villagers who claim to have seen the creature tend to classify it as an animal (that is, with the classifier 'ekor'), whereas urban people who know orang pendek only from stories talk about it as a kind of dwarf human being (pers. comm., November 2005).

European sightings in the nineteenth and twentieth centuries

Although Marsden was the first to write about orang pendek (or 'gugu'), the first European claiming to have seen one was evidently the French geographer and traveller G.L. Domeny de Rienzi (1836).[7] De Rienzi says he observed several beings locally named 'gougons' on the east coast of Sumatra near the Indragiri River. These he describes as humanlike but resembling four-legged animals by virtue of 'a body covered in long hair, very receding forehead, the form of the glottis and their limited mental capacity'. The creatures, he adds, derive from the state of 'Ménangkarbou' (Minangkabau in western Sumatra) and 'scarcely exceed monkeys in intelligence, but are nevertheless human' (1836: 24). In regard to the name and description, De Rienzi's report obviously corresponds to Marsden's gugu. Still, the report is not completely unoriginal: in addition to the information on the 'glottis' (the source of which may never be known), the Indragiri region, on Sumatra's east coast, is associated less with orang pendek (or 'gugu') than are areas to the south and west, including the Mingangkabau highlands.[8]

The next European to report seeing a 'gugu' (incorrectly written as 'gugur') or a 'kubu'—names he uses indiscriminately—was the Anglo-American adventurer, Walter M. Gibson (1856). Gibson's observations occurred on two separate occasions during his ill-fated visit to southern Sumatra in the mid-nineteenth century. The first concerned a group of five or six dark brown hairy creatures he detected in the tree-tops, in a jungle creek off the Musi River not far from Pulau Rimau. From their arboreal habit and Gibson's description of their size and hominoid form, these were almost certainly orang-utans. The interpretation is not contradicted by Gibson's Malay guide, who referred to the creatures as 'big monkeys (or apes)'. Yet the guide also called them 'orang utan' and later 'orang Kubu' (1856: 116, 117). Translated quite correctly by Gibson as 'wild man', the name 'orang utan', despite its connection with modern English 'orang-utan', is not the southern Sumatran word for the ape (Yule 1903: 643–44), which is mostly called by the Malay term 'mawas' or a cognate. Rather, as Gibson also noted, it is a general name for the Kubu people and 'other wild aboriginal races' (1856: 120). Nevertheless, by this time Indonesians already knew 'orang utan' as a European designation for the ape. The Malay guide's application of 'orang Kubu' is another matter and is likely symptomatic of a Malay stereotype of the jungle foragers as people living in trees.

Gibson's second encounter with what he evidently considered a creature of the same kind took place in Palembang, where a Malay nobleman directed his attention to a hairy 'kubu' he had put to work. Remarking on a gruff 'yuh' uttered by the creature (apparently in response to a foreman's order), Gibson's Malay host told him that these beings are able to speak only 'some short grunting words' and pronounce just one syllable of Malay bisyllables. Gibson describes the individual as 'a dark brown form, tall as a middle-sized man, covered with hair, that looked soft and flowing', with 'well formed' limbs, hands and legs and a 'straight' body well capable of bearing heavy loads (1856: 181) As regards the face, the nose was fuller and the lips thinner than a Malay's, while the mouth was wide, the lips protruded and there was no chin. Gibson provides a sketch of his hairy 'kubu' (ibid.: 180; see Plate 5.2). The illustration corresponds to his verbal account, except that the body hair, rather than 'flowing', appears short, even very short. The head hair, too, is depicted as quite short, except for what appears to be a mane covering the neck and falling to between the shoulders, a feature curiously reminiscent of some twentieth-century accounts of orang pendek.

Other than a band worn around the forehead and strips of material apparently covering the genitalia and the one visible wrist, Gibson's 'kubu' is shown naked. Besides the face, the only hairless part of the body is an area on the back of the one visible leg, which appears quite bare to about the middle of the buttock. Finally, although it may be a function of faulty perspective, the feet of the hairy being seem small in proportion to the rest of the body and the legs quite short in relation to the trunk.

Whatever it was that Gibson saw, it was quite different from what he observed in the jungle creek and, by the same token, was certainly not an orang-utan. Hagen (1908: 29) straightforwardly describes Gibson's subject as a 'Kubu slave', an interpretation consistent with the regular Malay practice of enslaving the forest-dwellers which continued until the end of the nineteenth century (Winter 1901: 217, 222–23; Persoon 1989: 71). Contradicting twentieth-century depictions of orang pendek, Gibson also described the hirsute labourer as being of average height; certainly he was not smaller than his Malay companion, shown in the same picture. If he were a real Kubu—a forest-dwelling hunter–gather—the individual should not only have been smooth-skinned but probably shorter than the Malay. At the same time, while Hagen describes Kubu as generally short (1908: 47), he distinguishes a taller minority 'type' with a shorter trunk and longer limbs and concludes that it was one of the latter that Gibson observed (Hagen 1908: 53).

The next putative observation by a European pertained not to an orang pendek as such, but to a person reputedly sired by such a creature. In January

1899, T. Lubeck, a sergeant major stationed in Sumatra's Padang Highlands (in Minangkabau territory), wrote a letter to the Batavian Society describing how sometime previously, in a village named Aer Tanang, he had encountered a 'full-grown female person' just 60 centimetres tall. She was 'well built' and had white teeth, black eyes and a head of 'rich black hair hanging to the ground'. The woman spoke 'barely intelligible Malay'. Villagers told Lubeck she had been born to a Malay girl who often went alone into the jungle and had had intercourse with an 'orang ratas'. The name refers to small, extraordinarily shy and 'lightly haired' forest people, living on roots, fruit and fish and speaking an unintelligible language (Notulen 1899: 55–56). Although cited by De Santy (1925) as relevant to later accounts of orang pendek, 'orang ratas' occurs nowhere else in the literature. Moreover, the subject of Lubeck's observation was, ironically, too short for an orang pendek and can reasonably be attributed to a human abnormality, probably a midget.[9]

What may be counted as the fourth European sighting of orang pendek involved a young man, evidently with a Dutch name but of part Indonesian descent. Hailing from Padang, he is identified only as 'L. v.d. W.' (Westenenk 1932: 40–41) and was employed as an overseer by a Mr. 'v.H.' (evidently not the Van Herwaarden discussed below). Occurring in April 1910, the overseer's encounter was reported by L.C. Westenenk, first in 1918 when Westenenk was Resident of Benkoelen (Bengkulu) and again in 1932 after his retirement (in 1924) as Governor of Sumatra's East Coast. According to the young man's own hand-written statement, he was clearing forest in the Barisan mountains near Mount Sugirik with several native labourers, when about 15 metres away they observed a 'large, short-legged creature' crossing their path. It had thick hair and walked like a man, but had a face somewhat different from an 'ordinary human' (1918: 109).

There are discrepancies between Westenenk's two accounts of the incident. While his 1918 article refers to a single specimen, the Padang overseer's written statement, published verbatim by Westenenk in 1932, describes a group of three 'large, broad-shouldered', hairy, naked and weaponless 'mountain men'. In addition, the written statement further records a sighting, four days afterwards and in the same region, of a group of about 25 small but similarly hairy manlike beings at a distance of 200 metres. A minor discrepancy concerns the date of the observation, given as 10th April 1910 in the 1918 account and 20th April in the 1932 version. Nevertheless, despite these inconsistencies, the physical appearance of the creature (or creatures) remains much the same. Unfortunately, exact details of size and hair colour are not provided. The Padang man states expressly that what he observed was not an orang-utan; Westenenk, by contrast, suggests that an orang-utan may be exactly what he saw.

The next two European eyewitness accounts are more detailed and rather more famous. In June 1917, on the east side of Mount Kabah in southwestern Sumatra, a coffee plantation administrator named Oostingh encountered a strange animal in daylight, in a jungle clearing.[10] A month later he reported his observation to Resident Westenenk (1918), who interpreted it as a possible orang pendek sighting. The generally hominoid creature, found sitting (or possibly squatting) on the ground with its face away from Oostingh and at a distance of some 10 metres, was as tall as himself (1.76 metres) and was covered in short hair—specified as noticeably shorter and thinner than that of a siamang. The neck was 'huge' and 'extremely leathery', the shoulders were broad and square and the arms 'ridiculously long' (Westenenk 1918: 110). Oostingh further described the creature as 'dusty greyish black', like 'black earth'; but it is not entirely clear whether this refers to the skin or the hair. When it became aware of his presence, the creature got up, 'walked calmly away' and then 'with giant strokes, swung into a tree' (ibid.).

Although he never saw the animal's face, Oostingh, a man quite familiar with Sumatran primates, was certain it was not an orang-utan; it looked, he said, more like an extremely large siamang, except that it had short rather than long hair. Westenenk therefore opined that his informant may indeed have encountered a very large siamang, particularly a very old, lone male that had lost much of its hair (ibid.: 110). A noteworthy detail of Oostingh's report is the size of the creature—much taller than a normal adult male siamang and well in excess of native estimates of orang pendek height. Another is the claim that it 'swung into a tree', suggesting an arboreal habit contradicting native representations of orang pendek as ground dwellers and, at the same time, pointing to an orang-utan or other known ape. As we shall see, however, nothing of Oostingh's observation contradicts late twentieth-century observations of orang pendek reported by Deborah Martyr and her associates.

By far the best known colonial eyewitness of orang pendek is J. van Herwaarden (1924), a Dutch timber prospector stationed in Palembang. According to Herwaarden, one afternoon in October 1923, while hunting wild pigs in Pulau Rimau, the large river delta in southeastern Sumatra, he encountered a solitary female specimen of what he identified as a 'sedapa', standing on a branch of a tree, about three metres from the ground (see Plate 5.3). Proceeding to the base of the trunk, he was thus able to observe it quite closely.

Herwaarden's creature was 1.5 metres tall and covered in 'dark' body hair, somewhat darker at the back than at the front. It was 'lightly haired' on the upper side of the hands and the face was 'almost hairless'. Herwaarden says nothing about the length or thickness of the body hair. The head hair, which was 'very dark' (but not, it seems, of an unequivocally different colour from the body hair), was considerably longer, reaching almost to the waist and was 'not particularly dishevelled' (1924: 105). When the creature later took flight, the long hair 'fluttered' in the wind. As the 'sedapa' turned to face him, Herwaarden got a good look at its 'brownish' face. He describes the head as 'more pointed' than a human's and the forehead as 'more high than low'.[11] The face, Herwaarden says, was not like a monkey's or ape's and the eyes, lips and small ears were similar to a human's. Nevertheless, he describes the 'eyebrows' (apparently the brow ridge) as 'very powerfully developed', the mouth as wide and the chin as slightly receding. The nose also was wide, with large nostrils and while the incisors were 'regular', the canine teeth, yellowish white in colour, were large and more developed than in a human.

While Herwaarden was definite that his creature was female, he mentions no basis for this determination, referring neither to breasts nor genitalia. Like Oostingh's creature, the arms of Herwaarden's subject were long and extended to just above the knee, but the legs were comparatively short. Although seen mostly at a distance of 30 metres as the creature fled, the feet appeared wide and short and the toes looked 'normal'.[12] As Herwaarden observed the hairy female, it made two sorts of sound. The first he reproduces as 'hu hu', a call that was immediately answered from nearby forest (implicitly by another creature in the vicinity). The second was a 'hissing sound', uttered as the creature fled. As noted, both vocalizations are also reported in native accounts. Unfortunately, Herwaarden does not specify the length of his observation, though it seems that the creature remained in the tree for some time while the armed hunter stood below. To effect its escape, it did not climb further up the tree, but ran to the end of the branch on which it was standing, dropped to the ground and fled quickly on two legs.

There is no way of establishing the veracity of Herwaarden's account. Sceptics have raised several points in criticism. These include the fact that Herwaarden did not immediately report his experience to Gustav Fischer, his host and the lease-holder of Pulau Rimau and that he did not publish his story until the following year (Dammerman 1929). It has also been asked why Herwaarden did not shoot the creature even while being aware of its scientific significance (he was prevented from doing so, he said, by its human appearance) and why a creature, reputedly shy and wary of humans, should remain so long in a tree in the face of an armed human observer (Anon. 1924a; De Santy 1925: 4). It is perhaps also curious that, despite his proximity, Herwaarden does not mention any body odour; but then neither does any other account of orang pendek. Given that Herwaarden's subject might have been an ape, it is also noteworthy that the extent to which known apes emit distinctive natural smells is controversial.[13]

During the early twentieth century and before Deborah Martyr's experiences in the 1990s, at most only three other Europeans claimed to have observed orang pendek. One may have been the aforementioned Gustav Fisher, on whose land Herwaarden's sighting of a 'sedapa' took place'. Yet if Fischer did see the creature (as is stated in Anon. 1926a), his observations are nowhere described. Much of Fischer's knowledge appears in any case to have come from native sources. The same goes for Fischer's Japanese brother-in-law, also described as having seen sedapa on Pulau Rimau (Westenenk 1921).

In May 1927, A.H.W. Cramer, a Dutch plantation employee in Kerinci, reported seeing, at a distance of 10 metres and in the company of a native woodcutter, a humanlike figure moving at 'incredible speed' and leaving behind small human footprints. His native companion, who had a better view, described it as a small, naked male person, weaponless, with 'very long hair' and almost black skin. Cramer (1927) adds that such creatures are locally called 'orang item' ('black, dark people') and hail from 'the other side of Kerinci', where they live in the jungles like animals.

Apparently the last Dutch claim to direct experience of a creature possibly corresponding to orang pendek was made by a sergeant major in the Topographical Service named Van Esch (Anon. 1927b). While Cramer's report does not mention 'orang pendek', Van Esch was convinced that this is precisely what he had seen. Working in Surulangun (see Map 5) and descending a cliff to collect water, Van Esch saw coming towards him a 'humanlike figure' covered in brown hair. The creature was about 1.3 metres tall, with a chest half a metre wide, a dark face and long head hair apparently darker than the body hair (a feature also reported by one native-informed source). The limbs were not apelike, but more like a human's. Van Esch also thought he saw two prominent canine teeth. When he first spotted the creature, the unarmed Dutchman took cover. It then approached to just a few metres from where he was hiding, but fled after being startled by a noise in the forest. Van Esch's experience drew a response from no less an authority than the discoverer of *Homo erectus*, Eugene Dubois. Stressing his conviction that no hominins other than known humans could possibly exist at present and referring to the creature's brown colour, Dubois concludes that Van Esch probably saw an orang-utan (Anon. 1927a), an interpretation echoed by Coomans (1929).

Discounting Lubeck and Fischer but including Westenenk's Padang informant, there were apparently just eight European sightings of what could be interpreted as orang pendek during the early twentieth century. The most detailed accounts are obviously those by Oostingh, Herwaarden and Van Esch. Herwaarden and Oostingh reveal considerable agreement. The main difference concerns the length of the head hair and perhaps the colour of the skin, as well as the fact that Herwaarden's 'sedapa' dropped from a tree and fled on the ground while Oostingh's creature apparently did the opposite. Interestingly, Herwaarden shows even more agreement with Van Esch, especially in regard to the difference in head and body hair, the prominent canines and perhaps the hair colour ('dark' versus 'brown') and height (1.5 versus 1.3 metres). Nevertheless, Oostingh's account appears the most credible, at least insofar as what he describes sounds very much like an ape and—assuming a possible overestimation of the size—particularly a very large and old male gibbon or siamang.[14] It has been suggested that Herwaarden as well may have encountered a gibbon (Hill 1945); however, given the long 'fluttering' hair, the size and the largely human face, a young adult female orang-utan would appear more likely (cf. Rijksen and Meijaard 1999: 60–64, 417). At the same time, if these accounts are to be taken at face value and in view of how close Herwaarden and Van Esch were to their subjects, it is difficult to conceive how, in the 1920s, these men could honestly have mistaken known apes for something which they perceived as considerably more humanlike, particularly in regard to the erect posture and evidently efficient bipedalism.

Alternatively proposed as the actual subjects of the European sightings have been Kubu foragers. It is virtually certain that Gibson's semi-articulate labourer was a Kubu. It has also been argued that what Herwaarden encountered was a Kubu woman who, presumably from fear of the armed European, had climbed a tree (Anon. 1924a). A former ethnologist in the Netherlands East Indies Government reports hearing frequently that jungle-dwelling Kubu will 'always run away or climb a tree whenever a Malay approaches them' (Adam 1944: 21). Also relevant is a Malay representation of Kubu as tree-dwellers (ibid.) and De Wals's description, presumably drawn from native sources, of Kubu as sleeping in nests constructed between the branches of tall trees (1937: 170).[15] The notion coexists with a local belief that orang pendek descend from a Kubu woman who mated with an orangutan (ibid.).

Recalling descriptions by De Santy and others of orang pendek hair falling over the forehead and eyes, another noteworthy comparison is Van Dongen's description of the hair of Kubu he encountered as regularly covering the face (1906: 235). Similarly relevant is a Sumatran conception of Kubu bodies as 'covered in hair' (Forbes 1885: 233), a notion conceivably reflected in the claims of Gibson and Herwaarden. Although the notion does not stand up to empirical scrutiny (Forbes 1885; Hagen 1908: 39, 41), an impression of hairiness might just be derived from the appearance of skin diseases (including tinea, herpes and ichtyosis, Hagen 1908: 36, 263) very widespread among forest-dwelling Kubu in the nineteenth and early twentieth centuries. Referring to a Kubu woman shown in his monograph, Hagen (1908; see Plate 5.4) remarks how no part of her body larger than a finger nail was free of skin disease, so that her skin looked 'more like cracked, rough, lichen-covered tree bark than human skin' (ibid.: 263; see also Anon. 1924a, who speaks of 'a chronic scurf of a greyish colour'). At a distance, such cracked and abrased skin could conceivably suggest body hair.

Widespread skin ailments among Kubu have been associated with an aversion to water and want of bathing, while avoidance of bathing has further been linked with unpleasant body odours attributed to the foresters (Forbes 1885: 236; Van Dongen 1906: 233, 240–41; Hagen 1908: 15, 36). Ironically, though, this feature argues against an identification with orang pendek, since the latter appear not to be associated with a particular smell. Also detracting from the hypothetical connection is the fact that the orang pendek, contrary to earlier suggestions (Brasser 1926: 256; Westenenk 1932), is a creature recognized by Kubu themselves (see Herwaarden 1924; and more recently Martyr 1995c, 1998, 1999a: 35; Rintjema 2001). According to Martyr, Kubu names for orang pendek include 'yiao' (1998) and 'orang gerunggan batang', or 'person who rolls over logs' (1995c). On the other hand, the category is far more closely associated with sedentary Malay cultivators, that is, people less familiar with forests than the jungle-dwelling Kubu. The 'Kubu hypothesis' of course errs in the other direction from the 'ape hypothesis', inasmuch as it assumes that Herwaarden and other Europeans who were familiar with the glabrous jungle dwellers could mistake them for hairy 'ape-men'.[16] Nevertheless, if Malay villagers represent human foresters as beings like orang pendek, then there is no reason to believe that Europeans might not sometimes have misconstrued encounters with lone Kubu as instances of ape-men, especially if they were familiar with the local stereotype.

What is more noteworthy in the present context is how similar the European accounts are to the majority of native descriptions of orang pendek. As demonstrated above, native representations do reveal variation. Yet this implicitly much larger corpus of images does not obviously display any more internal variation than do any two of the European accounts.

Orang pendek at the end of the millennium

In 1932, O.J. Rookmaker, a colonial officer stationed in the Rokan district, received the skeleton and hide of what Rokan hunters, operating on his request to obtain a specimen, claimed to be an infant orang pendek. These he sent to K.W. Dammerman, head of the Zoological Museum in Buitenzorg (now Bogor, in West Java). While the case was under investigation, the status of the creature remained uncertain and popular interest in orang pendek naturally peaked. Later, in July 1932, it was revealed that the remains belonged to a Leaf monkey, a kind locally known as 'simpai' (Tukang Hikayat 1932) which had been doctored to effect a more human appearance.[17] Public interest in orang pendek then rapidly waned. So too did more serious writing on the topic, although short news articles continued to appear sporadically until 1938.[18]

Not until the last decade of the twentieth century did the orang pendek come once again to command the serious attention of westerners, most particularly in the person of British journalist and amateur primatologist Deborah Martyr. Having heard about orang pendek during a visit to Sumatra in the 1980s, Martyr—evidently an intrepid person and quite at home in tropical jungle—began looking for the creature in earnest. By the mid-1990s she had gained the support of Cambridge primatologist David Chivers and the conservation body Flora and Fauna International (FFI) and was at various times assisted in the field by Achmad Yanuar (a Sumatran doctoral student at Cambridge), Jeremy Holden (a zoologist and wildlife photographer) and several others.[19] Much of the research was concentrated in the Kerinci region, particularly in Kerinci Seblat National Park at locations above 750 metres but also at lower elevations, on the northern slopes of Mount Tujuh and elsewhere. Numerous interviews with local people were conducted, both in Kerinci and in parts of Bengkulu province to the south (Martyr *et al.* n.d.: 2–5).

Just below, I summarize the findings of Martyr's team, drawn from a series of unpublished reports written between 1994 and 1999.[20] I draw other information from a master's thesis by Annelieke Rintjema (2001), a Dutch anthropologist who conducted ethnographic fieldwork, also in the 1990s, centred in the southern Sumatran village of Lempur, also in Kerinci. In this remarkable study, Rintjema (2001) identifies five local categories of beings which, sometimes at least, take the form of hairy hominoids. Of these, one, which is called 'orang hutan' ('people of the forest'), corresponds to what is elsewhere named 'orang pendek'. Not to be confused with the ape named 'orang-utan' (as I explain below), this creature is, as Rintjema shows, quite distinct from the other four categories, all of which by contrast suggest varieties of spiritual beings. The large majority of Lempur stories concerning orang hutan comprise eyewitness accounts, while just two reveal what Rintjema calls a 'mythical tinge' (2001: 20). Orang hutan, moreover, is the only one of the five categories regularly sighted near cultivated plots (ibid.: 156) and it is the only one deemed capturable (ibid.: 152).

My objective in what follows is not so much to assess the validity of Martyr's and Rintjema's findings, as to gauge how far they coincide with native and earlier European accounts and to consider what light they may shed on these. Much of what Martyr reports derives from local interviews. However, she also refers to material evidence in the form of footprints and hair, vocalizations, nests and remains of feeding not readily attributable to known species and even brief sightings of specimens claimed by herself, Yanuar, and Holden.[21]

Based on a combination of interview data and sightings, Martyr records height estimates for orang pendek of 1.0–1.5 metres (Martyr 1990), 0.75–1.2 metres (1994: 6), 1.0–1.25 metres with a median of 1.15 (based on field observations, 1994: 66) and 0.90–1.25 metres (based on informant accounts, Yanuar and Martyr 1995). Lower estimates (such as 0.75 and 0.90 metres) may reflect Martyr's discovery, part-way through her 1994 research, that local people tended to underestimate height verbally by 15 to 20 per cent in relation to visual height comparisons—comparing the height of a creature seen to that of a particular child or an object. Possibly, this factor could also account for the discrepancy between the lower figures and colonial era estimates, especially those given by Oostingh and Herwaarden. A larger size is also suggested by references to orang pendek as 'big' or 'very big' and alternative names which Martyr recorded, including 'uhang besar' and 'orang gedang', both meaning 'big person' (Martyr 1994: 17; 1995c). These, however, Martyr plausibly explains as reflections of the creature's broad build and muscularity rather than its stature (Martyr 1994).

Martyr provides various details of orang pendek hair. Sumatrans generally described short, fine and glossy body hair, with a minority indicating longer hair on the head and thicker hair on the arms. Also mentioned are a normally black mane extending to the base of the spine and short head hair growing to just above the eyes (1990: 61; 1994: 12, 68). Whereas Martyr mostly reports short body hair, her associate, Achmad Yanuar, cites informants' descriptions of long hair on the body, especially on the back, as well as long hair on, or around, the head (Yanuar 1998: 5; 1999). Yet long hair on the back is consistent with the 'mane', while long hair on the head recalls colonial references to hair falling over the forehead and eyes, as may Martyr's reference to short hair on the forehead. Also recalling early twentieth-century accounts, informants described hair colour mostly as dark ('black', 'dark grey') or reddish brown (1990; 1994: 12, 14; cf. Rintjema 2001: 125, who mentions 'red-brown' or 'dark brown' hair). However, largely from her own putative observations and a sighting claimed by Jeremy Holden, Martyr further records 'dark yellow or tan', 'light red gold', 'ivory', 'parchment' and 'fawn' (1990: 61; 1994: 50, 60; 1995b), colours partly comparable to the 'yellowish-red' reported by Yanuar (1999). Summarizing this information, Martyr and colleagues later describe hair colour as 'golden brown, varying from a yellowy phase through to a reddish colour to peak at a darker brown phase'—adding that this contrasts with orang-utan hair, which 'peaks in the middle of this spectrum' (Martyr *et al.* n.d.: 5). According to one informant sighting (Martyr 1995c), head hair is darker than body hair; but this difference is not mentioned again.

In one place, Martyr says orang pendek skin is invariably described as 'dark grey' (Martyr 1994: 12, 68). However, she later cites local eyewitness descriptions of facial skin as ranging from 'pinkish tan, through to tan, to grey' (1995c), thus partly recalling the 'pinkish brown' attributed by Rookmaker to the 'letjo' of the Rokan. Features of the head and face reported by Sumatrans included a high forehead (1990: 61), a heavy brow ridge, a bridgeless nose, lobeless ears, close set eyes, large and powerful jaws, a wide mouth, an upper lip shorter than an orang-utan's, large upper canines, a receding chin and a 'sagittal crest' (1994: 12, 13, 44, 67). A sagittal crest, a ridge of bone supporting powerful facial muscles in mature male gorillas and orang-utans, might also be suggested by the pointed head reported by Herwaarden. Yet, curiously, Herwaarden's subject is described as a female. Neither the crest nor any other of these features receives mention in reports appearing after 1994. Moreover, Martyr's own sightings and those of her colleagues were not of sufficient length to confirm them. Suggesting a generally apelike appearance, the details largely agree with more detailed colonial descriptions; yet in the absence of numerical data, most could equally apply to known non-human primates. One exception is the high forehead, which is at variance with earlier representations of orang pendek, but which on the other hand is suggestive of an orang-utan.

In regard to general body form, Martyr depicts the orang pendek as erect, broadly built and muscular and with shoulders and chest extremely wide relative to height (1994: 13, 67, 68). With a head 'carried low on the shoulders' and inclined forward, a virtually absent neck and shoulders that are 'almost square' rather than sloping (ibid.: 44, 67), the orang pendek sounds remarkably similar to the creature described by Oostingh. Referring to one of her claimed sightings, Martyr further describes the animal as 'extraordinarily *hominid* in body posture' (1995c; my emphasis). Citing a Kerinci headman, her published account (1990) mentions a large and prominent belly, a feature which Martyr correctly notes is not previously recorded in print. While the belly accords with the orang pendek's mainly vegetarian diet, Martyr's subsequent reports appear not to mention this again. However, drawing on over twenty sightings between 1993 and 1998 by residents of the Kerinci (southern Sumatran) village of Lempur , as well as information from Bukit Kasang, just north of Padang, Rintjema as well attributes a 'large belly' to the orang pendek (2001: 114, 205). Although sex is apparently readily distinguished, overt sexual features are not mentioned in Martyr's reports, nor in Rintjema's.

Among the most remarkable findings of Martyr's team are footprints quite different from the bear prints locally attributed to orang pendek. First observed in 1993, the prints reveal a curved foot apparently belonging to a non-human primate, with a large toe positioned well back of the others and pointing away from the foot by up to 80 degrees (1994: 68–70; 1995c; see Plate 5.5). The length of the foot averages more than 18 centimetres (1994; 1995c). The source of the prints is naturally controversial. Rijksen and Meijaard (1999: 63) have interpreted them as possible orang-utan hand prints, while other authorities remain puzzled.[22] Martyr herself, who registers a conviction that the prints were not 'manufactured' (1995d), suggests that their most troubling aspect is an absence of prints attributable to infants or even sub-adults (1995c). Some tracks discovered by her team comprised multiple prints which allowed measurement of the stride (1995d) but, oddly, no figures are provided.

From both informant interviews and observations claimed by herself and colleagues, Martyr records various details of orang pendek behaviour. While both Herwaarden (1924) and colonial native accounts attributed hooting calls and hissing to the creature, Martyr's team have identified a greater variety of vocalizations—including 'soft hooting', 'hissing' or 'hiss shrieking', 'muttering', 'cooing', 'laughing' and a coughing sound (1994: 26, 41, 49, 52). Later, Martyr refers to 'grunt' calls and a booming sound which, she infers, is produced with the chest rather than the throat and by a primate with an 'exceptionally large diaphragm' (1995c). A sometime assistant of Martyr's, zoologist Barney Long, similarly reports hearing a 'terrifying laugh' apparently made deep within the chest cavity by a large animal (Long 1995: 3). Implicitly not attributable to known species of animals, the various sounds are reduced to five types by Martyr. She concludes that these are produced only when the creature is alarmed or threatened and that none is a territorial call. Local people, Martyr suggests, recognize just three calls—apparently in addition to the 'hissing' mentioned by one informant (1995c)—but no one appears to have attributed language to the orang pendek.

On the basis of her own putative sightings, Martyr describes orang pendek locomotion as displaying an unexpected humanlike 'fluidity' and an 'accomplished, fluent bipedalism'. Nevertheless, she also suggests that the creature occasionally moves quadrupedally (1994: 39, 42–43; Martyr *et al.* n.d.: 5, 10). Perhaps consistent with this, Martyr further describes orang pendek as grasping branches on either side of a path while walking bipedally, or according to her own observation, as walking with arms reaching forward 'like a man wading through deep water' (1994: 15, 45). Similar descriptions are given by Jeremy Holden of the creature he saw (Martyr 1995c; Van Beurden 2002: 17, 18). Both sources furthermore recall early twentieth-century native descriptions of orang pendek moving bipedally while grasping onto nipah palms or pushing aside tall cane.

Most natives questioned denied orang pendek could climb trees; a few reported seeing them in trees but not travelling arboreally. According to another local idea, the creatures sleep in hollow trees and caves (Martyr et al. n.d.: 5) and Martyr mentions evidence of one having entered a natural tunnel (1994: 34). Martyr's team also found nests similar to orang-utan nests, but built low in trees or occasionally on the ground. These appeared not to have been made by sun-bears, tapirs, or wild pigs. Local people are either ignorant of the nest builders or attribute the structures to 'various mammals or birds', but not to orang pendek (Martyr *et al.* n.d.: 9–10). Reminiscent of the 'letjo mats' mentioned in the earlier literature, nests built on or close to the ground make little sense in Sumatra, where tigers are active at night. Orang pendek may also be active at night, though they are described as mainly diurnal (ibid.: 4, 9). Local sightings reported in cultivated fields mostly took place either at night or at dusk or dawn, when few people were about (Martyr 1994: 9; Martyr *et al.* n.d.: 9).

Characterizing orang pendek as generally shy and intolerant of humans, Martyr claims they can nevertheless behave aggressively when disturbed. She records an alleged killing of a man by an orang pendek (1994: 17) and another case where a young Kerinci woman collapsed from shock and later died after one had 'laughed at' her (1995b). An implicitly high intelligence (1994: 22) is cited as a reason the creatures are able to elude human trackers and are rarely encountered. Another factor is their evident scarcity. At present, orang pendek seem restricted to the Kerinci region and a few neighbouring areas in southwestern Sumatra. Owing to extensive deforestation since the 1930s and the spread of plantations, they are no longer found in the lowlands of southeastern Sumatra (where Herwaarden's sighting took place). In several locations—including the Minang Hills and the Siparuk region—Martyr was told that the creatures were once present but are no longer (1990: 59; 1994: 54, 66; 1995b; 1995c; 1999b). The primary habitat of orang pendek is lowland hill forest, but in Kerinci they are also reported at elevations up to 1200 metres (Martyr *et al.* n.d.: 4–6). Another description specifies the range as from 'close to sea level to about 2000 metres' (Martyr 1996). According to Martyr, there is evidence for a 'highly mobile' creature that requires a large home range. Generally, orang pendek appear to be rare, and possibly on the brink of extinction (Yanuar and Martyr 1995), with a population of less than 2000 in Kerinci Seblat National Park, where they appear to be most numerous at present (Martyr 1995c). In 1994, Martyr suggested there was a 'relatively thriving population' in the Kerinci region, but she later described the population estimate on which she based this assessment as too high owing to double counting (1994: 54, 66; 1995c).

Contradicting a 'general perception' of orang pendek as solitary animals, interviews suggested to Martyr that two or more animals can often occur together in the same location (Martyr *et al.* n.d.: 4). She further describes orang pendek as 'pair but not group forming' (1995b), thus qualifying an earlier impression that they either form pairs or 'live in single-sex pairs' or small groups (1994: 54). Sightings of orang pendek reported to Martyr and her team often concerned animals involved in feeding; also, the creature's existence is partly inferred from signs of feeding observed by the investigators themselves. Similar to what is described in the earlier literature, the diet appears to comprise various wild and domesticated plants (including cucumbers, durians, sugar cane and jackfruit), invertebrates, snails, fish and even stolen chickens and cooked food taken from huts. In mountain and hill forests, orang pendek roll over large logs to find invertebrates, a behaviour mentioned by several colonial authors (e.g., Maier 1923). Initially, Martyr suggested that ginger (genus *Achasma*) was a primary source of wild plant food, but she later revised her opinion after discovering that Pig-tailed macaques also feed on the plant (Martyr 1994; 1995a; 1995c; 1999b; Martyr *et al.* n.d.). She further acknowledges that much of the orang pendek diet is shared with sun-bears (Martyr 1994: 20).

Possibly linked with their reputed feeding habits are two other behaviours Martyr's team hypothetically attribute to orang pendek. One is a 'forceful tapping' or banging on trees, interpreted as a method of 'echo-location' when searching for larvae (Martyr 1995b; 1995c). Another concerns the occurrence of many local sightings during or just after light rain, when invertebrates are numerous (1995c). Also associated with orang pendek by local people, the banging sound reputedly led to two sightings by Martyr's principal Sumatran guide. In a comparative context, the meteorological association and tree-striking are curiously similar to behaviours attributed to putative hominoids in Flores and Sumba (see Chapters 3 and 4)—and in the case of tree-striking, even more curiously to a famous hominoid reported from North America (see Chapter 8).

In Bukit Kasang, north of Padang, Rintjema was told orang pendek use fire to roast fish (2001: 205). But this is the sole reference to any sort of technology and with a remarkable uniformity the creatures are depicted as entirely lacking in culture.

More fantastic aspects

In spite of some variation, it may be agreed that the picture of orang pendek constructed by Martyr—on the basis, it should be added, of more sustained and systematic investigation of the hypothetical creature—is quite comparable to images, drawn more exclusively from native sources, which are reported in the colonial literature. Insofar as Martyr and colleagues base their findings on direct empirical evidence, it is moreover quite conceivable that its source, whatever this might in fact be, is the same as that which has inspired and sustained the image of the orang pendek among rural Sumatrans from the time Marsden wrote up to the present. While sceptics will no doubt continue to reject orang pendek as a real creature, physical features and behaviours attributed to the entity mostly fall within the realm of empirical possibility, even though they may in fact reflect already attested species. Of quite a different order are apparently fantastic features. Such features as are reported in colonial accounts are, however, also broadly similar to ones recorded by Martyr's team and in the interests of parsimony, all sources can be treated together.

Even by comparison with hairy hominoids reported from other parts of the world, orang pendek possess relatively few features that appear simply incredible. Mentioned most often are feet that are 'inverted' or 'turned the wrong way' (Westenenk 1921; Maier 1923; Van den Pijl 1938). In a global view, such inversion is frequently associated with witches and malevolent spirits and these connections have been interpreted as a symbolic device expressing, in physical terms, profound moral difference from ordinary humans (Needham 1978: 35–36). In her study of Lempur categories, Rintjema claims that it is not the orang pendek (locally known as 'orang hutan') that possess reversed feet, but a distinct kind of spirit known as 'tirau' (2001: 140–41). Whoever the owners of such feet are thought to be, on Sumatra the representation is sustained by material evidence in the form of peculiarities of attested footprints and several commentators have explained it accordingly. Given that Sumatrans attribute bear tracks to the orang pendek, Malayan sun-bears standing erect with toes characteristically pointing inwards could well create the impression of a creature walking with the feet turned outwards (the normal position in hominins) but with the heel to the fore (Sody 1925: 59–60). Similarly, MacKinnon has suggested that the idea of 'back to front feet' may derive from prints of sun-bears, which have 'the big toes on the outside of the foot', or on the side opposite the arch (1974: 114–15).[23]

In contrast, Rijksen and Meijaard suggest the idea of 'backwards' orang pendek feet reflects the hand prints of a 'long-fingered type' of orang-utan, which walks on hands held flat with folded fingers and with 'wrists pointing in the walking direction' (1999: 62–63). But there is yet another possibility. Through constant walking over rough terrain or up and down inclines, barefoot humans can develop feet with the big toe and other toes turned dramatically inwards. Cases have been noted for the Igorot of Luzon (see Worcester 1912: 192–93; Hagen 1908: 45; see Plates 5.6 and 5.7). The condition seems also to have occurred among the Toala of Sulawesi (F. Sarasin 1906: 56, Figures 4a and 4b). But most relevant in the present context is Van Dongen's description of the Kubu he met along the Ridan River, as having 'inward turning' feet, resulting from a lifestyle involving 'much climbing and walking over fallen tree trunks and a constantly slippery forest floor' (1906: 232, 235); a photograph of a Kubu woman named Nor illustrates feet thus misshapen (see Plate 5.4). Prints of such feet could easily give the impression of a foot turned the wrong way, thus suggesting Sumatran forest-dwellers as one source—if not the most fundamental—of the 'inverted' feet of the orang pendek. Although himself unable to appreciate the difference, Forbes notes the claim by Malays that they can always distinguish Kubu footprints from those of their fellows (1885: 241).

Other notions difficult to accept as literal attributes of an empirical animal include: orang pendek not disturbing people if offered tobacco or dried moss (Westenenk 1918: 108); their attraction to the smell of burning incense (Martyr 1994: 26; Rintjema 2001); and their 'herding' of wild pigs (Westenenk 1918; 1932: 37).[24] Orang pendek taking tobacco left out for them does, however, recall the 'silent trade' conducted by Malays and Kubu, in which tobacco was one favoured item (Forbes 1885: 235, 242; Winter 1901: 231–32; Persoon 1989: 510). Reminiscent of a widespread representation of certain spirits as the owners of wild animals (Forth 1998a; Valeri 2000)—and as we shall later see, also of African representations of pig-herding hairy hominoids (Chapter 8)—the herding of wild pigs is a practice more specifically ascribed by Sumatrans to 'dwarfs' called 'bigau' or 'si bigau' (or 'gembala babi', pig herder, Martyr 1994). But only in part are these identified with orang pendek (see Westenenk 1918; 1932; Moussay 1995, s.v. 'bigau', 'léco').

Before he encountered the specimen on Pulau Rimau, Van Herwaarden had heard from people in Palembang that orang pendek had a single eye in the middle of the forehead and furthermore that they could turn into tigers (an attribute of some Sumatran humans), climb trees like a house lizard, make mats from stolen sheep's wool and help lost rattan collectors find their way home (1924: 104). None of these ideas is recorded anywhere else in the literature. But this may be explained by their urban source—Palembang was a multi-ethnic city of 108,000 residents in the 1920s and one of the largest in Indonesia—and more particularly by a tendency to exaggeration and fantasizing among people either unfamiliar with forested areas or residing at some spatial and cultural distance from the locus of a representation (see Chapters 2, 3 and 8). Rural Malays told Herwaarden of encountering a 'sedapa' in a body of water; it escaped by diving, re-emerging some distance away and then swimming off quickly. But this ability to swim also seems not to be recorded anywhere else.

A recent journalistic account (Van Beurden 2002: 13) describes orang pendek turning into humans and back again—an ability equally attributed to known animals, including Bornean orang-utans (Brooke 1866, cited in Roth 1896: I: 350). Sumatrans told Martyr that orang pendek were mystically powerful creatures (Malay and Indonesian 'gaib') as they could 'disappear', a notion she interprets as reflecting a natural ability to elude humans (1995c). Both disappearing and transformation are of course hallmarks of spiritual beings and indeed Martyr cites a local evaluation of orang pendek as 'half animal, half god/spirit' (1994: 16). Kerinci people refer to the orang pendek as 'hantu pirau', 'dark brown ghost', a category apparently bound up with the local belief that a person who is lost in the forest and cries out will be answered by an orang pendek and thus led further astray (Anon. 1995). Drawn from an Indonesian newspaper article, this last idea recalls similar ideas concerning hominoidal figures on Sulawesi (see Chapter 4). At the same time, it contradicts Herwaarden's report that orang pendek will, on the contrary, help lost rattan collectors find their way home. Similarly suggesting a spiritual representation is a Kubu term for the creature, 'hantu pandak', 'short spirit or ghost' (Rintjema 2001: 215). But, again, such designations could equally describe gibbons, considered by Sumatrans 'not as an animal but as a visible forest spirit' (Bakels 1995: 16). As equally fabulous ideas apply to known species the world over, the more general point is that none seriously compromises the orang pendek's empirical status, let alone proves it to be a completely imaginary being.

Stories of capture, mating and abduction

As with hominoids reported from other parts of Indonesia, stories concerning the capture of orang pendek have occurred from time to time, going back as far as Marsden (1783). The colonial period saw several failed attempts to capture a specimen, including a farcical hunt with dragnets in 1925 on Pulau Rimau (Anon. 1926a; Schebesta 1928) and another unsuccessful attempt by an administrator named Van Kan (Dammerman 1924: 177–78).[25] The previously mentioned hoax of 1932, perpetrated with a doctored monkey shot by Rokan hunters, was actually a killing but was nevertheless reported in the press as a 'capture'.

In southern Sumatra in 1927, what were suspected to be orang pendek were, in three cases, caught in tiger traps, some deliberately set for the purpose. In each instance, the creature somehow broke free of the trap and escaped (Coomans 1929: 6–7, 12, 28; Ridderhof 1929; Westenenk 1932), but the incidents nevertheless yielded samples of blood and hair and a tooth fragment. Tests proved generally inconclusive. According to one opinion, the blood belonged to a 'human or non-human primate' (Docters de Leeuwen, cited in Coomans 1929: 10). K.W. Dammerman, the zoologist, determined that it was not from a Malayan sun-bear, nor from various domestic animals, a siamang, or a human (letters cited in Coomans 1929: 39, 40). Later, however, the same authority stated that the blood 'pointed faintly to human origin' and he speculated that it might have come from a native who had injured himself while handling the trap (Dammerman 1929: 4). The hair may have belonged to a sun-bear as, according to a letter from Eugene Dubois, may have the tooth fragment (Coomans 1929: 34–36, 37). Apparently referring to the same hair, however, another source described it as 'reddish' (Captain Bor, cited in Rookmaker 1932: 219) and thus suggestive of an orang-utan rather than a sun-bear (which is black-haired).[26] Although not obtained by trapping, Martyr and her team also collected hair and faecal samples of suspected orang pendek (Martyr *et al.* n.d.: 7–8). Results of samples sent for analysis appear so far not to have been published. However, hair samples collected by a trio of British explorers in Kerinci in 2001 have been analysed by Dr Hans Brunner, an Australian authority on mammalian hair. Two samples, orange-brown in colour, did not match hair of any known Sumatran mammals, including orang-utan, siamang, or gibbon, even though their closest resemblance is to orang-utan hair (Brunner 2003).[27]

Also in the 1920s, a report from the Rokan district described another capture of an orang pendek, in a trap set by a resin collector named Amat. Amat had discovered the creature stealing food from his lean-to; it managed to escape from the trap by digging its way loose, but not before seriously injuring Amat's hand (Anon. 1933). About the same time, another Sumatran reputedly shot an orang pendek at some unspecified location (Vogelpoel 1930). Described as standing about 1.5 metres tall, with short legs and long arms, sparsely covered in red-brown hair and with a 10-centimetre beard, the victim sounds rather like an orang-utan. Much later, in March 1958, western presses carried a story relating the capture in southern Sumatra of a creature designated only with the name 'sindai', but evidently also identified as an orang pendek.[28] Captured by a method unspecified near Perabumulih (to the southwest of Pelambang), the creature—described as an 'anthropoid ape' but also as resembling a human—was a female 1.2 metres tall, with long nails and a body covered in short hair. The age of the 'sindai' was estimated at seventeen years, but how this was determined is nowhere indicated.

According to Sanderson (who does not specify a source), the captured creature was tailless, bipedal and behaved 'in every way like a tiny human being, but apparently had no speech and ate only raw foods'; its short body hair was 'pure white' and 'fine' (Sanderson 1961: 226). As the same author points out, 'sindai'—a name found nowhere else in the literature on orang pendek—is suspiciously similar to 'simpai', a local name for the Banded leaf-monkey (*Presbytis melalophos*). Although one news article concludes by stating that the creature was to be taken to Palembang, what happened to the animal after its capture is unknown. Sanderson simply says it was sent to Java for inspection by 'leading scientists'. Seemingly consistent with the possibly simian reference of the name, Heuvelmans (1995: 142) suggests it was a hoax comparable to the 1932 affair, which also involved a leaf-monkey. Further hinting at a connection with the 1932 case is the colour of the sindai, which recalls the whitish hair of the young monkey killed some 26 years previously by the Rokan hunters (Dammerman 1932: 127). However, 1.2 metres is very large for a monkey and in fact this detail and others besides correspond very closely to what has been described for orang pendek over the entire twentieth century. One exception is the hair colour, although a light colour is one of the possibilities recorded by Martyr for orang pendek. Another is the long nails, a feature commonly described for hairy hominoids of eastern Indonesia and Oceania (see Chapter 8) but not mentioned elsewhere for the orang pendek. At the same time, I am reliably informed by the Indonesian anthropologist Juara Ginting (pers. comm., March 2006) that the Karo of northern Sumatra are fearful of orang-utans, not only because of their great strength but also because they can seriously injure a person with their long, sharp nails.

Like similar reports from eastern Indonesia, the 'sindai' story is inconclusive. According to a much later tale reported in a Sumatran newspaper (Anon. 1995), a man from Kerinci had some years previously trapped an orang pendek in a natural cavity; while he attempted to bind its limbs, the creature managed to escape. Martyr has recorded several similar stories from the same region. In 1980, an infant orang pendek, just 35 centimetres in length and of a yellow or tan colour, was somehow captured, kept for several days and later released into forest on Mount Kerinci (Martyr 1995c). Interestingly, the creature is said to have 'cried' (or 'wept'), a detail reminiscent of the human crying reported by De Santy's informant (1925), as well as reports of similar captures from eastern Indonesia (Chapters 3 and 4). Rintjema, also, relates a capture story from Kerinci in which the creature 'cried very easily and was very shy' (2001: 124). In the seventeenth century, Bontius described a female 'ourang-outang' he observed in Java as crying and 'shedding tears' and—in a way paralleling quite remarkably Rintjema's capture tale—as displaying 'great shyness' (Bontius 1931: 285). Particularly if Bontius's beast was not a specimen of the ape *Pongo pygmaeus* (as has commonly been supposed), these details might suggest that Bontius's report derived from a Javanese representation and particularly one from western Java (see Chapter 6), comparable to ones encountered not only in Sumatra but also in more easterly parts of Indonesia.

Another story investigated by Martyr concerned an orang pendek caught in a pig snare in the 1990s; however, the putative remains appeared not to belong to a primate (1999a: 4). In 1998, Martyr heard of a reputed killing and burial of an orang pendek, said to have weighed 30–35 kilograms, near the south Sumatran village of Lempur. But in this case it turned out that, for a reason unspecified, someone had dug up the grave and burned the body two years previously, thus preventing further identification. It may be too fanciful to suggest a connection between this report of corpse burning and Flores stories regarding the destruction of local hominoids, including the extermination of the ebu gogo and the Lio creatures trapped in a cavity and burned alive around 2001. More reminiscent of the eastern Indonesian tales is an annual 'cave burning' in Pelompek, near Lake Gunung Tujuh in the Kerinci region—apparently a ritual performed to prevent orang pendek from reoccupying a particular location (Rintjema 2001: 209, citing a personal communication from D. Martyr). However, there is no evidence that fire was employed to remove the creatures in the first place. I have enquired about local stories of people using fire or smoke to attack or kill orang pendek, but everyone asked has replied that they have never heard of such tales (Rintjema,

Martyr and Jeremy Holden, pers. comm., November 2005); hence, any resemblance to the eastern Indonesian representations is implicit at best.

While capture stories apparently occur with some regularity, the only putative discovery of an orang pendek corpse was evidently the one reported by De Santy (1925) from Banjuasin. Orang pendek mating with humans is, similarly, mentioned much less often than capture of living creatures. Apart from the case reported by Marsden, successful mating is suggested by Lubeck's account of a diminutive woman reputedly derived from intercourse between a Malay girl and an 'orang ratas'; but this is only marginally connected with orang pendek. Under the name 'segugu', orang pendek in the far south of Sumatra are claimed to have sexually assaulted local women several hundred years ago and thus given rise to hybrid offspring (Coomans 1929: 5). Conversely, southern Kerinci (Lempur) people say that female orang pendek, unable to find a mate, will kidnap solitary hunters and force them into intercourse with the aim of becoming pregnant (Martyr 1994: 2). A Kubu folktale in which a resin collector is similarly forced into sex by a female hantu pandak ('short spirit', a term sometimes applied to orang pendek) relates how the man finally escapes by inserting burning eaglewood into the creature's vagina, an act that explains the hantu pandak's continuing aversion to the smell of this substance (Rintjema 2001: 215–16; pers. comm., October 2005). Whether this is an exception to the generalization about people using fire against orang pendek is moot. Yet another report of human–hominoid intercourse concerns unidentified descendants (possibly still living in Sorolungun in Jambi province) of a man who produced children by an orang pendek female (Rintjema 2001: 213). In an apparently more general vein—and recalling Coomans' 1929 report—the Besemah people of southern Sumatra recognize certain lineages as descending from ancestors who married with 'cultivated' orang pendek (or 'guguh'; Barendregt 2002: 304, note 29).

Although apparently of a less historical character, comparable stories concern orang-utans that abduct either human males or females, force them into intercourse and produce hybrid offspring. Folktales of this sort have been recorded from both Borneo and Sumatra, thus the two areas where the apes definitely survive to the present (for Borneo, see St. John 1862: I: 22; Roth 1896: I: 350, citing C. Brooke 1866; Rutter 1999: 60–65; Appell 1991: 76, 80–81; for Sumatra, see Voorhoeve 1927: 111; Juara Ginting pers. comm. March 2006). Also common, especially in Borneo, are reports, some apparently well founded, of orang-utans accosting and attempting intercourse with human females (Harrisson 1955: 602, 631; Appell 1991: 76–77; Galdikas 1995: 84, 293–95). Except for the notion of man-grabbing sexually aggressive females, however and a story in which a bride is carried off by an orang pendek and never seen again (Martyr 1994: 1), abducting humans apparently plays a lesser part in local representations of the hypothetical creature. Also, in contrast to hominoid representations from eastern Indonesia, orang pendek are apparently not thought to kidnap children. Whether referring to capture, mating, or human abduction, moreover, such stories are with very few exceptions presented as straightforward historical incidents and not as timeless folktales or passages from myth.

In fact, orang pendek receive no mention in the older ethnographic literature, nor in collections of southern Sumatran folk narratives (e.g. Helfrich 1927a; Van Dongen 1931), even while beliefs and stories concerning both zoologically documented animals and a variety of spiritual beings appear in these sources with great regularity. Why this should be is puzzling, especially as the authors must surely have heard of the orang pendek. Most likely, however, the lacuna reflects the fact that orang pendek, as described both locally and by Europeans, do not clearly fit the category of 'spiritual being' while, simultaneously, not matching, any animal known to western zoology. As regards traditional narrative, another factor may be the creature's rarity, since myths and folktales most often feature local species that are well known, including monkeys and gibbons (see Bakels 2000: 124, regarding the 'important role' of siamangs and gibbons in Sumatran folktales). Interestingly, orang-utans do appear in folk narratives of northern Sumatra, where the creatures are rare but definitely attested; but here too they are less common than other animal characters, including tigers, mousedeer and monkeys. Except for Schlegel and Müller's essay (1839–44), early zoological studies, also, do not mention the orang pendek. But this is more easily explained by the lack of evidence for the creature's existence as a distinct species. In the twentieth century, a strictly unilinear view of human evolution —nowhere more clearly expressed than in the pronouncements of Dubois— would also have weighed heavily against the existence of any primate that sounded too human.

Comparison and conclusions

The image of orang pendek presented by Martyr and associates generally reveals little difference from earlier European writers' accounts of native ideas. Martyr's animal seems somewhat smaller, possibly with shorter body hair and head hair and with a notably different foot shape. Hair colour is also rather more various than the generally dark colour mentioned by local people, although this may partly be a function of methodology—Martyr being more sensitive to problems of translation and interpretation in regard to local colour terminology (1995c). Furthermore, her creature appears more arboreal in its habits than native representation allows. Much of this can be summed up by saying that Martyr represents the orang pendek as rather more apelike than do Sumatrans who, as noted, often describe it as a sort of human. On the other hand, there are noteworthy points of agreement between the two images: the erect posture and humanlike bipedalism, the broad, powerful build and the attributed diet. Neither image clearly includes speech or language nor, with the exception of inverted feet, what would seem to be impossible physical features or behaviours (for example, breasts so long they can be slung over the shoulders). Even the pointed head mentioned by Martyr echoes both a native account (Vogelpoel 1930) and the reputed sighting by Van Herwaarden (1924).

As European eyewitness accounts show substantial agreement with native representations, it is hardly necessary to compare these further with Martyr's account. In several respects, the most discrepant native image derives from Rokan descriptions of the creature locally called 'letjo', which is characterized as being no hairier than a human, having curly head hair, being slim rather than robust and according to one native source, even possessing speech (Rookmaker 1932). It may be no coincidence that prominent among Rookmaker's Rokan informants was a raja who was evidently central to the 1932 hoax.

The 'pinkish brown' skin of the Rokan figure does accord with a detail of Martyr's description, as possibly does an apparently apelike footprint. Nevertheless, the Rokan 'letjo' seems a quite a different representation from the orang pendek (or gugu or sedapa) of more southerly parts of Sumatra. More than these, it suggests in several obvious respects a basis in human beings culturally if not physically distinct from Rokan villagers and accordingly less of a connection with experience of some sort of non-human primate. On the other hand, the latter is indeed suggested by such simian features as the letjo's long arms, heavy eyebrows, absent chin and flat nose and the 'letjo mats'.[29]

Several factors may explain noticeable similarities between earlier European descriptions and Martyr's more recent and more systematically derived, account of orang pendek. Since Martyr too relies considerably on local informants (including their claimed sightings of the creature), one factor of course is that both draw on the same native representations. At the same time, both could substantially reflect experience of the same primate species, whether scientifically known or unknown. Some sightings may reflect sun-bears standing erect. In fact, Martyr interprets many informant sightings as involving either bears, large macaques, Agile gibbons, or siamangs (1995c). But bears especially are incapable of the rapid and sustained bipedal locomotion described for orang pendek and in other respects as well are far too different to be the sole, or even a main, source of the image.[30]

As already suggested, another possible source, at least of some part of the representation, are distinct human populations like Kubu, more closely associated with the forest than are the sedentary Malays who—from Marsden to Martyr—have provided by far the bulk of information about orang pendek. How far the Malay representation (or its partial European derivative) can be traced to actual physical differences is questionable, for Kubu do not in fact appear to differ significantly—in height, complexion, hair colour, nor, most importantly, body hair—from Malay villagers. Although Hagen (1908: 47) characterizes Kubu as 'small people' (with an average male height of 1.588 metres), Forbes described twelve individuals with a similar average height as being as tall as Malays of the Malay Peninsula (1885: 241). Also,

Van Dongen (1906: 235) described the Ridan River Kubu as no shorter than other Sumatrans. While Kubu can be darker than some Malays, like those resident in Kerinci, they are apparently lighter than others (Rintjema, pers. comm., February 2006). Hagen describes Kubu complexion as a 'rich reddish brown', with females being somewhat lighter than males (1908: 33). This is perhaps reminiscent of the Rokan 'letjo', as is Hagen's description of Kubu head hair as 'dark black brown to dark brown with a reddish sheen' (ibid.: 40) and as more inclined to curliness than Malay hair (Forbes 1885: 236; Hagen 1908: 39; see also Van Dongen's report of an individual with somewhat frizzy hair, 1906: 233). As noted, however, the Kubu described by Hagen and Forbes do not live as far north as the Rokan. What is more, other descriptions of Kubu hair colour (including Forbes' specification of 'jet black', 1885: 236) suggest little if any difference from Malay hair.

Perhaps more noteworthy is the report that most Kubu have relatively short arms (Hagen 1908: 47, 51), a feature which of course weighs against depictions of them as physically apelike and particularly against one interpretation of the long-armed creature Herwaarden reputedly observed. Malays evidently maintain a stereotype of Kubu as hirsute, wild and cultureless creatures not entirely dissimilar from orang pendek. But, ironically, it is Kubu and explicitly not orang pendek, who, according to Malays, live in trees. And it should also be recalled that, despite apparent similarities, Marsden's 'Labun' informants nevertheless distinguished 'kubu' from 'gugu'.

Cramer's 'orang hitem' was very likely a dark-skinned human and possibly a Kubu. Despite qualifications already registered, it is also conceivable that Herwaarden, influenced by native images, could have mistaken a frightened Kubu hiding in a tree for an orang pendek—as indeed might a Malay in similar circumstances. Noteworthy here is Van Dongen's description of an old Kubu woman and a boy whom he and his companions suddenly encountered while travelling through jungle:

> They were searching for edible roots and did not know we were coming. When they saw us and we them (we were already very close to them), they huddled together like frightened game that knew not how to escape. They uttered not a sound. Fearfully, they leered up as if to see where the blow would fall. No matter how much the jenang [a Malay agent and trader known to these Kubu] reassured them, they remained sitting until we had gone and then they made off.
>
> (Van Dongen 1906: 242)

One needs, however, to distinguish between occasional claimed sightings and a pre-existing general representation, with Kubu foresters being more plausibly linked to the former than the latter. In addition, human subjects, Kubu or otherwise, would appear to be ruled out for observations claimed by Martyr and her associates (and likewise for the remarkably similar creature spotted by Oostingh).

Ever since Schlegel and Müller interpreted 'orang panda' as a southern Sumatran name for the orang-utan, the red ape has regularly been proposed as the subject of native and European sightings.[31] The most articulate advocate of this view is primatologist Herman Rijksen. Rijksen argues that orang pendek reflect occasional encounters with orang-utans which survive in small numbers in southern and central parts of Sumatra. These moreover belong to a 'delicate, long-fingered variety' (not a species or sub-species) of the ape, with dark rather than reddish hair and more given to crepuscular habits, ground-dwelling and bipedal locomotion than are their congeners in northern Sumatra (Rijksen and Meijaard 1999: 63–64, 263–65). Although Sumatran orang-utans are currently attested only in regions north of Lake Toba, there is indeed evidence for the occurrence of small populations in the southern part of the island until relatively recent times. Nineteenth-century evidence includes two orang-utan skulls from Jambi province, deposited in one of the Dutch national museums (Schlegel and Müller 1839–44: 6), as well as the skeleton of a creature shot somewhere near Palembang and placed in a Florence museum in 1888 (Rijksen and Meijaard 1999: 55, citing Beccari 1904: 205). There is also Gibson's observation (1856), almost certainly of orang-utans, in the treetops by a jungle creek along the Musi River, also near Palembang. Similarly, Bickmore (1868: 408) describes an orang-utan captured in the region of Padang and sent to the Dutch governor, sometime in the 1860s.[32] Based on occasional published reports since 1917, a map produced by Rijksen and Meijaard shows twentieth-century sightings of orang-utans as far south as Bengkulu (1999: 62–63), while from their own surveys, the authors report the occurrence of the apes in areas where they were not previously known, at least as far south as the equator.[33]

Circumstantial evidence pointing to southern orang-utans as the source of orang pendek includes the previously mentioned reddish hair (found in a trap), the creature's overall size and shape described in a variety of accounts, the high forehead (reported by Martyr) and perhaps even the 'pinkish' complexion (suggesting the skin of younger orang-utans). Still, several considerations cast doubt on the connection. To begin with, evidence for *Pongo pygmaeus* in places associated with orang pendek is based largely on unsubstantiated information that has yet to be confirmed by zoological study (see Rijksen and Meijaard 1999: 55–58). As the distinct physical and behavioural characteristics posited by Rijksen are similarly conjectural, interpreting orang pendek as reflecting a hypothetical southern population of *Pongo* is, therefore, ultimately to explain one unknown with reference to another. Second, during the past twenty years or so, even people living in the most isolated parts of southern Sumatra have increasingly become familiar with orang-utans from media (especially television), zoos and pictures on banknotes. Thus, Sumatrans encountering orang pendek during the same period should either have identified what they saw as orang-utans (or 'mawas', using the Malay term) or should at least have compared them to orang-utans. Yet, this seems rarely to have been done. Two local observers of orang pendek described a resemblance to orang-utans they had seen in a circus and on television (Yanuar 1999). But others, including people who had observed orang-utans in a Bukittinggi zoo, distinguished the two categories (Martyr 1994: 12). Similarly, while some people in the Kerinci village of Lempur identified orang pendek with zoo orang-utans, others claimed these were different creatures (Rintjema 2001: 167).

Relevant to such dissociation is a distinction of names. Martyr's initially arresting report that some Sumatrans actually use 'orang hutan' ('people of the forest') for the creature elsewhere called orang pendek (Martyr 1994: 76; Rintjema 2001) may appear to confirm Rijksen's identification of the mystery creature with the known ape. But actually it does not, for only recently have many Indonesians begun to call the ape 'orang-utan' (or 'orang hutan'), a designation probably originating in a European misunderstanding. Indeed, everywhere on Sumatra, the orang-utan (*Pongo pygmaeus*) is called by quite another name, either Malay 'mawas' or a term related to this. Contrary to what Schlegel and Müller seemed to argue, moreover, the cognate 'maweh' does occur in southern Sumatran languages and dialects for 'orang-utan', as these authors themselves appear to confirm (1839–44: 12; see also Anon. 1926b). Also and somewhat ironically, these include dialects in which the orang pendek is called 'orang hutan' (Rintjema 2001). Obviously, southern cognates of 'mawas' are quite different from 'orang pendek', 'sedapa', 'gugu' and 'letjo'. It is indeed possible for distinct names, relating to largely or partly different representations, to have a single zoological referent (see Forth 2004a: 42), but this is unusual and, in any particular case, cannot simply be assumed.

A final problem with Rijksen's interpretation concerns the fluent bipedalism that so impressed Martyr, which suggests something different from an orang-utan, even a darker-haired variety that spends more time on the ground. Nevertheless, an explanation of orang pendek as a representation grounded in experiences of orang-utans would appear the most plausible, particularly if these have been generated by people unfamiliar with the ape encountering specimens sporadically. This is particularly so if such experiences have sometimes been combined with encounters with siamangs, another dark-haired and occasionally bipedal primate, thereby producing and sustaining a composite image. The interpretation is also not entirely dissimilar from the one offered by Dammerman (1924: 182), although the colonial zoologist traced orang pendek not to an extant southern population of *Pongo pygmaeus* but to a folk memory of the locally extinct ape, triggered mostly by native sightings of sun-bears.

An explanation of orang pendek as a non-human primate, either one related to the orang-utan or something like a large gibbon, has also been advanced by Martyr and associates. In either case, this is a 'terrestrial, bipedal ape' (Martyr *et al.* n.d.: 11) and by contrast to Rijksen's interpretation, represents a new, as yet zoologically undiscovered species.[34] Positing an undiscovered animal is obviously a less parsimonious explanation than identifying orang pendek as a variety of the known species *Pongo pygmaeus*. Nevertheless, it is supported by factors counting against Rijksen's interpretation; and a large gibbon-like creature, especially, sounds remarkably like the animal encountered by Oostingh (or would do so if the Dutchman had overestimated the height). Also, the Oriental zoographic realm, which includes western Indonesia, has recently been the site of several remarkable mammalian discoveries, so a new primate on Sumatra, even a bipedal ape, is not ruled out.

With regard to their findings, it goes without saying that Martyr and associates do not consider the orang pendek as possibly hominin. In the Dutch colonial press, the orang pendek was sometimes speculatively identified with a 'missing link', or as Dubois' *Pithecanthropus* (see also Rookmaker 1932). But it was left to the British anatomist and primatologist W.C. Osman Hill to make this argument systematically. From the existing literature, Hill (1945) reviewed descriptions of orang pendek and concluded, with some justification, that these sound more like what is now called *Homo erectus* than they do either bears, non-human primates, or human beings. For all that might weigh against an argument for *erectus* surviving in Sumatra, with some qualifications (such as the creature's evident lack of fire and other technology), Hill's demonstration of the more human appearance of orang pendek and moreover its similarity to palaeoanthropological reconstructions of pre-sapiens hominins, do make a kind of sense of the essentially ethnographic evidence of his time. Martyr's primatological evidence of course points in a different direction. What is more, certain features of orang pendek—such as the small stature and long arms—while not clearly consistent with *erectus*, arguably fit better a more recent, albeit hypothetical hominin (and a possible descendant of *erectus*) found in another part of Indonesia: *Homo floresiensis*.

In view of the smaller size of *floresiensis* and its interpretation as an island endemic dwarf, Rookmaker's suggestion that the Sumatran hominoid might represent a form of *erectus* that has decreased in size (1932: 219) appears almost prophetic. However, in the zoological context of Sumatra, the orang pendek is obviously more readily accounted for as a non-human primate, either a known species or a new ape related to known species. Comparison with *floresiensis* inevitably invokes comparison with eastern Indonesian folk categories like ebu gogo. But while in many physical respects the smaller Florenese figures are generally similar to the Sumatran creature, they are also clearly more hominoid and particularly so in respect of the attribute of speech and, in the case of ebu gogo, the pendulous female breasts. What is more, the eastern Indonesian figures appear somewhat more fantastically conceived than orang pendek, for example, in regard to the extreme voracity (swallowing plates and dishes) and anthropophagy attributed to both ebu gogo and the Sumbanese mili mongga.

I therefore conclude that ebu gogo, to speak only of this category, refers to something quite different from orang pendek. Zoogeographically speaking, eastern Indonesia lacks the apes that are crucial to interpretations of the Sumatran entity as a non-human primate. Flores and Sumba also lack populations of forest-dwelling food-collectors like the Kubu, who may also have contributed to Malay (and European) representations of orang pendek. Thus, if the eastern Indonesian and Sumatran images are connected, they are more likely linked as variant expressions of a pan-human cognition motivated by different empirical or historical triggers, than as reflections of zoologically closely related kinds.

Quite different from essentially ethnozoological interpretations of orang pendek as grounded in local experience of known or unknown primates, would be a suggestion that the category is entirely imaginary and is to be understood as a reflection of social factors decisively informing a particular cultural symbolism. In the present case, however, such an approach, although the one most likely to find favour among cultural anthropologists, is arguably the least parsimonious. This is not to say that orang pendek might not participate in some sort of cultural symbolism; after all, the representation does comprise some fantastic components and may be symbolically deployed to the same extent as are attested natural species. On the other hand, the case that the image is largely or entirely non-empirical has yet to be made. Even Dammerman, whose disdain for native representations bordered on what nowadays would be termed 'racist' and who deemed these inherently fantastic and completely untrustworthy (see 1924: 178, 181), interpreted orang pendek as a folk memory of locally extinct orang-utans.

The possibility of explaining eastern Indonesian hominoids in mainly symbolic terms will be further explored in Chapter 10. What has been demonstrated in the present chapter is that the orang pendek has a very probable zoological basis—and in this lies the importance of the case. If it could be shown conclusively that the figure reflects an undiscovered species—a new, bipedal ape or something like a large ground-dwelling gibbon—then we should moreover have evidence for an empirically grounded 'wildman' comparable to palaeoanthropologically grounded interpretations of *Homo floresiensis* as a similarly diminutive hairy hominoid contemporary with *Homo sapiens*, on the island of Flores, several thousands of years ago.

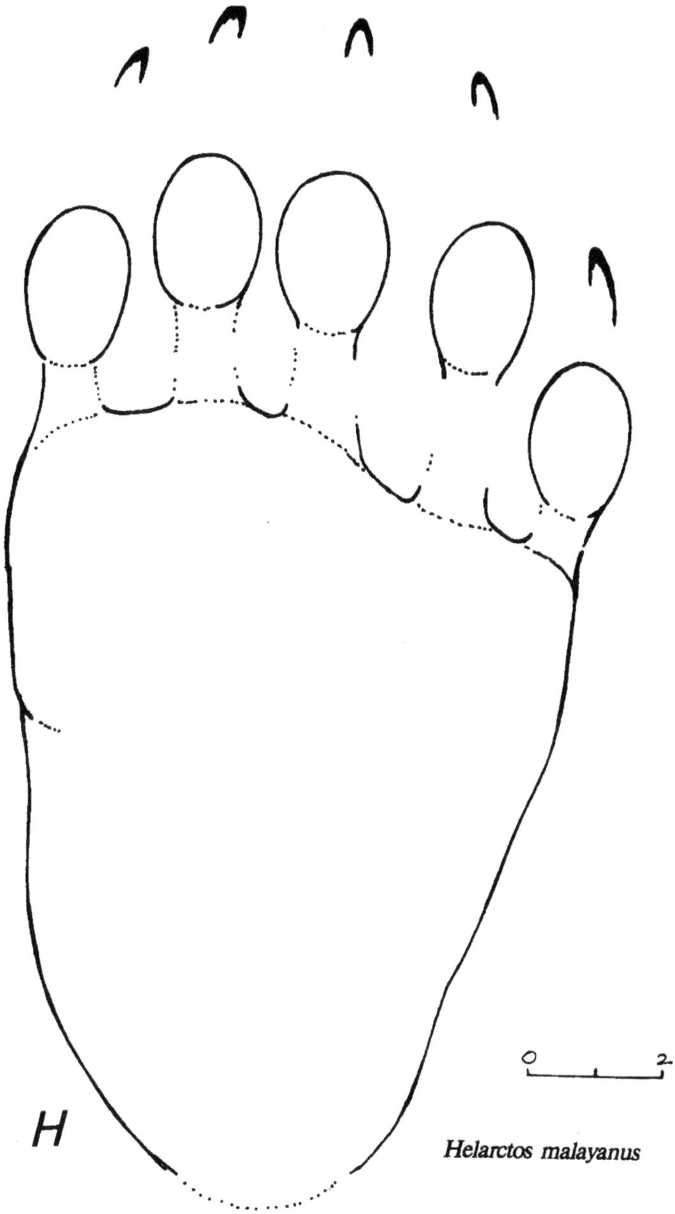

Plate 5.1 Sunbear print.

Plate 5.2 The 'kubu' reputedly seen by Walter Gibson in the mid-nineteenth century.
Source: Gibson (1856).

American Museum of Natural History

Plate 5.3 A sketch of 'Orang Pendek', probably based on Van Herwaarden's (1924) description.
Source: Heuvelmans (1995), with permission of Kegan Paul.

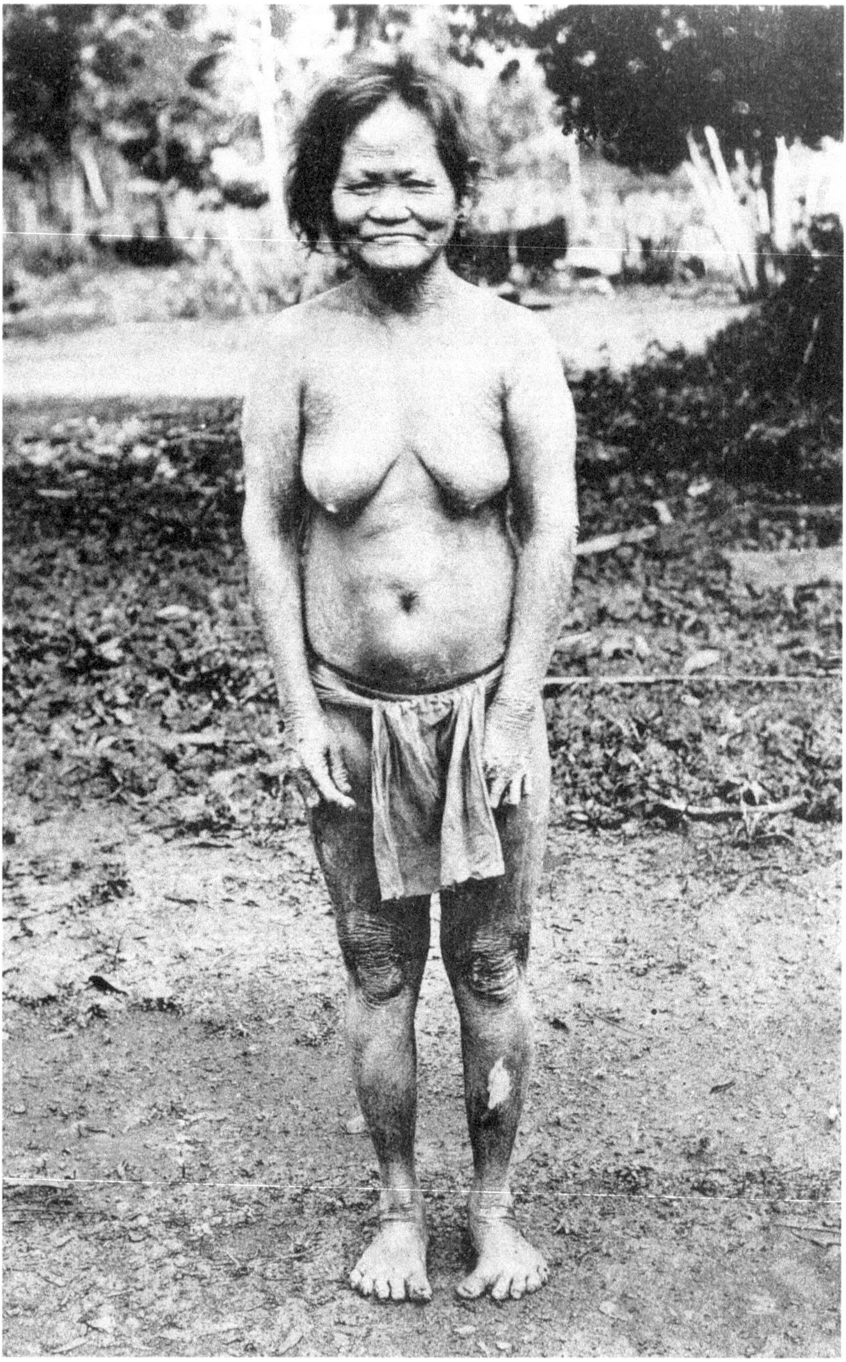

Plate 5.4 A Kubu woman (Nor) with skin disease and deformed feet.
Source: B. Hagen (1908).

Plate 5.5 Cast of a footprint discovered by Deborah Martyr and attributed to the orang pendek.
Source: G. Forth, with permission of the Natural History Museum, London.

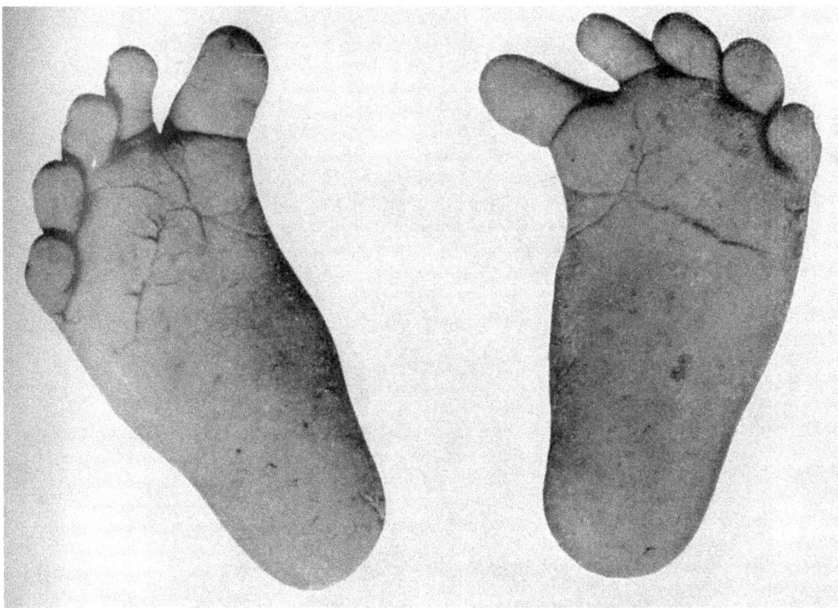

Plate 5.6 Deformed feet of an Igorot highlander, the Philippines.
Source: Worcester (1912).

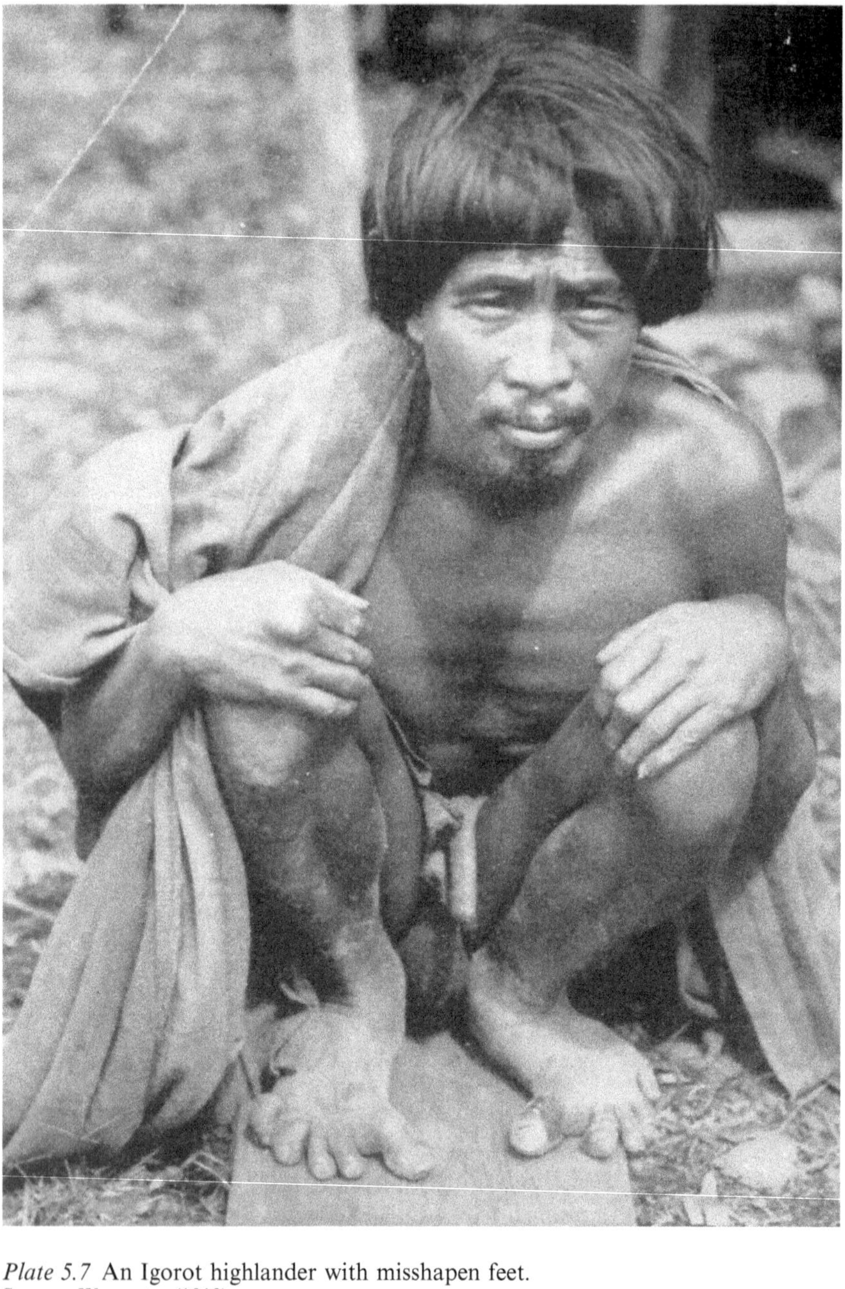

Plate 5.7 An Igorot highlander with misshapen feet.
Source: Worcester (1912).

6 Wildmen of western Indonesia and Mainland Southeast Asia

Biogeographically, the island of Sumatra is of a piece with Borneo and the Malay Peninsula, all of which in geologically recent times were joined by the Sunda shelf. Borneo is further linked with Sumatra in being the only other location where the orang-utan (*Pongo pygmaeus*) is known still to survive. Given the evidence for these apes playing a large part in the Sumatran representation of 'orang pendek', one might therefore expect to find comparable images both in northern Sumatra and in Borneo. But while the inference is supported for northern Sumatra, it is less so for Borneo. What is more, images of hairy hominoids appear to be almost as common in the Malay Peninsula and western Java, where modern populations of orang-utans are not attested.

Hominoids in northern Sumatra

In terms of language and culture, Sumatra is generally divisible into two halves. Southern and central Sumatra—the region of the orang pendek—is home to speakers of Malay, or rather 'Malayic' languages, also spoken in the Malay Peninsula and Borneo. By contrast, inhabitants of northern Sumatra speak non-Malayic languages, including the Batak languages (such as Toba and Karo), Gayo and Acehnese. Despite this ethno-linguistic divide, northern Sumatrans subscribe to representations of hairy hominoids as much as do southern Sumatrans. What is more, while southern and northern images are not identical, the differences pertain far less to their content—the way locals describe them—than to the way they have been interpreted by Europeans.

Immediately relevant to this contrast is the question of the geographical range of the Sumatran orang pendek. Several writers have claimed that orang pendek are not reported in northern Sumatra, or more particularly to the north of the Indragiri River (Anon. 1924a). This, however, is highly questionable, since very similar creatures have been described for Tapanuli and several more northerly regions, including Aceh and Gayo (see Map 5). The absence of orang pendek from northern Sumatra is pivotal to Dammerman's interpretation (1926: 196; 1929: 5–6); for according to the colonial zoologist, the image reflected a distorted folk memory of the orang-utan as a southern

Sumatran animal that was either very rare or recently extinct. In the same vein, Dammerman argued that where the animal survived in greater numbers (and was thus relatively familiar to local populations), as in northern Sumatra and Borneo, or had long disappeared, as in Java, the representation did not occur. As we shall see, he was wrong on both counts.

Generally similar to orang pendek are the 'homang' of the Toba Batak and the 'umang' (or 'omang' in an older transcription) of the Karo, who reside to the north of the Toba (see Map 5). By regular sound shifts, the two related terms are cognate with Malay 'kemang', often translated as 'dwarf' and generally referring to some sort of earth spirit (Joustra 1907: 30, 45; 1912: 38; Neumann 1951: 145). Nevertheless, Voorhoeve distinguishes the Toba homang from Malay kemang, remarking that only the latter are 'dwarfs' (Dutch, 'dwergen'; 1927: 115). Similarly, Hagen, referring mostly to the 'homang', states that people 'expressly consider [the creatures] not as spirits but as humans that possess supernatural power' (1883: 533, 534). Steedly, who has recently worked among the Karo (and who attributes Hagen's statement to the Karo umang), evidently agrees, noting that his 'characterization of the umang as a special kind of human (or rather quasi-human) beings is still current among Karo today' (1993: 259, note 2).

'Homang' and 'umang' denote largely identical images. Toba describe the homang as humanlike in both form and behaviour, resembling a 'wild cat with a human face' (Warneck 1909: 80) or a being half-human and half-animal which scratches and bites (ibid.). Heard in dense forests, their vocalizations comprise laughter and a nocturnal screaming recalling a small child (ibid.: 120). Similarly, the Karo umang, although unable to talk, can respond to human speech with sounds Steedly transcribes as 'hah, hah' (1993: 191–92; see also ibid.: 7). Describing homang in the past tense—as creatures killed off by expanding humans and which therefore had not been seen for several generations—American anthropologist Harley Bartlett, writing in the 1920s, focused largely on differences between homang and known apes. Although they lived much like orang-utans (Toba 'maos'), the homang looked very much like humans beings and had black rather than red hair. What is reputedly homang hair adorns a ceremonial spear at Silo Maradja. As regards size, the creatures, described by Bartlett alternatively as 'dwarfs' and 'spirits' inhabiting lowlands as well as mountains, were larger than the well-known siamang (Bartlett 1973: 50, note 34).

In an account differing from Bartlett's in several particulars, a Toba man, Guru Tinandangan, claimed to have seen a homang in the early twentieth century. This he described as having long red hair, large white teeth, long sharp yellow nails and inverted feet and as calling 'hu, hu' (Voorhoeve 1927: 115). It is obvious how, in most respects, the description suggests an orang-utan, but how far the alleged sighting report reflects widely held local images of the homang cannot be known.

Although similarly described as small, hirsute and reclusive 'wildmen' (Steedly 1993: 121), the Karo umang are represented as still extant. However, whether they are, for that reason, conceived any less fantastically than their Toba counterparts is moot. What are called 'umang houses' comprise cavities, caves, or rock shelters in cliff faces; these include a niche to the right of the entrance that umang reserve for their leaders (Westenberg 1892: 233; Neumann 1926: 4–5). Westenberg describes one such location as containing 'remarkable signs of decoration' and a 'roughly hewn human figure' (Westenberg 1892). Anyone entering umang territory should leave an offering, otherwise they will be pelted with stones. Like the Toba homang and like certain eastern Indonesian hominoids, umang will also abduct humans, who may disappear altogether or be absent for years. According to Westenberg, abductees especially include physically attractive people of both sexes and some abductions are thought to entail marriage with umang (1892: 232–33; Neumann 1926: 4, 5, 9). Karo also associate abduction by umang with obtaining the powers of a shaman or spirit medium, whose familiars the creatures then become (Steedly 1993: 5).

In contrast to the umang, the Toba homang appear more inclined to abduct children. These they detain only for a time, but afterwards such youngsters may suffer long illness and develop a fear of dogs (Warneck 1909: 80–81). Consistent with this last idea is the report that the homang themselves, like hairy hominoids elsewhere, are much afraid of dogs (Voorhoeve 1927: 115–16). According to the aforementioned Guru Tinandangan, homang cause people to become lost in the forest and then kill them (Voorhoeve ibid.). In the first respect at least, they recall certain eastern Indonesian representations as well as the southern Sumatran orang pendek.

In a more positive vein, the Karo umang provide knowledge of medicinal herbs (Steedly 1993: 5) and, according to a Karo legend (Westenberg 1892: 234), precious metals and agricultural fertility. Consistent with the last benefit is an apparently more general identification of umang as original occupants of Karo territory (Neumann 1926; 1927; Juara Ginting, pers. comm., July 2004). In this connection, Neumann argues that the representation of umang as ape-like creatures has become conflated with that of a fully human yet culturally distinct population which preceded the Karo ancestors and left behind such material traces as tombstones found in forests (1926: 4–5). 'Umang house' ('rumah umang'), it should be noted, also refers to these tombs (Anon. 1924c: 890).

Apparently the only fantastic physical peculiarity of both the umang and the homang is one also attributed to the southern orang pendek: feet 'turned the wrong way' (Bartlett 1973: 50; Westenberg 1892: 231). Referring to the Karo umang, however, Neumann (1926: 4; 1951: 210) denies this attribute, stating that the creatures simply walk with toes turned under. As noted in Chapter 5, Rijksen and Meijaard (1999: 62–63) ascribe such a method of walking, on the fore-limbs, to one variety of orang-utan and propose this as the source of the belief in the inverted feet of the orang pendek.

How exactly northern Sumatrans classify umang and homang is quite unclear. Several authors construe them as spirits. Warneck (1909: 80) even refers to the homang as 'begu homang', 'begu' being a general term for 'spiritual being'. But Westenberg appears to distinguish the Karo umang from begu (1892: 221). Hagen similarly asserts that northern Sumatrans regard homang not as spirits but as a kind of human being; yet he further characterizes the creatures as a variety of 'orang bunian', mysterious beings able to render themselves invisible (1883: 533, 534; and see Chapter 5, note 24). Pleyte (1894: 285), on the other hand, seems to distinguish 'bunians' from 'omang'. According to Steedly—who alternately designates the umang as 'spirits' and 'wild men'—what fascinated Europeans was the similarity of umang to the dwarfs and trolls of their own folk traditions (1993: 259 n2). If so, this fascination did not result in such a volume of colonial writings as did the generally comparable orang pendek. Equally remarkable is the virtual absence of reports of Europeans having sighted umang or homang creatures, or their footprints. Nor for that matter do reports of putative sightings by natives appear in the colonial press.

The absence of claimed experience of umang or homang would obviously accord with a European understanding of them as a kind of spiritual being, an evaluation for the most part not applied to orang pendek. Also striking a contrast with the orang pendek is the relative prominence of umang and homang as characters in Toba and Karo myths and folktales (Pleyte 1894; Joustra 1904; Voorhoeve 1927)—although their occurrence in these genres should not have distinguished them as non-empirical beings for the European recorders.[1] The narratives depict Toba homang as gambling with dice (Voorhoeve 1927: 43, 51) and having the power to cause continuous darkness (Pleyte 1894: 89–90). In the same genre, they are also described as occupying an underworld, even though in mundane discourse the creatures are 'usually represented as living on earth in the forests' (ibid.: 129). For the homang, another factor possibly affecting conceptions, or expectations, of their likely occurrence is their reputed extinction well before the twentieth century. At the same time, it is arguable that Europeans encountering something hominoid but apparently not human would likely have taken it for an orang-utan, a creature whose presence in the forests of northern Sumatra was, at the time, already well established.

Although occurring outside of Toba or Karo territory, a possible exception to the absence of reported sightings of northern Sumatran hominoids (apart, perhaps, from the one claimed by Guru Tinandangan) concerns an account from South Tapanuli, to the southwest of Lake Toba (Anon. 1932). From a hill overlooking a river bank, an unnamed native tax collector and a local penghulu (headman) observed what they described as a group of 150 to 200 naked creatures, apparently carrying heavy stones and wood to construct a dam for fishing. These 'little men' were covered in sparse hair, particularly around the neck, forearms, chest and lower legs. As 'many of the creatures looked like apes' (or monkeys; Dutch 'aap') and since they were 'good climbers' and able to jump from branch to branch, their activities would seem to have been partly arboreal. By the same token, it is probable that what the two men saw was indeed some kind of monkey or ape. Yet, other than fabrication or exaggeration, what would account for their large numbers, or their appearance of constructing a dam, is unclear. The report concludes by noting that Tapanuli terms for such creatures include—besides 'orang pendek'—'orang leso' or 'orang letjo'. As noted in the previous chapter, the last term is the name employed for orang pendek in the Rokan, a district to the south of Tapanuli. Whatever its source, therefore, the representation appears continuous with southern Sumatran images of orang pendek; hence the report is arguably the exception that proves the rule concerning the absence of sightings identified with the northern categories homang and umang.[2]

In the far north of Sumatra, creatures comparable to the umang and homang of the Karo and Toba are called 'mante' (or 'manti', 'manteu'). C. Snouck Hurgronje, in his famous work on the Acehnese (1906: I: 18–19), includes a brief (and sceptical) note on the 'mante', described as naked hominoids covered with thick hair and reputed to inhabit a particular mountainous region. Comparatively speaking, a more intriguing detail is Snouck's mention of a story concerning how, a couple of generations previously, a pair of mante—a 'husband and wife'—were captured and brought to the Sultan of Aceh. As neither would speak or eat food offered to them, the pair eventually died. Besides similar tales told in southern Sumatra, one is immediately reminded of capture stories from eastern Indonesia and particularly the 1996 story from Sumbawa (Chapter 4).

Another account comes from an elderly Acehnese hunter who claimed often to have seen such creatures in local forests. These he described as about a metre tall and covered in thick hair but lacking the long arms of an ape. In this report (Anon. 1935), the creature is referred to as an 'orang pendek'; but the local name is given as 'kukuman', a term not found in any dictionary of Acehnese. Describing it as a 'human being' and quite different from an orang-utan or a bear, the hunter further stated that 'kukuman' will steal rice and buffalo milk left in huts by collectors of forest products.

A variant of 'mante', the term 'manti' occurs in the language of the Gayo people, who reside immediately south of Aceh. The referent is similarly described as 'a sort of wild man' (Hazeu 1907, Djajadiningrat 1934), or a 'kind of human being' (Melalatoa and Sitanggang 1985), covered in hair and living in mountain forests. An entry in Melalatoa and Sitanggang's dictionary further suggests that the creatures are physically small. Like the Acehnese variant, Gayo 'manti' is employed metaphorically for humans, specifically for people who appear dumbfounded or startled (Djajadiningrat 1934). One might also insult someone by saying they 'descend' from manti (Hazeu 1907), but whether this entails a belief in mating and miscegenation is not specified.[3] The only colonial report of the Gayo manti appears to be a published letter from a police captain (A. Agerbeek, no date given, cited in Coomans 1929: 24) who identifies the creature as the 'so-called orang pendek' and describes it as having large hands with sharp nails and digging up worms, turning over fallen tree trunks in search of food and raiding remote gardens and 'potato fields'. Although he mentions only native Gayo sightings of manti, the captain himself claimed to have observed the creature's footprints, child-sized but wide and distinguishable from sun-bear tracks.

Despite his implicit denial, Agerbeek's description does indeed suggest bears. On the other hand, since orang-utans are found throughout this entire region (Aceh, Gayo and the Karo and Toba territories), the most obvious explanation is that all of the northern Sumatran representations are substantially grounded in experiences of these apes. Of course, the evidently more hominoid images are locally distinguished from apes, as shown, for example, by the Toba contrast of homang as black-haired and orang-utans as red-haired. Yet the colour contrast is consonant with Rijksen and Meijaard's (1999) identification of a dark-haired variety of *Pongo pygmaeus*, which these primatologists suggest is also present in southern Sumatra and which is there reflected in native reports of the orang pendek.

Wildmen in Borneo?

In view of the foregoing, it may appear odd that creatures like orang pendek are reported infrequently from Borneo, the only place outside of Sumatra where orang-utans are known to occur. Among the few accounts is John MacKinnon's well-known reference to a creature called 'batutut' (1974: 114), described by a native guide after MacKinnon had encountered small, wide footprints that appeared to have been made by a hominin. The similarity with native descriptions of Sumatran orang pendek is indeed striking: the batutut stands between 1.2 and 1.5 metres, walks erect and has a 'long black mane'. It is furthermore shy, nocturnal, averse to light and fire, inhabits deep woods and feeds on river snails. The name 'batutut' reproduces the creature's hooting call. From the guide's description, however, the batutut seems more aggressive than its possible Sumatran cousin, for it is said to kill people and feast on their livers. Also unlike the orang pendek—but similar to the northern Sumatran homang and umang—it will abduct children, although only temporarily and without harming them.

While noting differences between batutut and sun-bear tracks, MacKinnon suggests that the footprints he saw may have been made by a larger, undiscovered bear (1974: 114).[4] Yet in view of the evidently ursine origin of tracks locally attributed to the Sumatran orang pendek, it is probable that orang-utans play an equal or greater part in the total representation of batutut. In fact, some Dusunic-speaking groups in western coastal Sabah are reported to use 'batutut' (or 'batutud') for the orang-utan (Porodong, pers. comm., December 2005).[5] Also noteworthy is the Bornean category 'mayas rambei', recorded by Beccari (1904: 199) as the native (apparently Iban or Bidayuh) name for a 'form' of the orang-utan distinguished by its 'long hair'. 'Rambei' is evidently 'rambai', an Iban term for head or body hair that is long and 'flowing' (Richards 1981: 296). It thus recalls the 'mane' attributed to the batutut. But it is equally applicable to the long hair on the head, back and limbs of adult orang-utans.[6]

In Sabah, other groups of Dusun—including the Rungus, Kadazan and Orang Sungai—employ a completely different word, 'kogiu', for the orang-utan (cf. Rijksen and Meijaard 1999: 419; Evans 1953). Quite possibly, therefore, people who use this or yet another name for *Pongo pygmaeus* may reserve 'batutut' for a more fantastic representation. From casual enquiries I myself made along Sabah's Kinabatangan river in July 2005 (where I was fortunate enough to observe an orang-utan in the wild), Orang Sungai described 'batutut' as a very different sort of mysterious creature, resembling a cat rather than a primate, able to change its size and being in some unspecified way dangerous to humans. One man knew 'batutut' only as an ill-defined bogey invoked to frighten youngsters, a usage obviously consistent with the hominoidal creature's reputation for child abduction and homicidal attack.

Peninsular figures and the ape-men of Trolak

So far as is known, the orang-utan no longer occurs on the Malay Peninsula and may not have survived there for thousands of years. Possibly brought from Sumatra (from where peninsular Malays themselves derive), the ape's Malay name 'mawas' and cognates, however, continue to be used by both Malays and non-Malay speaking peninsular aborigines, as references to seemingly fantastic creatures.[7] These creatures reveal affinities to similar hominoid categories elsewhere—both further north on the Southeast Asian mainland and well to the south, in western Java.

According to Wilkinson, under the name 'mawas' the orang-utan is known by repute in Malaya 'as a semi-human monster with arms of iron and a pot-shaped head in which it cooks its food'; it is also believed to kill and eat humans (1932: 117). Although cooking with the head and anthropophagy seem not to be reported elsewhere as attributes of orang-utans, Wilkinson's characterization immediately illuminates several earlier accounts. The Malays of Kedah are thus said to apply 'mawas' to a 'race of wild people' closely resembling 'the Mawa or long armed gibbon' and who, in place of a bone in the forearm, possess a 'piece of sharp iron which serves the double purpose of an arm and a cleaver for cutting wood' (Anderson 1824: xxxi–xxxii). The last idea is also recorded by Skeat and Blagden (1906: II: 283), who further mention as synonyms of 'mawas' the Malay names 'hantu mawas' (roughly 'mawas spirit' or 'spirit mawas') and 'orang-outang' (here to be understood as 'wild people'). Similarly implicating a group of ethnic others, Begbie (1834: 5–6) suggests that the sharp forearm simply reflects a mistake by Malays, who confuse cleavers always carried by 'essentially migratory' people—apparently aborigines—for the arms that wield them. However that may be, Skeat's partial identification of the image as a 'spirit' (hantu) is particularly interesting, since it suggests another instance in which an empirical creature (human or animal) that is rarely or never seen has become 'spiritualized', or subject to a fantastic representation.

Designating the notionally sharp forearm is 'tulang mawas', or 'bone of the mawas', a phrase further defined as 'prehistoric iron tools' (Winstedt 1964). These are evidently the same as 'besi mawas' (mawas iron), described by Wilkinson as 'iron tools left in the forest by forgotten [sic] miners' which natives interpret as 'broken bits of the elbow-knife of the mawas' (1932: II: 117). Such a mawas bone is also characteristic of creatures named 'mawas' known to the Chewong (or Siwang) aborigines of the Malaysian state of Pahang (Needham 1956: 56). The Chewong 'mawas' is further depicted as a hominoid with inverted nostrils which subsists on an unknown diet. They are quite harmless, very shy and relatively scarce and formerly, when they were more numerous, the aborigines would kill them with blowpipes (Needham 1956; Howell 1984, who gives the name as 'mawes'). This last detail, especially, suggests an animal rather than a spirit. Indeed, consistent with her generally naturalistic portrayal (even the inverted nostrils suggest an exaggeration of a physical feature), Howell cites a local claim that the creatures do 'not really' belong to the general Chewong category of spiritual beings ('bas'). At the same time, neither Needham nor Howell describes these mawas as hairy. Nor does either mention orang-utans, or indicate awareness that 'mawas' is the Malay name for the ape.[8]

Also in the Malay Peninsula, Malay-speaking Jakun people in Selangor attribute sharp forearms, more particularly 'a sharp blade-like bone in their right forearm which they use in felling trees', to beings called 'hantu sakai' (Skeat and Blagden 1906: II: 283). 'Hantu' is 'spirit, ghost', whereas 'sakai' is a pejorative applied generally to Malaysian aborigines. These hantu sakai are otherwise described as small, hairy 'people', living in remote places and possessing a fine sense of smell. Essentially, this is the same representation as what other peninsular peoples call 'mawas'; but in a more general comparison, their specification as small and hairy provides a crucial link with Sumatran figures. Further noteworthy is an implicit parallel between, on the one hand, similar Sumatran Malay representations of the Kubu and orang pendek (or gugu) and, on the other, what would appear to be peninsular Malay views of both aborigines (called 'sakai' and 'orang hutan', or 'forest people') and creatures called 'mawas' or 'hantu sakai'. 'Hantu sakai' is a category attributed to the Jakun who, it is important to note, are distinct from non-Malay-speaking aboriginal groups like the Chewong. Of course, peninsular aborigines, too, subscribe to images like the 'mawas'. But then so do the Kubu in regard to orang pendek.

The Selangor hantu sakai are also comparable to hominoids that peninsular aborigines regard as the owners of wild dogs. David Labang, a native of Borneo at one time employed by the National Parks and Wildlife Department of Sarawak, reports having seen one of these creatures in the peninsular state of Pahang (see Map 5). He describes his subject as 'a tiny naked man about three feet high with short, thick seemingly curly hair, thin limbs, brownish skin, rather hairy body and looking exactly like a human being' (pers. comm., cited in McNeely and Wachtel 1988: 261). The individual was spotted in proximity to leaves of 'wild ginger', a plant which, interestingly, Martyr has claimed to be a major food source of both the Sumatran orang pendek and Pig-tailed macaques. Except for the body hair, Labang's creature might alternatively have been an exceptionally short aborigine. That the hominoid fled on seeing him is not surprising as Labang was armed and at one point contemplated shooting. McNeely and Wachtel suggest that Pahang aborigines designate these creatures as 'uyan'. This, however, is incorrect: 'uyan' is in fact a Kelabit (Bornean) name for 'pygmies' which Labang himself applied to what he saw (pers. comm., January 2006). Although I have been unable to find 'uyan' in any Kelabit dictionary, it resembles 'uyang', the term for 'human being' (cf. Malay 'orang') employed by the Kenyah and Apo Kayan of northeastern Borneo. It also appears in the same languages in names for the orang-utan (Rijksen and Meijaard 1999: 419).[9]

In recent years, Peninsular Malaysia has been the site of a rather different sort of hominoid representation. Since 2005, Southeast Asian media have been reporting sightings and footprints claimed to reflect large hirsute beings identified with the American category of 'bigfoot' (see Chapter 8). Concerning a giant creature, these apparently bear some relation to a traditional Malay category, 'gergasi' ('giant') or 'orang gergasi', denoting a 'tusked man-eating ogre' (Wilkinson 1932: II: 187). Comparable reports of large hirsute hominoids have, however, been appearing in Malaysian and Singaporean newspapers at regular intervals for over half a century. Evidently the first instance occurred in late 1953. As reported in the *Malay Mail* (1, 2, 4 and 8 January 1954) and the *Straits Times* (1, 2, 6, 13 and 16 January 1954), between December 24th and 28th 1953, a young Chinese woman, a little Malay girl, a Tamil man and a Malay corporal each separately encountered a trio of ape-like creatures comprising 'two men and a woman' in or near the Trolak Rubber Estate in the Malaysian peninsular state of Perak (see Map 5). Generally characterized as 'humans' or 'people', albeit with features of apes, the three beings were described as 'fairly tall' (over six feet according to one account) and light-skinned, with long jet black hair and long teeth, 'fangs', or 'tusks'. Although references to hair mostly concerned their long locks, the creatures were said also to have hairy bodies; the forearms were especially hairy and the two 'men' had moustaches extending to the waist. One news report further described the creatures as 'well built' and another mentioned 'jutting brows'. Yet another twice remarked how they exuded an 'animal smell'.

Vocalizations attributed to the three beings included grunting, laughter (they are said to have laughed when the Tamil, fleeing in fear, fell over and rolled down a hill) and speech which, according to the Chinese girl, sounded like a bird croaking. On one occasion, they were seen digging tapioca roots. Consistent with their implicit classification as some sort of human being, the three figures wore bark loincloths and carried knives. In two of the four encounters, the hairy female approached the human witness while her male companions remained in the background. According to opinions of Europeans cited in the news reports, the creatures did not behave aggressively, but appeared to want to establish some sort of contact with local people.

What these figures were is anything but obvious. The clothing and knives rule out any sort of animal. Also, the relatively large size excludes Malaysian aborigines, or any local people, as does the light skin. In fact, both size and complexion and perhaps the hairiness as well, suggest Europeans. Given the time of year and the fact that the four eye witnesses comprised a curiously comprehensive cross-section of local ethnicity, it is not impossible that the incidents reflected a prank played on unsuspecting natives by saturnalian white colonials.[10]

Whatever the explanation, the Trolak creatures do not particulary resemble the sharp-armed 'mawas', or for that matter the Sumatran orang pendek. On the other hand, two similar stories related by Wavell (1958: 161–62), one from Malaya and another from Borneo, arguably lend further support to the idea of tales of hairy hominoids reflecting orang-utans or other primates. Particularly the bark loincloths and perhaps the weapons might suggest a connection with an image from Vietnam, which I presently describe, but not to such an extent as to render other interpretations less preferable.

More reports from the Southeast Asian mainland

Sharp-armed figures comparable to the Peninsular Malaysian mawas and hantu sakai are also encountered further north on the Southeast Asian mainland. Among animals native to the Nam-Noung region of Vietnam, local Moï people describe nomadic and vegetarian 'wild men' distinguished by a 'sharp membrane like the blade of a knife which they use for cutting away the bush to make a path through the forest' (Maître 1912: 62). Standing about 1.5 metres tall, the creatures also possess a thick coat of reddish hair, small humanlike feet and a tail like a monkey's. They are further said to lack elbows and knees and therefore have inflexible limbs. Partly it seems because they were once hunted and eaten by villagers, these 'wild men' (for whom Maître gives no local name) have since become so rare they are no longer encountered.

Were it not for the tail and perhaps the inflexible limbs (possibly a reflection of primate bipedalism), the Vietnamese creatures might well be taken for orang-utans. Other features of Maître's account recall associations of 'wild men' elsewhere, particularly in Indonesia. Certain Rhade villagers on Vietnam's Darlac plateau claim descent from one such creature, identified with the proper name of 'Kjhat'. According to the legend, one day, while stealing fish from the villagers' weirs, Kjhat brought along his infant daughter, whom he placed on a rock. When the weir owner appeared, the wildman fled, abandoning his child, whom the owner subsequently raised. After having the sharp membrane excised from her arm, the wild female was married to the weir owner. Their descendants are distinguished by a 'remarkable beauty' to the present day (Maître 1912: 62–63). Comparable to traditions from several parts of Indonesia concerning wildman changelings, this tale is also reminiscent of the Sumbanese story of Apu Kalita, similarly described, despite her relatively hairy condition, as a beautiful woman. The Rhade, it should be noted, speak a Malayo-Polynesian language as do the Sumbanese and most other Indonesians; hence one cannot rule out a cultural historical explanation of these narrative similarities.

Evidently not beautiful, but implicitly also distinguished by sharp forelimbs, are hairy hominoids inhabiting the mountainous Dak-Mil region of Vietnam, near the Cambodian border. These were described by the Australian journalist Wilfred Burchett, who recounts a tale in which one specimen was captured (Burchett 1965: 161–67). Able to cut down palm trees 'using the side of their hands' (ibid.: 165), the creatures are described, somewhat differently from those sketched by Maître, as covered in thick black (rather than red) hair, with shoulder-length head hair and proportionally long arms. They walk erect and can run very fast. Their humanlike footprints indicate that they move backwards as well as forwards. This attribute arguably recalls the Sumatran orang pendek; yet unlike the Sumatran creature, the hominoids described to Burchett do not possess inverted feet, but only move backwards—an inversion of quite another sort. No indication is given of the wildmen's height or size; nor, unlike Maître's creatures, is there any mention of a tail. The Dak-Mil hominoids are, however, characterized both as very strong and extremely timid.

Unlike virtually all figures described so far, the creatures of Dak-Mil are not entirely lacking in culture. Burchett relates a Vietnamese eyewitness account of a male specimen found 'cowering' inside a mountain cave containing a hearth of three stones, the remains of a fire and scattered animal bones, a sort of bed of bark and creepers and several sharp stones. M'nong highlanders say the wildmen make fire by rubbing together pieces of dry bamboo. But they use fire not to cook, but merely to warm their food, which comprises freshwater crabs, eels, small mammals and birds, as well as fronds that grow at the tops of palm trees. Further revealing technological ability, a small garment of bark covered the cave-dweller's genitals. On the other hand, the creatures appear not to be able to speak, producing only a sort of non-linguistic babbling.[11]

Although they are apparently well known to M'nong highlanders, Burchett gives no local name for the Dak-Mil creatures. It is therefore impossible to say how they might be related to Vietnamese, Thai and Laotian hominoids mentioned in recent cryptozoological writings (mostly on the internet) and variously named as 'nguoi rung' (Vietnamese for 'forest man), 'da nhan' ('wildman', 'savage' in Vietnamese) and 'briau' (a Laotian name). A Thai name given by McNeely and Wachtel (1988) is 'tua yeua'. Inferring that it must be both rare and shy and thus evidently like the creature described to Burchett, these authors describe the tua yeua as 'something very much like an orang-utan'.

Karen people, they add, know the animals very well but are afraid to shoot them (1988: 257).

Serving as his guide in the 1960s, Burchett's Vietnamese informant claimed to have encountered the Dak-Mil wildman in the cave in 1949, thus some fifteen years before recounting the story to Burchett. The Vietnamese together with several M'nong companions seized and bound the creature and took it to their camp. But as it would eat very little, they decided to return it to its cave. The hairy troglodyte did not, however, survive the journey and so was buried where it expired (1965: 165). A curiosity of this account is that the chapter in which it appeared was deleted from the English original of Burchett's book, but not from the French translation (from which the foregoing is taken). One might speculate that Burchett's editors thought his credibility as a journalist and the value of his previous publications on the Vietnam War, might suffer from his apparently credulous treatment of a story which, to many westerners, would appear thoroughly incredible.

What Burchett's Vietnamese companion actually experienced cannot be known. Both information provided by M'nong highlanders generally familiar with such creatures and the relatively detailed description of the specimen in the cave give the impression of a non-sapiens hominin. On the other hand, it is curious that no details are given of the head and face. What is more, implicitly sharp fore-limbs and inverted locomotion respectively link the Vietnamese figure with Malayan and Sumatran categories—and therefore possibly also with orang-utans.

There have been other reputed eyewitness accounts of hairy hominoids from the Southeast Asian mainland. Like David Labang's 'uyan' and the Dak-Mil wildman taken captive by Burchett's Vietnamese guide, these mostly concern creatures encountered by outsiders on a single occasion. Unlike these, however, two reports come from Europeans. The first is by A.D. Frederickson, a Fellow of the Royal Geographic Society. In an account of his Asian travels, Frederickson includes a sketch of 'the wild man of Johore' (1889: xii, 276), a creature he had observed in the Malaysian state of that name sometime in the 1870s. All Frederickson says about the evidently male 'wild man' is that he had been found living deep in the forests of Johore (now Johor) subsisting on fruits and roots and was about to be 'conveyed to the coast for shipment to some society at Calcutta' (ibid.: 276–77). Showing an individual with a head and face covered in very long hair and with shorter hair on the body, Frederickson's accompanying sketch (see Plate 6.1) gives every impression of someone with hypertrichosis.[12] In fact, it seems likely that the sketch was inspired by the Burmese Moung-Phoset (born in 1850 or 1854) or another member of his family afflicted with this condition (see Plates 6.2 and 6.3), who were much photographed about this time and who travelled abroad as far as Europe and America (Bondeson and Miles 1996). It may well be significant, then, that an unspecified 'two or three' of the illustrations included in Frederickson's book are described as 'facsimiles of photographs' (Frederickson 1889: vii).

Also dating from the nineteenth century, the other European observation concerns what was taken to be a large unidentified ape inhabiting the mountains of Tenasserim in southern Burma (Blanford 1891: 9–10). Two Englishmen were collecting birds on Mount Muleyit, when one of them, a man named Davison, encountered what he thought were monkeys in 'a very dense part of the forest'. From a distance of 10 feet, a closer look revealed one specimen standing erect with a height of about 1.2 metres; the creature was tailless and its hair was a 'deep ferruginous colour'. As Davison's gun was loaded with just a half-charge of the smallest shot, he did not fire at the creature. Not far from the same spot, a Captain Bingham later came across the carcass of a female tailless ape, over a metre in length and with 'long grizzled red hair on the outside of the limbs'. So decomposed was the carcass that only the skeleton could be salvaged and this too was subsequently lost. On another occasion, Bingham reported seeing a group of four or five 'large tailless monkeys' at the foot of Mount Muleyit, but these appeared to be black rather than red. Remarking that these creatures were too large to be gibbons, Blanford (1891: 10) suggests that the only animal corresponding to the descriptions of these two 'excellent observers' is the orang-utan. But this he dismisses on the grounds that the orang-utan is so well known that the observers would have recognized it and he suggests that a more probable candidate 'may have been a tailless, or nearly tailless' macaque monkey. Whether any sufficiently large, tailless Macaque occurs in Burma, or elsewhere in Southeast Asia, is another matter. If it does, then previously discussed and similarly red-haired creatures from Vietnam and Malaya might have an empirical primate referent other than *Pongo pygmaeus*.[13]

Back to the islands: Java and Bali

Images of sharp-limbed hominoids occur again among the Sundanese people of western Java. While linking the current absence of orang-utans on Java with a supposed paucity of 'stories of ape men' there, the zoologist Dammerman, at the time posted in Jakarta, nevertheless cited reports of 'large apes (or monkeys)' named 'aul' (Dutch 'aoel') that possessed 'forearms so sharp that they can cut a banana trunk in two' (1926: 196). Rather uncharacteristically in view of his tendency to dismiss the orang pendek and similar representations as 'native fantasies', the zoologist moreover left open the possibility that this image may reflect an unknown Javanese primate.

Only Dammerman appears to ascribe sharp forearms to the aul. But information on the category is generally scarce and, in a way recalling literature on the Sumatran orang pendek, ethnographers and folklorists have had little to say about it. In a book on Sundanese custom and folk religion, Hidding, for example, simply lists 'aul' as one of a variety of named 'ghosts and spirits which people sometimes perceive' (1935). More recent sources (for example, R. Satjadibrata 1954), however, indicate that the category is the subject of Sundanese folktales. Usually referred to as a kind of 'animal', albeit a 'fabulous' one, the aul is characterized as possessing either a human form or that of an ape or monkey (Rigg 1862: 27; Coolsma 1930: 27). While otherwise comparing it to a human, Coolsma says it walks on four limbs. The aul occurs in old forests and mountains and is in the habit of spitting. It also utters a nocturnal cry (specified by Coolsma as a shrill 'ul, ul') which gives the creature its onomatopoeic name (ibid.; Erlinga 1984: 40).

Another attribute of the aul is inverted feet (Rigg 1862). Such feet may be what Coolsma alludes to when he says the aul's 'legs are pulled upwards'. Back-to-front feet obviously link the creature with the orang pendek, known just across the Sunda Strait from western Java (see Map 5). Yet other information, including the reputation for spitting, connects the aul with the orang-utan (on apes spitting see Yerkes and Yerkes 1929; Maple 1980; Erin Vogel, pers. comm., January 2007). Indeed, Raffles (1817: II: lxxxix) listed 'áwul' as the Sundanese name for the orang-utan and while this cannot of course be taken as primatological confirmation, it is noteworthy that Coolsma (1910; 1930) provisionally makes the same identification.[14]

Insofar as central and eastern Javanese recognize hominoidal beings in any way comparable to images found elsewhere in Southeast Asia, these are quite different from the apelike and mostly naturalistically conceived aul, let alone the orang pendek and mawas as represented by Sumatran and peninsular Malays. The same applies to a legendary figure from Bali. This makes a certain ecological sense, for although both Bali and Java still contain tracts of mountain forest, the two islands are among the most densely populated parts of Southeast Asia. By the same token, western Java, the territory of the aul, is less populated and generally more rugged and mountainous than are more easterly regions.

What one does find in Bali and Java are representations of monstrous females with long, pendulous breasts. Probably the best known of these is the dramatic character named Rangda, the Balinese queen of witches, depicted as possessing, in addition to sagging breasts, long dishevelled hair, long fangs and long nails or claws. Although sometimes described as 'hairy', the adjective evidently refers to Rangda's long head hair; when represented in Balinese performances, her (clothed) body is 'striped black and white' (Covarrubias 1937: 326; Belo 1949: 12). Linked with an historical personage (the widow of a Balinese king and mother of Prince Erlangga) Rangda exists for Balinese largely as a mask and costume. In this respect, Rangda suggests a comparison with the bogey masks of Flores and particularly the Nage bogey named 'gogo meo' and 'ebu gogo'—as she also does by virtue of her antipathy to children and cannibalistic habit, a feature of all Balinese witches and indeed witches almost everywhere.

Very similar to Rangda, although evidently pertaining more to a category of spirits than an individual character, is the Javanese 'wewe' or 'wewe gombél', manifest as an ugly, shrivelled old woman, tall and thin, with loose hanging hair and breasts hanging to the ground. The malevolent beings are further described as abducting people, especially children, whom they carry away under their pendulous paps and hide in some 'sequestered nook' (Van Hien 1994: 129; cf. the Sundanese 'wewe', specified as a 'malevolent female forest spirit', Erlinga 1984: 840). Variants of this long-breasted female demon are the 'hantu kopek' of the Malays (McHugh 1955: 101) and the 'rogon gaiyoh' of the Tempasuk Dusun of Borneo (Evans 1953: 31–32). All of these images find further counterparts among eastern Indonesian spiritual beings described in Chapter 3. By the same token, all are obviously more comparable to the widespread figure of the 'pontianak' than to apparently more naturalistic representations of hairy wildmen.[15]

In regard to the foregoing, it is a point of considerable analytical importance that nowhere in Sumatra, Borneo, or the Southeast Asian mainland does one find long, pendulous breasts attributed to hirsute hominoids that people claim occasionally to encounter in mountain jungles and other remote locations. In other words, wherever in this region one finds both hairy hominoids and malevolent female spirits (in other words, instances or variants of pontianak), the two representations and the particular features of long breasts and hairy bodies and also sharp limbs, reveal a thoroughly complementary distribution. A similar contrast applies to the deployment of the two sorts of figures as bogeys, female instances of which appear often to feature the long breasts (although one interpretation of the Bornean 'batutut' specifies this as a bogey). Given their reputation as abductors, the northern Sumatra umang or homang could conceivably assume a bogey function; yet the southern Sumatran orang pendek by all accounts does not. By contrast, as shown in Chapter 3, in parts of Flores the two images—hairy wildmen and sometimes long-breasted female bogeys—appear to have been subject to a partial conflation.[16]

'Forest people' ('orang-utan') reconsidered

From pendulous breasts we return to sharp limbs. If not quite all, then most images that include sharp limbs point to a source in the orang-utan. This interpretation could even encompass the Sumatran orang pendek, for southern Sumatrans describe both the 'orang hutan' (the local name for orang pendek) and the orang-utan (*Pongo pygmaeus*, called 'maweh' in the Lempur dialect) as possessing 'bones of iron', a reference to their common reputation as creatures endowed with extreme strength (Rintjema 2001: 136). In a similar vein, inhabitants of East and Central Sumatra describe hairy, cave-dwelling 'giants' called 'orang gedang' ('gedang', or 'gadang' means 'big') as possessing forearms so sharp they can fell a tree in a single blow (Moszkowski 1909: 997). Although sharp forearms are not explicitly attributed to orang pendek, it is relevant nonetheless that Martyr has recorded 'orang gedang' as an alternative name for the normally smaller hominoid (Chapter 5).

The Sumatran phrase 'bones of iron' is of course virtually identical to the 'arms of iron' attributed to the 'mawas' of the Malay Peninsula, understood by Wilkinson (1932) as a fantastic representation of a barely known ape.

Citing similar ideas from the peninsula, Moszkowski (1909) infers that the referent of the Sumatran 'orang gedang', also, may be the orang-utan. As to the possible empirical basis of this idea for the orang pendek, we should recall reports of this creature holding its arms outwards and forwards while moving bipedally through forest (see Chapter 5), a form of locomotion that could indeed suggest a hairy manlike creature slashing or cutting vegetation with sharp or 'iron' arms or forearms. As an alternative primate reference, one might consider gibbons, the metallic limbs then possibly reflecting the way these smaller apes hold their extremely long arms aloft while moving bipedally. Found in Java and the Southeast Asian mainland as well as on Sumatra and Borneo (though the largest species, the dark-haired Siamang, occurs only in Sumatra and the Malay Peninsula), gibbons, however, are significantly more familiar to local people than are orang-utans, even in areas where the red apes still occur.[17]

Interpreting Mainland Southeast Asian and Javanese hominoids as ultimately grounded in experience of orang-utans naturally raises questions about the disappearance of the apes from regions where they are no longer zoologically attested. Some contemporary writers (e.g. Jardine 1833: 68) specified the Malay Peninsula as one of the few areas where orang-utans survived in the nineteenth century, but the basis of their claims is not specified.[18] Fossil evidence indicates the presence of orang-utans in most lowland forest areas of continental Southeast Asia until the end of the Pleistocene (about 12,500 years ago) and in Vietnam and the Malay Peninsula in post-Pleistocene times (Chuan *et al.* 1995: 260, 261). At least two modern primatologists, however, have suggested that the apes may have survived in some interior locations as recently as the sixteenth or seventeenth century (Rijksen and Meijaard 1999: 40–41).

Other questions concern the name 'mawas' and cognates, applied both to the empirical orang-utan in Sumatra and in a few Bornean languages and to a rather more fantastical representation in Peninsular Malaysia. From the fact that both peninsular Malays and Malaysian aborigines speak of sharp-limbed 'mawas' as a contemporary presence, one cannot of course infer that orang-utans are still extant on the peninsula.[19] What is far more likely is that Malays brought the term 'mawas' to the peninsula, together with knowledge of the apes, when they began migrating from Sumatra some two thousand years ago. With equal probability, the term was then adopted by Malaysian aborigines (like the Chewong and Batek Negrito). The occurrence of cognates of 'mawas' as names for *Pongo pygmaeus* in a few Malayic languages of Borneo (for example, Iban 'mayas') has also been interpreted as reflecting borrowing from Malay (Sutline and Sutlive 1994; A. Adelaar pers comm. February 2006).

It is perhaps less certain that names for the orang-utan in Batak languages, like Karo 'mawas', Simalungun 'mawas', Toba Batak 'maos', Acehnese 'maweh' or 'mawaih' and Gayo 'mawas' (Saragih 1989; Djajadiningrat 1934: 42; Hazeu 1907; Melalatoa and Sitanggang 1985)—none of which is a Malayic language—also derive from Malay. Nevertheless, comparative evidence suggests that ‘mawas’, as a reference to an empirical primate, is not a particularly old term in western Indonesia and Malaysia and moreover that it has a Sumatran origin.[20] If this is correct, then important implications follow. If *Pongo pygmaeus* has long been extinct on the Malay Peninsula—as per the current zoological consensus—then in the fantastic image of thesharp-limbed creatures to which the imported name ‘mawas’ is now applied, we see clear evidence of a process of mythification accompanying an ever diminishing familiarity with an empirical animal: as it were, the ‘real mawas’. Something different may have occurred in southern parts of Sumatra, where the orang-utan, if still surviving, has become extremely rare. Here, rather than a referential change, the name may have undergone a conceptual expansion. In West Sumatra and South Tapanuli, ‘mawas’ can refer not only to the orang-utan but to primates in general, including monkeys and even human beings (Ian Singleton, pers. comm., February 2006). According to Singleton, a primatologist working in Sumatra, monkeys are distinguishable as ‘mawas monyet’ (‘monyet’ is Malay for ‘monkey’ or ‘ape’) and humans as ‘mawas orang’. ‘Mawas orang’ (roughly ‘human mawas’), he adds, sometimes includes the orang pendek and probably Kubu people as well. Interestingly, ‘mawas orang’ also occurs as an apparent reference to the orang pendek among Malays of northern Pasaman, in central Sumatra (Rintjema 2001: 135–36).[21]

If north Sumatran names for the orang-utan like Karo ‘mawas’, Toba Batak ‘maos’ and Acehnese ‘maweh’ reflect a Malay loan, a question arises as to how the attested apes may have been designated prior to the borrowing. It is not impossible that terms like ‘umang’, ‘homang’ and ‘mante’, now denoting seemingly more fabulous hominoids, may once have been applied to the red apes, or to representations based on these. If so, then the introduction of ‘mawas’ may lexically and conceptually have disjoined the empirical primates and similar yet more fantastic images nowadays associated with the distinct and presumably older, northern Sumatran terms. The disjunction could even have promoted a development of these categories in the direction of fantasy. Such a development would then account for the initially curious circumstance that, contra Dammerman, hominoidal images on Sumatra are no less fantastic and are in certain respects arguably more so, where orang-utans are definitely present than where the animals are either currently absent or far rarer.

At the same time, the Karo umang’s more fantastic impression could be ascribed to a conflation, such as Neumann claimed, of an ape-like image with an originally different image of an aboriginal population. In both Malaya and northern Sumatra, one also cannot rule out an ideologically mediated experience of contemporary forest-dwelling peoples like peninsular aborigines as a factor shaping representations maintained by dominant sedentary populations (even though the sharp-limbed ‘mawas’ appears to be recognized by this minority as much as by majority Malays). Among Sumatran Batak peoples, including the Toba and Karo, a comparable referent may be found in the Lubu, a group of foresters living in the territory of the Mandailing, the southernmost of the Batak populations. Noteworthy in this connection are representations, maintained at least partly by Mandailing, of Lubu as dirty, foul-smelling nomads afraid of water and susceptible to skin disease, who are despised but also feared owing to reputed spiritual powers (Lekkerkerker 1916: 117–18, citing Kreemer 1910; Loeb 1935: 295–96). In regard to physical if not spiritual attributes, the description is obviously redolent of nineteenth- and early twentieth-century accounts of the Kubu, one possible source of the southern Sumatran orang pendek (see Chapter 5).

Returning to the question of orang-utan survival: slightly more substantial than indications of recent presence of the apes in the Malay Peninsula are reports from Java and particularly western Java, the territory of the Sundanese 'aul'. In 1697, the navigator Francois Leguat reported seeing a tall, female specimen of an 'extraordinary Ape' somewhere near Batavia, now Jakarta (Leguat 1891: II: 234–35). Although he says he observed it several times, Leguat does not give a name for this ape. He represents the 'Animal' in a particularly anthropomorphic way, describing it as residing in a small 'house', neatly making a 'bed' with 'pillow' and 'coverlet' every day and binding its head with a cloth whenever suffering from headache. However, since Leguat later refers to the creature being sent to Europe, it would seem that what he observed was a captive animal which, as he remarks, 'might have been taught all these tricks'. (It died in one of the ships of Leguat's fleet off the Cape of Good Hope.) Another curiosity is the resemblance between Leguat's reference to the female creature covering its genitals with one of its hands and the earlier report by Bontius—to which I return just below—of another female, apparently of the same kind, doing much the same. Leguat was told that this 'particular Species of Apes' was 'to be found only in the Island of Java, but all were not of this Opinion'.

Half a century later, the English navigator Charles Noble reported experiences of a young, tailless, particularly humanlike and apparently captive ape, somewhere near Batavia or 'Bantam' (Banten, near Java's western end). The creature, for which Noble does not provide a name, is further described as bipedal (walking 'upright with great ease'), just over a metre tall and possessing a face almost bare but with 'head, eyebrows and lower part of the chin very rough'. It was fond of grasping Noble and made a 'low, pensive sound as if whining and crying'. Noble's statement that the creature 'was very often found in the woods' might be taken to apply to the entire species rather than this particular specimen, but this is unclear. Like both Leguat (1891: II: 235) and Bontius (1931: 285) before him, who recorded local opinions that the animals they observed were ape–human hybrids, Noble suggested that the creature (or its kind?) might derive from intercourse between forest animals and 'Malayan female slaves' who had fled into the woods to escape the cruelties of their Dutch masters (1762: 61–62).

In the latter part of the eighteenth century, Louis Relian, a surgeon stationed in Batavia, reported that 'orang outang' were found 'in the mountains of Java' but were more numerous in Borneo (Dobson 1953–54: 8, citing Buffon 1789). Better known than all these reports, however, is the earlier account penned by another physician, the seventeenth-century Dutchman Jacob de Bondt (called Bontius). Also stationed in Batavia, Bontius (1931) claimed to have observed several specimens of a humanlike ape, both males and females, and, like the other authors just mentioned, he described these as walking erect. Bontius records the local name of the creatures as 'ourang outang', or 'woodmen', a name that is actually Malay rather than Javanese or Sundanese. In so doing, he may have given rise to a misunderstanding reflected in European vernacular naming of *Pongo pygmaeus* to the present day (Yule 1903: 643–44), since 'orang utan (or hutan)'—'forest person'—mostly refers to human beings. Nevertheless, the Dutchman's report has commonly been understood as a reference to the red ape (see, for example, Dobson 1953–54: 2). Bontius does not state explicitly that he saw the creatures in Java. Yet he specifies as a Javanese belief the notion that they are able to speak but refuse to do so for fear of being put to work. According to Moszkowski (1909: 997), in Peninsular Malaysia the same idea is applied to the orang-utan, represented as a human that is clever enough to remain silent in order to avoid corvee and paying taxes (1909: 997).[22] In addition, Bontius's evaluation of the Javanese idea as 'really ridiculous' suggests that the attribution of language was inconsistent with the generally animal-like character of what he saw—thus tending to contradict any interpretation of his subjects as being, simply, *Homo sapiens*.[23]

There are, then, some grounds for supposing the orang-utan, though decidedly rare, was still extant in western Java during part of the second millennium. Even though orang-utans usually move quadrapedally, the early reports of their walking erect are of no great concern, as they do sometimes do so; and it may be supposed that it was to this occasional and for them remarkable, habit that these early writers referred. If any one of them had actually observed a human being (as has sometimes been suggested of Bontius), the bipedal gait, as a trait expectable in humans, would not likely have been remarked. On the other hand, Javanese knowledge of the apes in the seventeenth century, as in the twenty-first, could have derived either from imported specimens or tales of travellers returned from Sumatra or Borneo. In either case, the sharp-armed western Javanese aul, like its counterparts in the Malay Peninsula, is credibly accounted for as an image grounded in some experience of *Pongo pygmaeus*. And if the experience has not been direct, then as we have seen from Moszkowski's description of the 'orang gedang' as a creature with sharp forearms (1909), the aul could reflect a similarly equipped image that has crossed the Sunda Strait from Sumatra.

Whether local human minorities in Java may also have contributed to the aul is doubtful, or at any rate less obvious than in the case of Sumatran or peninsular Malay figures. For Java, one might think of the Kalang, originally a derogated group of forest-dwelling humans reputed to possess tails and hair on the lower halves of their bodies (Van Hien 1912: 111, cited in Wieringa 1998; Kalff 1925; Dammerman 1926; Seltmann 1987). Kalang, however, reside (or resided) in central Java rather than the western region associated with the aul and their reputed physical attributes appear mainly to reflect their myth of origin, which specifies derivation from a woman who mated with a dog (rather than a primate). Moreover, while Kalang are said to possess tails, such an appendage is attributed neither to the aul nor, of course, to the orang-utan.

Viewed in relation to the aul, the 'mawas' of the Malay Peninsula, various Sumatran figures and even the tailed hominoids described by the Moï of Vietnam, the cave-dwelling creature captured by Burchett's Vietnamese informant seems of a different order and less obviously attributable to experience of either apes or culturally distinct humans. Unless one supposes the account to have been fabricated or grossly exaggerated, it is difficult to avoid the conclusion that the Vietnamese encountered some kind of pre-sapiens hominin. Parallels with Van Herwaarden's orang pendek immediately come to mind, except that the Dutchman's report is more explicable in relation to a body of local lore pertaining to the category in question. Indeed, Burchett's story concerns a single incident and though the creature apparently instanced a category known to M'nong highlanders, details of this category perhaps point in somewhat less of a hominin direction than does the account given by Burchett's Vietnamese guide. By the same token, this particular capture tale suggests a circumstantial fantasy comparable to the Malaysian media story concerning the creatures that reputedly appeared in Trolak, in the Malaysian state of Perak, in December 1953.

However they might be explained, like the orang pendek of southern Sumatra, all the figures reviewed in this chapter are described as rarely encountered or no longer extant. With the possible exception of the Karo umang, all moreover appear decidedly marginal to traditional social and religious life. Available ethnography reveals no particular cultural value attaching to the images, nor therefore any reason why local communities should be inclined to fabricate or perpetuate them. Clearly, a degree of fantasy is involved in their representation. But this does not contradict an interpretation of the images as elaborations of more basic, zoological precepts, or cultural memories of these.

Plate 6.1 The 'wildman of Johore'.
Source: Frederickson (1889).

Plate 6.2 Nineteenth-century Burmese afflicted with hypertrichosis.
Source: With permission of the Natural History Museum, London.

Plate 6.3 Nineteenth-century Burmese family afflicted with hypertrichosis.
Source: With permission of the Natural History Museum, London.

7 Other Asian hominoids

Similar images of hominoid creatures in western Indonesia and proximate parts of Southeast Asia might be explained either by experience of common primate fauna or, in the case of the Malay Peninsula at least, cultural relatedness. As regards the wider distribution of such images in Indonesia and Malaysia and also in the southwestern Pacific (see Chapter 9), one might further consider a common Malayo-Polynesian linguistic and cultural heritage. By contrast, the occurrence of representations similar to the orang pendek or ebu gogo in more distant parts of the Asian continent challenges both cultural interpretations and explanations based on recent historical experience of known animals.

The 'nittaewo' of Sri Lanka

According to reports originating in the nineteenth century, Sri Lanka was once home to mostly small hairy hominoids called 'nittaewo'. For students of Flores hominoids, one of the more remarkable aspects of the Sri Lankan wildmen is an interpretation proposed some sixty years ago by the British anatomist and primatologist W.C. Osman Hill (1945). Not only did Hill construe the nittaeweo as a species closely related to the Sumatran orang pendek; he further argued that both could be understood as dwarf descendants of 'Pithecanthropus' (*Homo erectus*), made small by evolution in similar island environments. This of course replicates the recent interpretation of *Homo floresiensis* advanced by the new hominin's discoverers. As will become clear, there are other possible interpretations of the Sri Lankan creature. But the nittaewo assume yet another significance in relation to Flores discoveries. For not only are the former reputedly extinct, but the story of how local humans accomplished their extermination bears a striking resemblance to the Nage legend of ebu gogo.

Knowledge of nittaewo derives from the Vedda people, traditional hunters once linguistically and phenotypically distinct from the majority Sinhalese population. The specific source is Vedda inhabiting the Lenama (or Leanama) district of Panama Pattu in eastern Sri Lanka.[1] Especially in regard to nittaewo physical appearance, published information is limited and mostly comprises accounts by English explorer Hugh Nevill (1886a), naturalist Frederick Lewis (1914, 1915) and A. de Zylva (1886), a district headman (mudaliyar) and Nevill's initial informant. Each of the European authors draws on several local accounts. While some were provided by Sinhalese, the ultimate source in each case appears to be Vedda.

According to Nevill's Sinhalese informants, nittaewo were covered in shaggy red hair and possessed long and powerful 'claws'. Two Vedda hunters rejected this description as reflecting confusion with a rare and locally extinct sun-bear, claiming instead that nittaewo were 'rather dark'. The Sinhalese De Zylva's account is somewhat fuller. Nittaewo were covered in long, thick hair 'like that of the sloth-bear' and were 'of a dark or monkey colour'; sloth-bears, it should be noted, are generally black, but are 'very occasionally brown' (Gurung and Singh 1996: 16). Like humans, the nittaewo walked erect and their hands, legs and feet were 'almost human', being distinguished mostly by elongated 'claws'. The head was large in relation to the body and 'bull-shaped' (possibly indicating a thick or short neck). Summarizing reports of people 'who profess to have seen them', De Zylva describes nittaewo height as varying from 1.2 to 1.8 metres. Lewis, by contrast, says less than a metre and adds that nittaewo females were shorter than males (1914: 289).

Lewis further describes nittaewo as tailless and walking erect. Only their legs were hairy, resembling in this respect Wanduroo monkeys (apparently the Wandura, or Purple-faced langur, *Presbytis senex vetulas*); the upper body was like a human's. Although nittaewo had short arms, they possessed strong hands and long, powerful nails which enabled them to tear apart small animals (ibid.). Employed also to kill human beings, these nails (or claws) are the most consistently described physical attribute of the creatures. There are no further indications of the size and general form of the nittaewo. Other than the 'bull shape' mentioned by De Zylva, no account describes the head or face; thus there is nothing to indicate whether these features were hominoid, simian, or neither. Lending an apparently human aspect to the nittaewo, however, Lewis's principal informant described them as 'a little people' (1914: 289). Nevill's Vedda sources similarly characterized nittaewo as a 'race of men', while according to his Sinhalese informants, they were 'half man and half brute' (1886a: 67, 66). Indeed, Nevill himself interprets the reports as reflecting a category of human beings, more specifically an impoverished group of food-collectors or a former population of slaves.

In regard to vocalization, Nevill's Sinhalese characterized the nittaewo as uttering only 'brutish noises'. In contrast, Vedda told him that they possessed a language which sounded like Telugu, spoken in southern India (1886a: 66–67). Lewis, too, reports a 'language', intelligible to some Vedda but sounding 'like the twittering of birds' (1914: 289); De Zylva, on the other hand, compares it to the chatter of monkeys. Some accounts describe nittaeweo as residing in trees, or tree-top platforms covered in thatch (Nevill 1886a) and as good climbers (De Zylva 1886: 68). However, the creatures are also depicted as inhabiting 'rocks' (Nevill 1886a) or 'caves, hollow trees and crevices' (Lewis

1915: 128). According to Sri Lankan legend, it was in a cave that the creatures met their end.

If in some respects, such as cave-dwelling and possibly red hair and bipedalism, nittaewo appear similar to images reviewed previously, they are distinguished from these by greater arborealism, extreme aggression, man-killing and a carnivorous diet. The beings also composed relatively large groups—'gangs' of ten to twenty individuals (Lewis 1914), or 'troops' of ten to forty (Lewis 1915: 128). Only collectively were nittaewo able to kill their prey; these included mouse-deer, hares, squirrels, iguana and tortoises, which they would surround and seize with their bare hands (Lewis 1914: 289). As this should suggest, nittaewo lacked technology and had no knowledge of fire. One account (Noone 1945: 263) alludes to an earlier report of the creatures using implements of stone, but this identifies no source. In addition to predation, nittaewo would take meat left out to dry by Vedda hunters. Some authors construe this as stealing meat (Hill 1945; Rambukwella 1963), but Nevill, the original source of this information, does not actually mention theft. Moreover, the scene he describes somewhat recalls a former silent trade, in which Vedda hunters would leave meat for Sinhalese smiths, who were then expected to reciprocate with arrowheads or tobacco (Knox [1681] 1983: 62; Lewis 1914: 286; Spittel 1933: 73). Nittaewo were fearful of Vedda, particularly their dogs and weapons (principally, bows and arrows); they were also afraid of water buffalo. Nevertheless, the creatures would attack any sleeping Vedda they encountered, falling upon him 'in a mass' and disembowelling him with their 'talons' (Lewis 1914: 289).

Continually fearing attack, the Vedda finally resolved to exterminate the nittaewo. This they accomplished in much the same way as the Nage wiped out their local wildmen, the ebu gogo: they drove them into a cave, piled wood in front of the entrance and set all alight, creating a fire which burned for three days (Nevill 1886a: 66–67). The location of the cave seems no longer to be known. As to when this is supposed to have occurred, several accounts suggest a date in the late eighteenth or early nineteenth century. At least two reports, however, locate the nittaewo holocaust in the time of informants' 'grandfathers', which might suggest a genealogical compression reflecting an association of the tale with the generation of story-tellers.[2] Speaking in the early twentieth century, one man described his grandfather joining in the 'burning out of one of [the nittaewo] encampments' (Lewis 1915: 129), a specification which suggests that the creatures inhabited various locations. Nevertheless, like the Nage legend of ebu gogo, the Sri Lankan representation is focused on a single place (albeit one no longer known) and the nittaewo are securely located in the past.

Such similarities are obviously important for determining the status of Florenese narratives, which it will be recalled concern not only the ebu gogo, but also the ana ula and lae ho'a of other parts of Flores (Chapter 3). Occurring in places as far apart as Sri Lanka and eastern Indonesia, they might suggest a widespread mythological theme of the sort commonly registered by folklorists. Yet with the possible exception of one or two stories from Oceania (Chapter 9), tales in which hairy hominoids are destroyed by fire are not apparent in the folklore of other parts of the world and within Southeast Asia the theme does not definitely occur anywhere outside of Flores. In Chapter 3, I suggested that, whatever the ontological status of the Nage ebu gogo, the story of their entrapment in a cave and destruction by burning may not be original to this particular representation and could have been adopted from another part of Flores. However, as cultural or historical connections between eastern Indonesia and Sri Lanka are wanting, similarities between the legend of the nittaewo and Florenese tales of extermination by fire are almost certainly coincidental. In this light, their resemblance challenges the historical veracity of the ebu gogo legend considerably less than does the occurrence of similar stories of wildman destruction in different parts of Flores.

At the same time, a certain ecological coherence attaching to the Sri Lankan and eastern Indonesian legends suggests that the two traditions may reflect a convergence. If the nittaewo were based on an empirical species, the fear of attack which led Vedda to wipe them out may have been connected with increased stealing of meat resulting from new territorial competition comparable to the Nage encounter with ebu gogo and the wildmen's theft of Nage crops. Fifty years before Nevill wrote, thus about 1835, large game was still plentiful in the Lenama region, where the last nittaewo resided; however, by the 1880s the game was much reduced owing to the introduction of firearms (Nevill 1886a: 68). As 1835 may not be too long after the nittaewo extinction, it is therefore conceivable that, owing to a reduction of game elsewhere, Vedda hunters in the late eighteenth or early nineteenth century had become more attracted than previously to the plentiful wildlife of Lenama and hence had come more and more into contact with nittaewo. Noteworthy in this connection is Nevill's remark that, after the Vedda had killed all the nittaewo, their territory became a favourite Vedda hunting ground.[3]

If nittaewo were real creatures, however, this ecological context does not preclude their existence as bears, a possibility considered by several authors (Spittel 1936; Hill 1945; Rambukwella 1963). Apart from the dark or reddish shaggy coat, bears are obviously suggested by the sharp claws, carnivorousness, arboreal habits and association with caves. The 'twittering' attributed to nittaewo may similarly be attributed to the ursine habit of 'sucking and nibbling their forepaws' (Hill 1945: 261). Rather than the less carnivorous, mainly nocturnal and almost exclusively quadrupedal sloth-bear (*Melursus ursinus*), the nittaewo image specifically suggests the more aggressive Ceylonese sun-bear (possibly *Ursus inornatus*; Hill 1945: 260), an animal more given to standing erect and which, owing to its inclination to attack humans, was exterminated sometime before the early nineteenth century (ibid.; Nevill 1886b). It hardly needs mentioning that humans killing off bears owing to their savage temperament provides a direct parallel to the legendary Vedda extermination of the nittaewo, an event which seems moreover to have occurred about the same time. Vedda were certainly skilled in uses of fire. Apart from smoking meat and fish and smoking out the hives of a venomous species of honey-bee (Bailey 1863: 282, 290), Vedda fired parklands in the dry season to attract deer to fresh grazing (Deraniyagala 1992: 384, 423, 424). Yet I have found no evidence for Vedda hunting or killing bears, let alone using fire and smoke for this purpose. More directly arguing against bears as a main source of the nittaewo are the creatures' bipedal locomotion and their occurrence in relatively large groups, both obviously suggesting a hominoid character. While bears generally are largely solitary animals, an editor's note in Hill (1945: 260) refers to a report of a Vedda being killed and eaten by 'a mob of six bears'. But these were apparently sloth-bears, not sun-bears, the more likely model for the nittaewo.

An interpretation of nittaewo as creatures of generally hominoid form need not be attributed solely to the presence of monkeys in Sri Lanka, or even the especially humanlike 'forest apes' reported in the eighteenth century by Francois Valentijn (1978: 181).[4] For in several respects the nittaewo bear a palpable resemblance to the Vedda themselves. Not only were both 'hunters', sharing similar items of diet (such as tortoise and monitor lizard), but until the late nineteenth century Vedda too dwelt in caves (see e.g. Rütimeyer 1903). Also like the nittaewo, Vedda would shelter in hollow trees (Sugathapala 1972: 4) and sleep in 'tree platforms', particularly when elephants were about (Deraniyagala 1992: 389). Finally, references to nittaewo having large heads echo a fifth-century account of Vedda by Palladius, who described the Sri Lankan hunters as 'small-made, shaggy, [and] large-headed' (Sugathapala 1972: 1).

These parallels tend to suggest that nittaewo reflect, in part at least, an image of the Vedda perpetrated by the Sinhalese. The problem with this interpretation is that the nittaewo derive not from Sinhalese but from Vedda tradition. Yet there is another way of viewing the evidence. The extent to which the representation of nittaewo recorded by Nevill was a Vedda one might be doubted, since by the time he wrote, 'pure' Vedda were reckoned to be virtually extinct in southeastern Sri Lanka. In addition, both Spittel (1963) and Kadirgamar (1963) question the Vedda identity of the two informants Nevill describes as Vedda. Nevill himself is quite clear that these men were Vedda; both were related to Korayela, a Vedda who died around 1870, and who is described as having been a storehouse of traditional Vedda knowledge. Nevertheless, it is noteworthy that, as Buddhists and nominal vegetarians, Sinhalese cultivators disparaged Vedda practices of hunting and flesh-eating. Although abduction is evidently not a charge laid against nittaewo, Sinhalese claim Vedda once kidnapped non-Vedda children (Bailey 1863: 291). Nineteenth-century Kandayan Sinhalese also compared Veddas to 'wanderoo' monkeys; they did so apparently not with reference to their appearance, but to similarities between simian mating habits and Vedda monogamy and life-long marriage, of which Sinhalese disapproved (ibid.).

Whatever the details, as a Vedda way of accommodating to Sinhalese culture, the nittaewo can reasonably be understood as a Vedda deflection of their own negative image—a means, in other words, of transferring features of their lifestyle known to be disapproved by the Sinhalese majority onto another, partly imaginary creature. More specifically, the nittaewo could have originated in an image of so-called 'wild Veddas' who maintained a hunting economy, a representation propagated by more sedentized Vedda at a particular point in Sri Lankan history (Brow 1978: 13–15).

An origin of the nittaewo in ethnic differences does not of course preclude influence from experience of non-human animals. In fact, besides bears and primates, the image has another possible zoological inspiration. Although the derivation of the name 'nittaewo' is unclear, it may be significant that Vedda apply a similar term, 'nittawe' (cf. Sinhalese 'ittaewe') or 'nittaewâ', to porcupines (Bailey 1863: 317; Nevill 1886a: 67), evidently the Indian porcupine (*Hystrix indica*). Since porcupines are mostly nocturnal quadrupedal herbivores, a connection with a bipedal hairy hominoid may seem far- fetched. Nevertheless, *Hystrix indica* can attain a length of up to a metre, possess large claws and, with their quills, can inflict fatal injuries even on tigers and leopards. When threatened, moreover, the animals rattle the hollow spines in their tails, producing a sound conceivably similar to vocalizations reported for the nittaewo (Gurung and Singh 1996: 48). Obviously, one cannot reduce the Sri Lankan wildmen simply to experience of porcupines. Yet some influence on the hominoid representation cannot be ruled out, particularly in view of the similarity of names, which suggests a metaphorical connection at least.

I conclude by registering a further comparative significance of the nittaewo. Not only does the story of their extermination, like the legend of the ebu gogo, refer to an apparently quite recent period of local history; like the Nage wildmen and other Indonesian representations of hairy hominoids (for example, the ana ula and lae ho'a of other parts of Flores and the Sumatran orang pendek), the Sri Lankan category leaves no other trace in the culture of either the Vedda or their Sinhalese neighbours. The nittaewo are apparently not a more general subject of folktales, nor do they possess any religious or cosmological significance. And it is presumably for this reason that they receive no mention in any professional ethnography (such as the Seligmans' work on the Vedda).[5] Equally relevant is Lewis's observation (1915: 129) that the story of the nittaewo extermination is 'told devoid of the usual fantastic embellishments that characterize the history of mythical creatures such as Yakko', a reference to 'yaku', or 'spirits' (Lewis 1915: 129). In this respect, too, the nittaewo reveal a representation significantly resembling the hominoids of Flores. But by the same token they differ, for example, from the mythically elaborated mili mongga of Sumba (Chapter 4)—and, as we shall presently see, from the Himalayan yeti as well.

Varieties of 'yeti'

So much has been written about the Himalayan creature called 'yeti'—or, more misleadingly, 'the abominable snowman'—that it is among the best known of the world's hairy hominoids. In popular western imagination, the yeti is a giant creature, a view mostly traceable to the huge footprints with which it is identified. Found in snow and very probably enlarged by processes of melting and sublimation (Napier 1972: 127, 140), the prints have produced not only a distorted picture of the yeti's size but also an inaccurate association of the being with snow-swept mountains. In regard to images reviewed in previous chapters, the main relevance of the yeti lies in the fact that most reports from the Himalayas pertain to a forest-dwelling hominoid or ape-like creature no bigger than a human or even significantly smaller. The resemblance to Southeast Asian hominoids is thus obvious.

Although the yeti is one of the few hairy hominoids allegedly observed by westerners, evidence for yeti sightings, by local people as well as Europeans, is contradictory. Anthropologist Fürer-Haimendorf says that 'most Sherpas have seen yeti at some time or other' (1954: 477); however, Napier cites a personal communication from the same author stating that he had 'never met a Sherpa who has actually seen a Yeti' (1972: 63). Sanderson lists eleven instances of Europeans claiming to have sighted the creature, dating between 1906 and 1957 (1961: 260–63). However, after reviewing the evidence for several of these, Napier concludes that 'only two Europeans have ever seen what might conceivably be a Yeti' (ibid.: 61) and he is inclined to dismiss even their reports as circumstantially unreliable.

Most authors have recognized at least two kinds, or two representations, of yeti. One is the creature known to the Sherpa people of Nepal and Tibet as 'dzu-teh', a huge beast of tremendous height with long shaggy hair, a flat head, very large feet and partly quadrupedal locomotion (Stonor 1955: 63, Sanderson 1961: 268). Classification of smaller yeti is more various. Stonor, who interprets the cover term 'yeti' as comprising Sherpa 'yeh' (referring to an association with rocky habitats) and 'teh' (denoting an animal), simply contrasts the giant dzu-teh with a single, smaller kind called 'mih-teh'. Evidently alluding to a more humanlike appearance, 'mih', meaning 'man, human', also occurs in 'mi go', 'or wild man', which Nebesky-Wojkowitz (1956: 156) suggests is the commonest name for yeti in general. On the other hand, Sanderson speaks not only of a small, human-sized yeti ('meh-teh') but also of an even smaller 'pigmy type', designated as 'teh-lma' (Sherpa for 'man-like being'). Inhabiting lower, warmer valleys and distinguished by tiny footprints and thick red fur with a slight mane, this smallest yeti, Sanderson claims, is the least known variety, but is nevertheless 'probably the commonest of all' (1961: 268).

Based largely on accounts by Sherpas, the height of smaller yeti is given as: 1.2–1.5 metres (for Sanderson's 'pigmy'); 1.4 metres (Tishkov 1993: 64, describing a claimed observation by a Russian member of a glaciological team); the size of a small man or youth, or a 'twelve-year-old boy' (Stonor 1955: 38, 69); or simply 'about the size of a man' (Heuvelmans 1995: 107). A creature allegedly observed in 1949 by a Sherpa, Sonam Tensing, stood about 1.65 metres (Izzard 1955: 43). Accordingly, Napier (1972: 63) notes that the height most frequently mentioned is between 1.5 and 1.8 metres. Nebesky-Wojkowitz, who describes a creature possessing very large arms, gives a height of between two and 2.3 metres. But this author does not distinguish varieties and evidently refers only to the larger kind of yeti (1956: 160; all heights are converted to metric).

Since most descriptions appear to apply to images other than the giant variety, hereafter I refer simply to 'yeti', distinguishing named categories only when necessary. Known mostly from tracks, yeti feet are generally described as humanlike. Corresponding with relatively short stature and of course contrasting with better known huge prints, foot size has often been specified as small—13 centimetres long (Sanderson 1961: 267–68) or just 10 centimetres in the case of a native sighting of a creature 1.4 metres tall (Byrne 1958). Sanderson also speaks of 'short broad feet' with a second toe larger than the big toe, or 'widely separated first and second toes' (1961: 267, 254). A combination of relatively small body size and extremely large feet is suggested by an early eyewitness account by a British forest officer in Sikkim, who described a creature about 1.2 metres high but leaving tracks 45–60 centimetres in length and 15 centimetres wide (Elwes 1915: 294). Further describing the toes as pointing in the 'wrong direction', the officer inferred that the animal walked 'on his knees and shins instead of on the sole of his foot' (ibid.). Since the tracks were found in snow, however, this impression is more plausibly ascribed to the melting and sublimation mentioned by Napier as probable causes of giant prints.

Most authors describe yeti hair as reddish, including reddish-brown (Izzard 1955: 43), reddish-grey, yellowish brown (Elwes 1915), or simply 'red' (Stonor 1955, Sanderson 1961: 267–69). Length is variously and vaguely, specified as thick and long or 'longish' (Elwes 1915), but also as 'short' (Sanderson 1961: 267). This at any rate applies to smaller sorts of yeti. The giant kind is described as covered in long or shaggy, dark grey, dark brown, or black hair (Nebesky-Wojkowitz 1956: 160; Sanderson 1961: 268), although for this variety as well, Stonor (1955: 63) says 'reddish-brown'. Specifying the smaller yeti ('mih-teh'), Stonor further describes the hair as less abundant on the chest than on the legs and waist, while Nebesky-Wojkowitz (who speaks simply of 'yeti') says the face and head are only sparsely covered with hair (1956: 160). Local reports further depict the hair on the yeti's upper body pointing upwards and that on the lower part hanging down (Stonor 1955: 152).

Unlike the Sumatran orang pendek and many other Indonesian hominoids, the yeti is not usually described as having head hair longer or otherwise distinct from the body hair. Referring specifically to the 'miti', however, one source mentions a long mane of hair falling in strands over the eyes (Oppitz 1968: 139), a feature also recorded in early twentieth-century accounts of the orang pendek and as we shall see, hominoids reported from Central Africa. Probably the most commonly mentioned feature of the yeti's head is its 'conical' or 'pointed' shape (e.g. Stonor 1955: 69, 82, 138), a feature echoed in Van Herwaarden's famous description of the orang pendek. Of eighteen Sherpa accounts reviewed by Napier (1972: 63–64), as many as a third mention a conical head. Other features of the head include a sharply receding forehead (Sanderson 1961: 270), a bull neck (ibid.: 267) and a face like an ape (Nebesky-Wojkowitz 1956: 160) or a monkey (Stonor 1955: 45, 64, 152).

Descriptions of yeti bodies generally suggest a robust build, a correlate of the great strength commonly attributed to the creature. A generally hominoid appearance is of course a correlate of erect posture. Yet, interestingly enough, the yeti is not consistently portrayed as bipedal. Nebesky-Wojkowitz describes a creature which, when travelling in forest, 'moves on all fours or by swinging from tree to tree', but 'generally walks upright with an unsteady, rolling gait' when moving in open country (1956: 160). Whatever it was the British forest officer saw in Sikkim, Elwes (1915) describes this creature, also, as going about 'chiefly on the ground', thus similarly suggesting a partly arboreal habit.

The foregoing appears to confirm the association of certain yeti at least with forests. By contrast, some accounts depict the largest variety as the one more given to quadrupedal movement and also to occupation of snowfields (Stonor 1955: 63–64). Apparently clarifying this discrepancy, Nebesky-Wojkowitz describes all yeti as residing in 'the highest tracts of Himalayan forest' rather than in the 'zone of snow', but moving into snowfields—and then leaving behind the huge footprints—in order to obtain salt from 'saline moss' growing beneath the snow (1956: 160).[6] Other accounts have yeti residing, at least during the winter, at even lower altitudes, including warmer valleys and low-lying regions of tropical or sub-tropical forest (Sanderson 1961: 254; Napier 1972: 53). The creatures are further described as inhabiting caves and, in accordance with the previously mentioned interpretation of the name 'yeti' (or 'yeh-teh'), rocky areas (Napier 1972: 41).

If forest and cave-dwelling and the roughly human size of some yeti sound rather like the Sumatran orang pendek, as regards their evident preference for meat, the creatures are more reminiscent of the Sri Lankan nittaewo. Generally omnivorous, the reputed diet mostly comprises small animals, including mammals, insects, fledglings and snails (Stonor 1955: 81; Sanderson 1961: 267). Yeti also have a reputation for raiding cultivated fields and stealing from winter food stores (Fürer-Haimendorf 1954: 477), while in folk stories they often meet their downfall owing to a legendary taste for rice or millet beer (Stonor 1955: 137–38; Nebesky-Wojkowitz 1956: 159; Oppitz 1968). Consistent with their carnivorous habit, the Himalayan creatures are further accused of preying on yaks and cattle. But this accusation appears to apply specifically to the giant kind of yeti (Stonor 1955: 63; Fürer-Haimendorf 1954).

Generally, the literature suggests that Himalayans are fearful or at least wary of yeti. The sight of one can be inauspicious, as can its cries. On the other hand, the creatures themselves are characterized as 'very shy' and afraid of fire and accordingly are not often seen (Nebesky-Wojkowitz 1956: 156, 160; Stonor 1955: 97–98; Oppitz 1968: 138). Images presented in folktales reveal an animal that is violent and dangerous, given not only to killing humans but also to consuming their flesh (see Oppitz 1968: 140, who records a story in which 'bloodthirsty' yeti kill several Sherpas). At the same time, citing conversations with Sherpas, Tibetans and Lepchas, Nebesky-Wojkowitz claims that 'less superstitious inhabitants of the Himalayas' regard the yeti as 'a harmless creature that would attack a man only if wounded' (1956: 160). While Sherpas and other Himalayan natives know yeti largely from visual encounters (with footprints if not actual specimens), they also identify them vocally. Vocalizations mostly comprise a 'whistling' sound but also cries described as yelping or mewing (Nebesky-Wojkowitz 1956: 158), which Napier (1972: 153) suggests may be traced to choughs, a species of alpine crow. More important for our comparative purpose, however, no account of yeti depicts them as speaking.

Like several physical features already mentioned, this lack of language reveals the yeti's resemblance to the orang pendek of Sumatra. By the same token, both are distinguished from eastern Indonesian representations of hominoids that are able to speak. In regard to another sensory attribute, however, the yeti resembles eastern Indonesian hominoids while differing from Sumatran and other western Southeast Asian creatures and also from the nittaewo of Sri Lanka. For according to a common Sherpa report, the yeti possesses a 'vile, pungent smell' (Napier 1972: 64). Like eastern Indonesian figures, the image of the yeti can be used as a bogey (Napier 1972: 39), thus signalling another contrast with the orang pendek.

Whatever is to be made of Nebesky-Wojkowitz's distinction of less or more 'superstitious' Himalayan natives, it nevertheless recalls the contrast of naturalistic and mythical representations of legendary hominoids evidenced in previous chapters. Sherpas may very well regard the yeti as just another animal and 'speak of them in much the same way as Indian aboriginals speak of tigers' (Fürer-Haimendorf 1954: 477). However, in addition to many features suggestive of a real animal, the image incorporates several obviously fantastic elements. Among these are inverted feet, or footprints (Stonor 1955: 109; Napier 1972: 64; see also Elwes 1915, cited above), another attribute suggesting a link with the Sumatran orang pendek. Other extraordinary features are more reminiscent of eastern Indonesian figures. Prominent among these are long, pendulous female breasts, so large they are 'thrown over the shoulders' (Napier 1972: 65) or, sometimes, 'dragged along the ground' and even employed as toboggans (Oppitz 1968: 139). It is important, however, that neither the breasts nor the back-to-front feet receive mention in reputed eyewitness accounts by Sherpas (Napier 1972: 64, 65). And it is perhaps for this reason that the fabulous appendages also do not figure in several otherwise detailed reviews of yeti physical form (for example, those by Stonor, Nebesky-Wojkowitz and Sanderson). Pendulous breasts and inverted feet may occur in general descriptions of the creatures, that is, ones unconnected with putative experiences, but they are largely confined to folktales (see especially Oppitz 1968, who records stories featuring both).

Another evidently fantastic characteristic of yeti revealed in Sherpa folk stories is their desire for human clothing, or more particularly shoes (Oppitz 1968: 139–40). This might recall the ultimately fatal liking for palm fibre by the Florenese ebu gogo, who similarly desired to use this material as clothing. In one Sherpa tale, a gift of shoes from an abducted human female proves to be a male yeti's undoing, not unlike the fibre which becomes the medium of the wildmen's destruction in the Nage legend. On the other hand, the theme is strangely reminiscent of reports of captive Bornean orang-utans liking to dress in human clothing (St. John 1862: I: 23; McDougall 1992: 230).

As Sherpa tales also demonstrate, yeti are further depicted in mythical discourse as abductors of human beings, with whom they sometimes mate and produce offspring (Stonor 1955: 136–37). According to one idea, yeti kidnap humans of both sexes, killing the men but keeping the women as mates. Resulting children then resemble their yeti fathers (Oppitz 1968: 139; Stonor 1955: 136). Whereas abduction is a common attribute of Southeast Asian wildmen, a more specific resemblance to traditions elsewhere lies in stories of extermination. According to another Sherpa narrative, villagers tired of yeti raiding once put out jars of beer to lure the creatures from crags they occupied high above villagers' fields. Once the creatures got into the habit of consuming the beer, the Sherpas added a poisonous root, which killed all except one pregnant female. Thus it is that yeti, now scarce but once much commoner in Sherpa country, 'were not quite exterminated and still survive to this day' (Stonor 1955: 137–38).[7] Yet another extermination tale describes how beer-drinking Sherpas once engaged in a mock battle with wooden swords. Following their 'imitative instinct' (Oppitz 1968: 140), yeti came down to the village on the following night, consumed beer left out for them and began similarly to engage in combat. The Sherpas, however, had deviously substituted weapons of steel for their wooden swords. Thus gulled, the yeti inadvertently killed one another, being survived, again, by just one pregnant female (Oppitz 1968: 140).

As versions of the legend of ebu gogo's extermination sometimes describe how the creatures were first made drunk with palm wine, it is clear how these Sherpa stories resemble Florenese tales of hominoid extermination. Also comparable are the motives, particularly the threat posed by wildmen raiding fields and food stores. Yet another common element is the surviving pregnant female, a widespread narrative device which also occurs, for example, in the Poma tradition of the ana ula (Chapter 3). Nevertheless, the means of killing are quite different in the two cases, as are the locations. Thus, while the ebu gogo were trapped and burned inside their cave, in the yeti tales the creatures are lured to human habitations and killed either by poisoning or by self-immolation following intoxication.

Yet another similarity between the Himalayan and Southeast Asian hominoids concerns stories (or rumours) of local people capturing yeti. One story recorded by Nebesky-Wojkowitz relates to a time when Europeans were present in the region. Also predicated on the creature's taste for beer, the story describes how men specified as 'Tibetans' contrived to catch a yeti (a 'mi go', or 'wild man') by making it drunk and tying it to a carrying pole. The captors then set off 'to the nearest town, believing that the European sahibs would give them a rich reward for their strange capture'. Before long, however, the yeti regained its senses, broke its bonds with a 'single effortless wrench' and escaped (1956: 159–60). Although no date is given for this supposed occurrence, the details suggest a time by which westerners had become interested in the 'abominable snowman', thus in the late nineteenth or early twentieth century.[8] Without specifying any source, Heuvelmans (1995: 156) states that the captured animal was a large monkey.

While some commentators have interpreted the giant yeti (or dzu-teh) as a remnant *Gigantopithecus* (Heuvelmans 1995: 181; Reynolds 1967: 102),[9] others have more credibly ascribed this particular representation to experience of one or more kinds of bear (Napier 1972: 151; Messner 2000), including the Brown bear (*Ursus arctos isabellinus*, sometimes called the 'Red bear' or 'Snow bear') and the Himalayan black bear (*Selenarctos thibetanus*). Positing a basis in linguistic confusion, Oppitz too offers a version of the bear theory (1968: 137). Bears could be responsible for elements of images of smaller yeti, including broad, humanlike footprints (Napier 1972: 147, 150–51). However, human-sized or smaller yeti are, for the most part, far more suggestive of some kind of primate. Since yeti territory is also home to langurs and macaques, an empirical basis for the creature might more particularly be found in one or more species of monkey (see Sanderson 1961: 71–75; Elwes 1915). Yet as Napier has demonstrated at some length (1972: 55–61), local descriptions of yeti appearance reveal considerable correspondence with the orang-utan (*Pongo pygmaeus*), as does certain circumstantial evidence. In fact, an apparent absence of head hair significantly longer than body hair and the reddish pelage generally attributed to smaller forms of yeti, ironically suggest a creature more like an orang-utan than do corresponding features of the Sumatran orang pendek. Oppitz's reference to hair falling over the eyes is also not inconsistent with an orang-utan.

This is surprising, as orang-utans disappear from the fossil record of mainland Asia by the end of the Pleistocene. Nevertheless, referring to a possible survival of the species on the mainland until just several hundred years ago, or even to the present, Napier posits a memory of *Pongo pygmaeus* as the primary source of the yeti representation. Coming from a respected primate biologist, Napier's interpretation is compelling. Similarly noteworthy is primatologist Vernon Reynolds' suggestion that the yeti may be a 'largely unknown type of ape' (1967: 58). Since Napier wrote, his 'orang theory' has arguably gained support from evidence favouring orang-utans as the empirical basis of comparable images in Sumatra and mainland Southeast Asia (Chapter 6). But clearly, if the yeti does reflect the orang-utan, the legendary creature's local representation also responds to other factors. Among these is a universal tendency to think and talk about manlike yet non-human animals in remarkably similar albeit often quite fantastic ways, a tendency I further explore in Chapter 8 with reference to African images.

Wildmen of China

If Napier is right about smaller yeti reflecting the orang-utan, then clearly the red ape, whether contemporary or historical, has played a substantial role in the production of wildman images over a large part of Asia—including western Indonesia if one interpretation of the orang pendek is to be accepted. Images of 'wildmen' from China point in the same direction. Indeed, for a large part, Himalayan lore concerning the yeti appears to be of a piece with what is found in China.

Although Chinese figures have been designated by several names, among the best known is 'yeren', a term that translates as 'wildman'. And it is as 'wildmen' that the putative Chinese creatures have come to be known in English. Besides references in Chinese folklore and classical literature, reputed sightings of hairy manlike creatures have appeared from time to time, mostly from uninhabited or sparsely populated mountainous regions of China. Since the 1950s, these have attracted the attention not only of world media but also of Chinese scientists.[10] Probably the best summary of the Chinese evidence is still a 1982 article by Zhou Guoxing, a physical anthropologist with the Beijing Natural History Museum. Zhou bases his overview on twentieth-century informant accounts, including putative eyewitness reports by 'scientific workers'.

The first thing to note about Chinese wildman images is the way they reveal the same bimodal representation as does the yeti. Simply stated, the wildmen come in two varieties, large and small, with heights of over two metres and just over one metre respectively. Standing 1.4 metres, a creature reported in the 1980s by a Hubei cowherd (Jaivin 1985: 26) falls within the shorter range. Footprint lengths vary accordingly. Otherwise appearing 'remarkably similar' to a human's, the larger variety are 30–40 centimetres long, while a shorter, more simian type measures about 20 centimetres. In both varieties, the big toe points outward (Zhou 1982: 20).

In other respects as well, larger and smaller wildmen appear not to differ significantly. Both are bipedal but become quadrupedal when moving fast or climbing slopes. Wildman faces and bodies combine features of humans and apes, although the hands, ears and external male genitalia are more like a human's. Just one account mentions 'long ears' (Poirier *et al.* 1983). According to a Chinese biology graduate who claimed to have seen a wildman that had been shot in 1940, the creature's face resembled reconstructions of 'Peking Man' (*Homo erectus pekinensis*). This particular specimen, a female, also reputedly possessed 'a large pair of breasts' (Zhou 1982; Yuan and Huang 1981: 6–7). Large or pendulous breasts are in fact a general attribute of larger and smaller female wildmen alike (Zhou 1982: 20; Jaivin 1985: 30). But in contrast to wildmen almost everywhere else, whose male genitalia rarely receive mention, one account credits the Chinese variety with a 'pendulous penis' (Poirier *et al.* 1983).

Chinese usually describe wildmen as covered in short hair, three to four centimetres in length. The head hair is longer. Evidently not varying between head and body, hair colour is mostly described as red (see e.g. Greenwell and Poirier 1989: 48–49), although shades of brown, grey and yellow, black and occasionally white have also been reported. Analysis of suspected wildman hair has indicated the source as an 'unknown higher primate (ape or human)' (ibid.: 50–52; see also Poirier and Greenwell 1992: 73–77). As noted in Chapter 5, similar conclusions have been reached with reference to hair attributed to orang pendek.

Apart from hair samples, putative physical evidence of Chinese wildmen includes 'nests' and preserved hands and feet locally attributed to a 'man-bear' (a translation of one Chinese name for the wildman) killed in 1957 (Zhou 1982: 19). The hands are shown in Plate 7.1. Resembling arboreal structures built by orang-utans (Poirier and Greenwell 1992: 78), the nests have alternatively been ascribed to Giant Pandas (Poirier *et al.* 1983: 35). On the other hand, Zhou interprets the 'man-bear' remains as probably reflecting a previously unrecorded 'enormous monkey'; in any event, their source has yet to be fully explained. Drawing on both generalized accounts and reputed eyewitness reports, Zhou (1982: 20) lists several behavioural and ecological attributes of Chinese wildmen. Living in thick forests far from human habitations, wildmen lack 'syllabic language', fire and tools. Their vocalizations comprise various non-linguistic sounds, including yells (ibid.) and a sound likened by an eyewitness, a male forestry protection officer, to 'a sparrow chirping' (Yuan and Huang 1981: 11). Observers usually report the creatures as occurring singly or occasionally in pairs. As wildmen have been observed in winter, they evidently do not hibernate. Like the yeti (Napier 1972: 60), they have been described as moving about at night, but in contrast to truly nocturnal animals, their eyes do not reflect light. Wildman diet comprises various plant foods, including ripe maize stolen from fields in the autumn. They are generally good at avoiding detection by people and when encountered, mostly by lone humans, they are not aggressive.

If these descriptions seem tediously familiar, particular similarities between the Chinese and other Asian hominoids are nonetheless worthy of mention. Not least because of the reddish hair and pendulous breasts, the two varieties of Chinese wildmen obviously resemble the yeti—as they do in regard to partly quadrupedal locomotion and lack of speech. Yet another similarity concerns the Chinese creatures' resemblance to large bipedal macaques, which are given to capturing male travellers of the opposite sex with whom they then mate and produce offspring (Van Gulik 1967: 26). According to a further idea, mating between male wildmen and human females results in 'monkey babies', a reference to certain birth defects (Poirier *et al.* 1983: 29–30). Arguably the only noteworthy differences between the Chinese and Himalayan wildmen are the former's evidently more vegetarian diet and less aggressive nature. Yet, as remarked earlier, aggression and carnivorous habits as attributes of the yeti are components of a quite distinctively mythical representation, as indeed is abduction and mating with humans.

To the extent that recently encountered wildmen are identifiable with fabulous figures from older Chinese literature, the former, too, are subject to a fantastic representation quite different from the naturalistic images summarized by Zhou. The 'fei-fei', mentioned in documents dating to the first millennium AD, resembles a long and red-haired, 'long-lipped' human (or, according to another account, a macaque) which, unlike Zhou's unaggressive creatures, devours people and walks on reversed feet. Large exemplars, moreover, can attain a height of over 3 metres, thus far in excess of the range Zhou describes for the larger kind of wildman (Van Gulik 1967: 27–28). Fei-fei are also described as being fond of wine and capable of speech. Their speech, however, resembles the twittering of birds, a comparison which immediately recalls descriptions of the Sri Lankan nittaewo and the Vietnamese creature described to Burchett (Chapter 6), as well as the recent report of a 'chirping' Chinese wildman. In regard to the inverted feet and the preference for alcohol, one is of course reminded of the yeti. Also from ancient Chinese literature, a distinctly named creature, the sing-sing (or hsing-hsing), is similarly credited with speech and a fondness for wine (Van Gulik 1967: 27, 55). The second feature possibly links both the sing-sing and fei-fei with an anthropophagous, red-haired 'wild man' described in a Chinese legend as the 'king' of 'wild beasts'; this mount-dwelling 'giant' is killed by herb gatherers, who first render the creature drunk by pouring wine down its throat (Werner 1922: 392–93).

An interest of a different kind is the selection of 'hsing-hsing' by modern Chinese zoologists as the vernacular name for the orang-utan (Van Gulik 1967: 29). In the same context, modified forms of 'fei-fei' have been chosen to designate baboons, gorillas and chimpanzees. Nevertheless, it is the fei-fei specifically which Van Gulik (ibid.: 28–29) proposes as a category possibly reflecting Chinese experience of orang-utans some two thousand years ago.

In accordance with their general resemblance, the Chinese wildman and yeti differ from the Sumatra orang pendek in identical ways, for example in regard to the latter's dark hair, absence of breasts and habit of bipedal flight. (As noted, the Chinese creatures use all fours when moving quickly.) At the same time, red hair and pendulous breasts are attributes that both mainland Asian representations share with certain eastern Indonesian wildmen—most notably the Sumbanese mili mongga, which is credited with both features. On Flores, the ana ula of Poma are sometimes described as red-haired, while the distinctive mammaries are of course ascribed to the Nage ebu gogo. Given the absence of non-human primates other than long-tailed monkeys in eastern Indonesia, this sporadic distribution provides yet a further indication of the fantastic quality of the breasts, a feature which, as we shall later see, reappears in images of the wildman of Europe and the North American sasquatch.

Consistent with physical resemblances between the Himalayan and Chinese figures, primatologists and anthropologists have interpreted the two equally bimodal representations in similar ways. Owing in part to the giant fossil ape's association with China, Zhou suggests a descendant of *Gigantopithecus* as a plausible source of the image of the larger Chinese wildman. So too do Greenwell and Poirier (1989: 56), who further conclude that wildman representations reflect 'two distinct animal species'. In regard to the smaller kind of wildman, all authors review possible referents among the known primate fauna of China. These include the White-browed gibbon (*Hylobates hoolock*), two species of macaque (*Macaca assemensis* and *M. thibetana*) and the rare Golden monkey (*Rhinopithecus roxellanae*). Another species suggested is the Stump-tailed macaque (Jaivin 1985: 32). In accordance with his interpretation of the 'man-bear' remains, Zhou favours observations of 'an enormous species of monkey still unknown to science' as the origin of the smaller wildman (1982: 22; 1984). Also positing a large primate, Greenwell and Poirier suggest that the representation may reflect a remnant population of orang-utans or a related species of the genus *Pongo* (1989: 56). The conclusion is not substantially different from that of Napier in regard to the yeti and Rijksen and others in respect of the orang pendek of Sumatra.[11] Nor is this a particularly new interpretation. In a discussion of the ancient Chinese category labelled 'sing-sing', Schlegel and Müller (1839–44: 4) considered a type of orang-utan as its possible referent; in the end, though, they rejected this, suggesting that some descriptions of sing-sing sound more like a gibbon.

That Chinese wildmen reflect orang-utans is no less credible than the idea that smaller yeti do so. However, in the Chinese case it is less necessary to posit a 'second species' to account for larger instances of the category. Since the larger Chinese wildman differs from the smaller mainly in height, instead of positing surviving *Gigantopithecus* (or for that matter a bear), the size disparity is more plausibly explained as a partial exaggeration of the dimensions of smaller primates. A less likely source is the giant orang-utans which renowned palaeontologist Ralph von Koenigswald suggested may still be living in higher and therefore cooler, regions of China (1981: 260).[12] Nevertheless, reports of very large orang-utans elsewhere could have some relevance for the issue of wildman size. An example is Clarke Abel's (1825) account of a Sumatran specimen killed in the early nineteenth century and estimated by natives and Englishmen who inspected the carcass as between 2.2 and 2.4 metres in length. This may well be an exaggeration. However, if Europeans have been prone to substantially overestimate the size of orang-utans, as they have the size of other known apes (see Jardine 1833: 58–59, regarding chimpanzees), then there is no reason why rural Chinese—or, for that matter, Himalayans—should not sometimes do so as well.

In any case, from the available evidence, one can readily agree with Zhou that the Chinese wildman is 'not of the human type' (1982: 21–22), which is to say, not a hominin. Rather, the image suggests some sort of animal. Whether it is an imaginary animal remains a question; so does the motivation for its representation if it is completely imaginary. Poirier *et al.* (1983: 38–39) have suggested several reasons why modern Chinese might be inclined falsely to report wildman sightings, including personal 'fame', dispensation from work (since such sightings are said to leave observers bed-ridden from anxiety) and a desire to promote tourism. However, such factors (which nowadays could equally apply in places like as Sumatra and Nepal) can hardly explain the origin of the physical image of the wildman, or its survival from ancient times to the present.

Central Asian exemplars

Asian regions to the north and west of China have produced a variety of reports of hairy hominoids. Given the geographical contiguity of this vast area with the domains of both the Chinese wildman and the yeti, it is not surprising that some Central Asian representations resemble these particularly closely. Among the earliest accounts is Roger Bacon's thirteenth-century description of hairy creatures 'with a human form in every respect but covered all over with hair' (1928: II: 387). Encountered in mountainous regions apparently in Mongolia, the beings were just a 'cubit' (about half a metre) high, had inflexible legs and moved by leaping.[13] Bacon further relates how they could be captured by first intoxicating them with beer poured into holes in rocks and how this is done so that their blood may be obtained as a dye. Except for the diminutive stature (which suggests a species of monkey), it would appear that the mediaeval Englishman may have been referring to creatures Chinese call 'fei-fei' or 'hsing-hsing'.

In contrast to this evident continuity with Chinese images, in west central Asia and particularly in the Caucasus, one encounters a distinctive representation of creatures noticeably more humanlike than Chinese and Himalayan exemplars. Despite this regional variation, several authors have treated wildmen of Central Asia (including large parts of the former Soviet Union) as a single category, often employing as a general label the Mongolian name 'almas' (singular and plural). Also known by a variety of local names, these almas, like comparable figures elsewhere, are mainly associated with mountainous areas, including the Pamirs, the Tien Shan, the Altai and the Caucasus. Other reports derive from the Gobi Desert and Siberia.

Including creatures both larger and smaller than local humans, Central Asian images reveal the same bimodalism as do Chinese and Himalayan representations. On the basis of 200 sightings drawn partly from the Tien Shan mountains, Supanov (1988) has analysed this bimodalism mathematically, isolating a 'tall stature mode' of 2.04 metres and a small mode of 1.56 metres. This variation not only concerns differences between widely separated regions, but also obtains within single regions, such as the Pamirs and the Caucasus. At the same time, very large creatures reported from areas bordering the Himalayas and China appear to be of a piece with the larger yeti. In connection with my point of departure in Southeast Asia, the interest of this section lies mostly in smaller central Asian wildmen. These in any case appear rather better represented in published accounts than do larger varieties. Drawn in part from reputed eyewitness accounts recorded in Tchernine (1971) and Shackley (1983), the following reports seem representative.[14]

Kazakhs describe a hairy bipedal creature from the Iren'-Kabyrga mountains, of less than average human height, with humanlike feet but broader and with a big toe set apart from the others. A captured specimen had long arms and a 'narrow, hollow chest', was stooped and ran awkwardly on legs 'far apart and bent at the knees'. The generally apelike head and face featured a massive jaw, sloping forehead, jutting brows, absent chin and ears that were large, somewhat pointed and lacked lobes (Tchernine 1971: 43).

From the Sanglakh mountains in Afghanistan, putative eyewitnesses have described a hominoid locally known as 'deva' or 'dev' (apparently related to the Sanskrit term reflected in European words for 'spirit'). The creature resembles a small, erect and very thick-set man, about 1.5 metres tall and covered in brown or reddish-brown hair. These dev are extremely rare and only occur singly or in pairs. Sightings are interpreted as an 'evil omen' (Tchernine 1971: 47–48).

Reports of almas from the Gobi Desert similarly depict manlike creatures covered in sparse reddish or 'reddish black' hair, though one specimen, with hair on the head longer than on the body, had grey hair. They also possess powerful jaws, low foreheads and prominent brow ridges and walk bipedally with a stooped gait (ibid.: 54, 55, 57). Other accounts describe the creatures as walking with 'knees semi-flexed' and feet 'bent slightly inwards' (Shackley 1983: 104). Mongolian almas are between 1.6 and 1.7 metres in height, thus about as tall as modern Mongols (ibid.). A dead specimen with short, broad feet is supposed to have stood between 1.65 and 1.7 metres (Tchernine 1971: 104). Also reported for the Mongolian creatures are long breasts, further characterized in by now familiar language, as being so long that they can be tossed over the shoulders. Female almas do this when feeding infants that crouch behind their mothers or cling to their backs. Mongolian stories also feature almas suckling human infants (ibid.: 55, 56; Shackley 1983: 104). As this might suggest, the creatures are implicated in tales of human abduction and mating with humans; they are also described as anthropophagous (Tchernine 1971: 59–60). Additional attributes of Mongolian almas include their ignorance of weapons and fire and inability to speak. As in one report of the Sumbanese mili mongga (Chapter 4), however, they seem not to be afraid of fire and may even be aware of its benefits (Napier 1972: 71; Colarusso 1980: 261); a story from the Gobi Desert thus describes almas warming themselves by a deserted camp fire (Tchernine 1971: 54–55). Finally, recalling the Himalayan yeti and some eastern Indonesian figures (but apparently not the wildmen of China), the Mongolian creatures are said to give off an unpleasant odour (ibid.: 54).

If reports of Mongolian almas are relatively detailed, accounts of Caucasian wildmen appear even more so. Although the Caucasian figures are known by various local names, the one mentioned most often in print is 'kaptar', a name of uncertain derivation used in Azerbaijan and Dagestan. An important source for the Caucasus is Colarusso's essay on North Caucasian (especially Circassian) exemplars; this is also one of very few articles on wildman figures written by a cultural anthropologist. Caucasian wildmen are generally described as thick-set, strong and agile and of medium human height, even as 'tall'—but also as short, standing just over 1.5 metres; Colarusso 1980: 256; Shackley 1983: 311). Footprints attributed to Caucasian wildmen reveal broad 'inward-turning feet, characteristic of bow legs', a feature reminiscent of the inverted feet of Sumatran and Himalayan hominoids (Tchernine 1971: 21). Covered in hair, possessing ugly protruding faces and large, dangling breasts and emitting a foul odour, the Caucasian creatures obviously resemble more easterly 'almas' (Colarusso 1980: 256, 259; see also Tchernine 1971: 161, especially regarding a Karbardinian account that specifies a 'stifling, sour smell'). A further resemblance, of course, are the inward-turning feet.

Recalling the 'silent trade' attributed to small-scale food-collecting societies (for example, the Vedda of Sri Lanka, the Kubu of Sumatra and the To Ala of southwestern Sulawesi), Colarusso describes Caucasian wildmen as exchanging 'vegetables and things' for trinkets offered by human neighbours (1980: 257). Involvement in a silent trade is also ascribed to small Mongolian almas, who leave 'skins' at 'appointed places' in return for 'simple basic articles' left by nomadic tribesmen (Sanderson 1961: 318). In one particular respect, however, Caucasian wildmen are quite different. For North Caucasian varieties are reputedly able to learn local languages and speak a 'crude, broken Circassian' whose shortcomings appear 'consistent with modern theories of early hominid speech' (Colarusso 1980: 257, 260). Rural Caucasians also describe wildmen as stealing from cultivated fields (ibid.; Tchernine 1971: 20–21). And while more easterly almas also are occasionally reported as living and mating with humans, such behaviour appears more commonly ascribed to wildmen of the Caucasus.

The most famous Caucasian case of co-residence and mating concerns the female named 'Zana', considered (apparently by her contemporaries) as an 'abnauayu', a category of indigenous 'pre-hominids' driven out by immigrating Caucasians. Zana was tall, extremely strong, heavily-built and covered in reddish-black hair; she was also dark-skinned and possessed 'negroid' features, a huge bosom and enormous teeth (Tchernine 1971: 155–59).[15] Reputedly captured and tamed in the Georgian province of Abkhazia in the mid-nineteenth century, Zana died in the 1880s or 1890s. She bore children by local men and these produced grandchildren still living in the 1960s. Especially recalling the Sumbanese Apu Kalita, Zana inadvertently killed her newborns by immediately washing them in a cold river. Villagers therefore took subsequent offspring from her and raised them themselves. Further recalling Apu Kalita, Zana was apparently sold as a slave and changed hands several times before arriving in her final place of residence. In contrast to Circassian wildmen, Zana never learned to speak (ibid.: 156), but then her age at the time of her capture is nowhere recorded. Whatever one is to make of this extraordinary story, the subject's large size appears to attest further to the variability of this feature. Yet since Zana's height is described only in the most general terms, it is not at all clear how much taller than human Caucasians she is supposed to have been.

Echoing points raised in regard to various other Asian representations, Colarusso remarks how Caucasian wildmen do not figure as characters in fables or fairy tales and how people speak of them as 'mundane beasts' (1980: 258). Similarly, Napier describes Central Asian hominoids as both less apelike and lacking the 'mythological overtones' of larger creatures like the yeti (1972: 67). In accordance with their apparently more human profile, several authors, including a number of Russian scientists, have interpreted wildmen not just from the Caucasus, but from Central Asia generally, as relict Neanderthals. Deemed 'startling but not wholly outrageous' by Napier (ibid.: 67), the hypothesis involves at least two major difficulties. First, whereas *Homo neanderthalensis* were skilled tool-makers and certainly possessed fire, Central Asian wildmen are usually described as lacking all technology. Second, depictions of the creatures as stooped and awkwardly bipedal strongly recall older and now outdated representations of Neanderthals as bent-kneed and shambling (Trinkhaus and Shipman 1993). One wonders, therefore, whether such accounts might somehow reflect influence from this now discredited palaeoanthropological view. Many parts of the region and certainly the Caucasus, seem too far from Southeast Asia for orang-utans or other non-human primates to provide a plausible source for the representations, notwithstanding the reddish hair mentioned in several accounts. In these respects, then, the Central Asian figures could point more in the direction either of anthropological or primatological knowledge emanating from Europe, or a widespread fantasy possibly reflecting a universal cognition.

An Asian summary

The most important finding of this chapter is the occurrence over a large part of Asia of images of hairy, tailless and largely humanlike creatures, either about the same size as local humans (thus, generally smaller than Europeans) or indeed shorter. Especially images from adjacent parts of mainland Asia, such as the smaller yeti and Chinese wildman, moreover suggest the same explanatory possibilities as do figures from mainland Southeast Asia and Sumatra. The most plausible interpretation is a figure substantially grounded in a non-human primate and, more particularly, contemporary populations of orang-utans or culturally retained memories of the apes. By contrast, the Sri Lankan nittaewo suggest a rather different image and accordingly a distinct empirical basis—namely, experience of bears further shaped by Sinhalese evaluations of Vedda hunters.

Bears likely play a major part in representations of the giant kind of yeti and possibly large varieties of wildmen elsewhere. To that extent, bears may illuminate the bimodalism common to Himalayan, Chinese and other Asian representations. (One estimate of nittaewo height, it may be recalled, was 1.8 metres.) Yet if other Asian images can be ascribed to experiences or memories of orang-utans, the ape is a less plausible source of Central Asian wildman images, especially ones from the Caucasus. And so are bears. While the Caucasus is home to black or brown bears, bears are ruled out by the noticeably more hominin, even human, physical and behavioural features of Caucasian wildmen—features comprised in apparently naturalistic or historical accounts rather than the timeless and fantastic imagery of myths (as, for example, characterizes some Sherpa tales about yeti). To a degree, the Caucasian material thus poses the same problem as do small hairy hominoids in eastern Indonesia, where bears and apes of all kinds are completely absent. The Caucasian town of Dmanisi in Georgia is the discovery site of a small-brained dwarf fossil hominin, 1.8 million years old, variously identified with *Homo erectus* and *Homo habilis* and more recently compared to *Homo floresiensis*. But whatever the similarities between Caucasian wildmen and Florenese hominoids like ebu gogo, it would surely be too fanciful to ascribe any significance to this coincidence.

Materials reviewed in the last three chapters suggest that a variety of wildman images have their basis in contemporary experience or cultural memories of zoologically recorded primates—and one in particular. Yet in every instance we find the representations include a residue, apparently (although not always certainly) fantastic, which cannot simply be derived from experience of known animals. This renders all such empiricist interpretations provisional. Yet if the images are partly, largely, or even entirely imaginary, such fantasy equally requires explanation, especially as the hypothetical tendency to fantastic elaboration evidently results in remarkably similar images in very diverse cultural settings. Like the accompanying cultural diversity, this similarity of images extends well beyond Asia, as I show in the next two chapters.

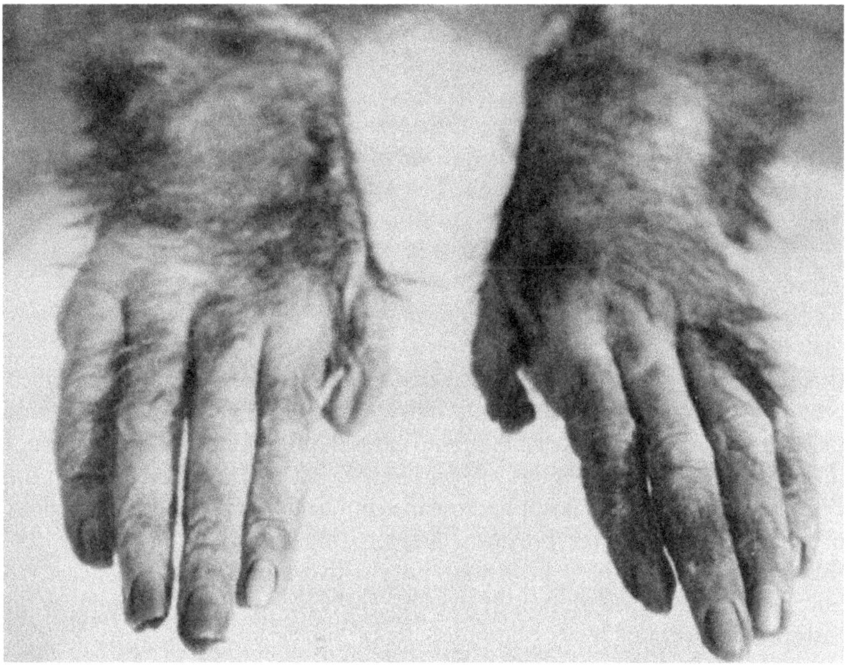

Plate 7.1 Hands of a 'man-bear'.
Source: Zhou (1982).

8 Outside Asia

Images of hairy hominoids, at least partly comparable to figures described previously, occur on all continents outside of Asia. In this chapter I must therefore be selective and attend only to cases that illuminate particular issues raised in preceding discussions of Southeast Asian wildmen. I begin with the European wildman. This is not because the continent is contiguous with regions of Asia discussed in the previous chapter—indeed, the European representation reveals relatively little continuity with wildmen of the Caucasus—but to facilitate consideration of possible influence of European images on European reports of hairy hominoids from North America, Australia and elsewhere. If not direct, this influence can be traced through an emerging palaeoanthropology and primatology, themselves influenced by figures of pre-Enlightenment European thought.

The wildman of Europe

As I have discussed in greater detail elsewhere (Forth 2007), in the European tradition 'wildman' (or 'wild man') refers to two related images: a figure (or figures) of ultimately pre-Christian European folklore and an image popular in the literature, art and iconography of the late mediaeval period. While informing the mediaeval representation, the wildman of folklore not only preceded but also succeeded the wildman of the Middle Ages, surviving in rural folktales and performances until the twentieth century. Nevertheless, the two representations are not always easily distinguishable, even in the comprehensive discussion of the images provided by art historian Richard Bernheimer (1952)—still the single most useful source on the topic.

Like some Asian counterparts, the European wildman comes in two sizes. One is much larger than humans (for example, the Alpine wild woman named 'faengge' or 'fanki') and the other 'of moderate physical proportions' or significantly smaller (Bernheimer 1952: 23, 45). Nowhere, however, does one find the numerical estimates of height or size reported for Asian exemplars. Illuminating this lack of specificity, reputed sightings of wildmen, from mediaeval or more recent times, are rare or non-existent. Indeed, it is difficult to say how far European folk, or participants in later mediaeval urban culture, ever considered the wildman as a contemporary creature inhabiting desolate mountains or woodlands. Descriptions of particular figures reveal features already familiar from previous chapters, including a body covered in hair, pendulous breasts, dark complexion, great strength and an absence of articulate speech (an attribute which of course *is* granted to some Asian wildmen). As an abductor of ordinary humans, lustful and sexually drawn to the opposite sex but also given to murder and anthropophagy, the wildman of Europe recalls more fantastic representations of Himalayan and other Asian figures. Equally comparable are apparently ancient tales from the Italian and German Grisons recounting the capture of wildmen by first making them drunk (ibid.: 25) and other European stories and visual portrayals of the hunting and capture of wildmen. On the other hand, depictions of the head, face or limbs that might recall non-human primates or pre-sapiens hominins are mostly lacking in representations of European wildmen and unlike the yeti or hominoid images from Southeast Asia, for example, they are not expressly described as physically resembling monkeys or apes. Perhaps the closest one comes to something like a simian or archaic hominin are descriptions of wildman faces, or masks that depict these, as 'ugly', 'horrible', or 'creased and oldish'. A particular female wild figure from thirteenth-century Germany is similarly described as possessing a hunchback, a huge black head, a flat nose and big teeth (Bernheimer 1952: 32, 38, 39, 82), while Sanderson (1967) has interpreted the feet of wildmen as depicted in mediaeval art as resembling those of Neanderthals.

Supporting a view of the wildman as essentially human, the figure of mediaeval literature and iconography was explicitly conceived not as a separate species of sub-sapiens hominin—a concept non-existent at the time—but as a feral man; that is, a human who, fallen from God's grace and removed from civilized society, had as it were 'reverted' to an animal condition and had in the process *acquired* the body hair and appetites of an animal. Yet the figure's degeneration had not produced a creature as low as an 'ape' (Bernheimer 1952: 1) and, indeed, being capable of redemption, the wildman's return to a fully human condition was a definite possibility. Accordingly, mediaeval art frequently depicts wild people as quite human in face and form and sometimes 'moderately attractive' (ibid.: 37), even though they were typically shown as naked and hirsute and, sometimes, as assuming a stooped or quadrupedal posture.

Perhaps not surprisingly, even cryptozoological writers (for example, Shackley 1983) have treated the European wildman as purely imaginary. If the figure is indeed completely fantastic, then, especially in view of resemblances with images from other parts of the world, one should have here the best evidence for the wildman as a universal archetype of human thought existing quite independently of empirical referents. Yet such referents are not entirely lacking. Bernheimer convincingly identifies a possible source of the mediaeval representation in sizeable populations of outcasts and vagabonds occupying forests beyond the bounds of European towns and cities (1952: 16). Despite the lack of explicitly simian morphological features in verbal and visual portrayals, also relevant to a representation of a hairy manlike being is European familiarity with imported monkeys (or a local monkey in the case of the Barbary 'ape', a species of macaque). Yet another possible source is knowledge, albeit irregular, discontinuous and indirect, of anthropoid apes. One thinks, for example, of ancient reports, such as those by Pliny and Hanno, of creatures whose ultimate derivation may have been orang-utans or gorillas. A similar inspiration may have been Asian reports of creatures possibly reflecting variants of current folk representations of the yeti, Chinese wildman and perhaps even the Sumatran orang pendek, as well as partial or exaggerated descriptions of technologically simple African or Asian peoples, most notably pygmies, Vedda and other forest folk. The irregularity and vagueness of such reports are important, as is their distance in time and space. For these factors would account for the depiction of their hypothetical objects not as something closely resembling a chimpanzee or orang-utan —or for that matter a real pygmy or Vedda—but rather as little more than a thoroughly human figure with a hirsute body.

As with giant wildmen of Asia and their possible connection with orang-utans, a question arises concerning the large size of some European wildmen by comparison with the small proportions of chimps and pygmies. Yet such disparity finds a reasonable explanation in the same tendency to exaggeration posited earlier for Asian instances. Depictions of the European wildman as little more than a hairy human bear a striking resemblance to representations of orang-utans and other apes, for example in the seventeenth-century representations by Bontius (1931: 285) and Beeckman (1973; see Plates 8.1 and 8.2). Such likeness arguably reflects the influence of the mediaeval figure on later understandings of anthropoid apes, an influence further reflected in European conceptions of non-western peoples. Yet it is also attributable to artistic expediency, where an illustrator followed a verbal description referring to a figure looking very much like a human but covered in hair. (In the case of Beeckman's relatively hairless orang-utan, one would have to add 'with feet like hands'.)

By the seventeenth century, the wildman of the late Middle Ages was little more than a vague memory among proponents of an emerging scientific urban culture. Noteworthy in this connection is a reference to what were probably apes, observed off the coast of Sierra Leone, by sixteenth-century Portuguese navigators who described them not as wildmen but as hairy bodied 'negroes' (La Roncière 1925, cited in Heuvelmans 1980: 507). Certainly, by the nineteenth century, few if any of the Europeans, mostly of British stock, supposedly encountering hairy hominoids in British Columbia or Australia would have had any direct knowledge of the mediaeval image. How far this wildman may nevertheless have contributed indirectly to such figures as the North American 'sasquatch' and the Australian 'yahoo' is addressed below.

Apes in North America

Probably the best known among the world's hairy hominoids, the sasquatch has also been the one most subject to empirical investigation, although mostly by amateurs rather than professional zoologists or anthropologists. The sasquatch is usually described as a very large, mostly dark-haired, cone-headed and speechless ape (Napier 1972: 85–87), quite unlike anything closely resembling a member of the genus *Homo* (Green 1980: 237). At the same time, like the yeti, the figure is associated with huge footprints of a decidedly hominoid form; thus it has come to be called 'bigfoot', the name by which it is mostly known in the United States. As a non-human primate of any sort, the sasquatch, a creature mostly identified with the mountain forests of western Canada and the northwest United States, is a zoogeographical impossibility, not susceptible to possible explanation with reference to orang-utans or other known primates in the same way as are comparable figures from East and Southeast Asia. Sasquatch territory does overlap that of Black and Grizzly bears; yet the common suggestion that sasquatch simply reflect mistaken sightings of bears is less plausible in this case than in Asian instances. Certainly, the giant footprints do not resemble those of the Ursidae.

Standing 2–3 metres tall and leaving prints as long as 45 centimetres, the giant sasquatch is mostly a Euro-American image developed during the twentieth century.[1] For the most part, the creature suggests an amalgam of large and small hominoids recognized by Amerindian cultures, recast in the light of older European wildman images, in which larger exemplars have come to predominate. This does not completely exclude a possible empirical basis for the category, but if there ever was such a basis, this is likely to have been something rather smaller. Supporting this suggestion is evidence from reputed sightings beginning in the nineteenth century, which indicate that the sasquatch has, as it were, grown larger with the passage of time. Many of the earliest Euro-American accounts of images corresponding morphologically to sasquatch thus relate to figures varying between about 1.2 and 1.5 metres in height. In addition, some Amerindian representations refer to hairy hominoids smaller than European or native humans.[2]

Such disparity is one factor complicating the relationship between apparently pre-European indigenous images—categories which Suttles (1972, 1980) collectively designates as 'sasquatch-like beings'—and the figure known to modern North Americans. At the same time, certain native figures themselves vary in size. A mid-nineteenth-century report by George Gibbs on the Upper Chehalis tribes, Salishan-speakers of southwestern Washington state, describes a 'race of beings' called 'tsiatko' or 'steh-tatl' (Clark 1955: 313–14) who inhabit 'holes in the ground' in mountainous regions, smell bad and are seen only at night. Members of some tribes describe the creatures as 'giants' covered in hair like a dog; yet others depict them as 'natural' or human-sized and resembling people except that they 'gibber and chatter' excessively. In regard to distinctions articulated in other parts of the world, another interest of these images is evidence that Chehalis do not include the tsiatko within the class of 'demons' named 'elip tilikum'. In other words, they consider them as something separate from spiritual beings and even as partly human. Also worth noting are stories of tsiatko abducting native women (ibid.: 314).[3]

Among the most detailed descriptions of indigenous hominoidal images and probably the most naturalistic, is Canadian ethnologist T.F. McIlwraith's account of the relatively small ape-like creatures which the Bella Coola (or Nuxalk) Indians of the central coast of British Columbia designate as 'boqs'. McIlwraith's description is worth quoting in full:

> This beast somewhat resembles a man, its hands especially and the region around the eyes being distinctly human. It walks on its hind legs, in a stooping posture, its long arms swinging below the knees; in height it is rather less than the average man. The entire body, except the face, is covered with long hair, the growth being most profuse on the chest which is large, corresponding to the great strength of the animal.
>
> (McIlwraith 1992: 60)

Further features include a variety of vocalizations (whistling, shrieking, grunting and roaring) but not, apparently, articulate language. McIlwraith classifies the boqs as one of several 'zoomorphic beings' recognized by the Bella Coola. He also refers to them as 'animals' thought still to be present in Bella Coola territory and records the last encounter as occurring in 1924 (ibid.: 63). According to native stories, boqs sometimes appear in groups and are most often found near beaches, where they gather and feed on shellfish. The attribute would appear significant in relation to numerous European sightings of sasquatch, or their tracks, near bodies of water, as well as indications that sasquatch are able to swim (Green 1980: 239).

At the same time, boqs possess several features consistent with McIlwraith's further interpretation of the creatures as 'supernatural'. The males are credited with extraordinarily long penises—so long that, like legendary characters from various parts of the world, they can surreptitiously engage human females in intercourse from a considerable distance. Boqs are also able to cause areas of land to rise and sea or river water thereby to drain away, a power they use against human intruders. However, their usual method of terrifying humans is striking tree trunks and breaking branches, activities in which they further deploy their enormous phalluses. Destroying trees is an attribute of similar figures, collectively designated in English as 'tree-strikers', that are recognized by other northwestern North American groups (including the Lummi, Salishan speakers like the Bella Coola and Chehalis). As noted, similar behaviours are ascribed to the Sumbanese mili mongga (Chapter 4) and Sumatran orang pendek (Chapter 5). Accordingly, one cryptozoologist has interpreted 'tree-striking' by both 'sasquatch' and orang pendek as reflecting unidentified 'hominoids' knocking over dead trees in order to feed on larvae (Bayanov 1982: 47).

Whatever powers boqs are thought to possess and despite their great strength, the creatures are not invulnerable. They can be killed with a bullet smeared with the human killer's blood. Less fantastically, one native report of an encounter with humans describes a boqs as acting frightened and taking flight (McIlwraith 1992: 60, 62).

Although the Bella Coola speak a Salishan language, the name 'boqs' (or puq'ws, Davis and Saunders 1980) is evidently cognate with terms from the non-Salishan Kwakiutl language, like 'bekwes' or 'bukwus', translatable as 'man of the woods' (Rigsby 1971; Suttles 1980). These terms thus have the same meaning as 'orang-utan', nowadays used by Europeans for the Southeast Asian ape, as well as several other Indonesian names referring either to hairy hominoids or human forest dwellers (see Chapter 4). By the second half of the twentieth century, moreover, English-speaking Amerindians were glossing Kwakiutl 'bekwes' and related terms as 'ape'. Reflecting another borrowing from Kwakiutl, 'baw'is', a term employed by the Tsimshian who reside to the north of the Kwakiutl and Bella Coola, is similarly equated with English 'monkey' or 'ape'—and also with 'sasquatch'. In addition, the Tsimshian variants 'pa'gwus' and 'pi'kis' respectively denote a 'monkey' and a sort of 'monkey mask' (Halpin 1980a: 20, 1980b: 214; see Plate 8.3). Revealing noticeably apelike features (a sloping forehead, heavy brow ridges, flat nose, prognathous lower face, wide mouth and receding chin), these masks have been interpreted by Halpin as a 'post-contact invention' reflecting introduced knowledge of non-human primates (1980b: 212). Be that as it may, the indigenous names comprise elements that originally designated a hominoidal forest creature, or 'man of the woods'. What is more, an indigenous representation of a being with simian features has been identified from carved stone heads, generally hominoid in form and displaying what could be construed as pronounced brow ridges, found in Washington and Oregon and dating from 1500 to 2500 years ago (Sprague 1980: 229–34, Plates 33–39).

Besides the boqs, other northwestern North American figures smaller than humans include the hairy dwarf hominoids of the Salishan-speaking Lummi of Washington state. Referred to as 'tree-strikers', these beings are in fact barely distinguishable from the Bella Coola creatures. Similarly described as small and also as very strong, is a female figure, or 'ogress', recognized by the Lummi and the Cowichan Indians of Vancouver Island (Suttles 1972: 74). According to a Lummi Indian living in Cowichan, one such 'little woman' was captured on Vancouver Island early in the twentieth century by railwaymen repairing tracks between Esquimalt and Nanaimo. (The further significance of this report and its possible relation to putative European experiences, is discussed just below.)

The Euro-American sasquatch has also been linked with much larger native figures. One is the image whose Salishan (specifically Halkomelem) name is the source of English 'sasquatch' (Suttles 1972: 69). Another is the giant being that Kwakiutl call 'zonokwa' (Lévi-Strauss 1982: 59, 65–66). Usually represented in indigenous narratives as females with pendulous breasts who abduct and consume human children, the zonokwa bear an arresting resemblance to images of hirsute female wildmen encountered in several parts of Asia. Prominent breasts also figure in a few reputed sasquatch sightings, as well as in the famous film produced by sasquatch-hunter Roger Patterson (see Napier 1972: 89–94). The zonokwa are further described as pronouncing words 'in such a way that every syllable of ordinary speech is repeated with initial "h" substituted for the consonantal beginning of the syllable, or with "h" introduced before the initial vowel' in words beginning with vowels (Boas 1935: 145). Suggesting an extraordinary coincidence with speech peculiarities ascribed to the Sumbanese mili mongga (Chapter 4), this specification provides obvious support for a view of wildman images as realizations of a universal proclivity of the human imagination that finds expression in culturally quite disparate places.

Amerindian classification, however, distinguishes large ogresses like zonokwa from smaller and in some ways more sasquatch-like creatures such as the Bella Coola 'boqs', at least in part. Like these smaller beings, the zonokwa possess great strength and characteristically tear down trees (Boas 1935: 145). Conversely, smaller figures (such as the short and stout 'cannibal woman' of the Salishan Halkomelem speakers of the southwestern British Columbian mainland) are equally depicted as anthropophagous child abductors (Suttles 1972: 69). Nevertheless, the zonokwa are not unequivocally described as covered in hair; according to one source, only their hands are hairy (Boas 1935: 144–45). 'Zonokwa', moreover, seems usually to designate a single, female figure, whereas the boqs and similarly small creatures are represented as a plurality. Partly in accord with these differences, indigenous terms for ogresses, both large and small, differ consistently from names applied to what Suttles calls 'sasquatch-like beings', not only in Kwakiutl but in all Salishan languages for which he provides 'ogress' terms (1980: 251–52).

In regard to attributes like anthropophagy, abduction of infants and engaging humans in sexual intercourse, neither the figure of zonokwa nor smaller hirsute beings such as the boqs bear a particular resemblance to the modern sasquatch, a creature usually described as a shy, retiring and non-aggressive. An exception is perhaps the story of Albert Ostman, a British Columbian lumberman who claimed to have been held captive by a family of sasquatch for several days in 1924; even so, Ostman suffered no harm at the hands of his captors and was able to escape with relative ease (Green 1978: 97–112). Whatever might explain these differences, it is clear that indigenous cultures of northwestern North America reveal no single figure corresponding to the representation of the sasquatch that has developed among twentieth-century Euro-Americans. Of several features attributed to the modern sasquatch, only two ('giant size' and residence in mountains or woods) are associated with seven named categories drawn from a sample of Northwest coast languages compiled by Suttles (1972: 87). Conversely, only two of the seven categories pertain to creatures characterized as 'hairy'.

Apparently corresponding closely to Amerindian representations of small hominoids like the boqs was a hairy being reputedly witnessed by several Europeans in the late nineteenth century. Dubbed 'Jacko', a name deriving from a European misinterpretation of a West African term for the chimpanzee ('engeco', see *Shorter Oxford English Dictionary*, s.v. 'jocko'), the lone creature was found lying near railway tracks in the Fraser Canyon, 20 miles from Yale, British Columbia, in 1884. After his subsequent capture, Jacko became the subject of a remarkable newspaper story which, by all indications, concerned a member of a real (although unidentified) species. The report described Jacko as 'something of the gorilla type', but also as 'half man and half beast' (*Daily Colonist*, Victoria, British Columbia, 4th July, 1884).

Based on the accounts of at least five railway employees who observed the creature simultaneously and a local physician named Dr Hannington, Jacko's physical description is quite precise. He stood 1.4 metres and weighed 58 kilograms. His form was generally human, except that his forearms were 'much longer than a man's'. His body, except for the hands, was covered in glossy hair about 2.5 centimetres long; and his hair (apparently head hair) was 'long, black' and 'strong' (sic). Jacko's favourite food was berries and he enjoyed drinking milk. Although no mention is made of external genitalia, the news report's consistent use of masculine rather than neuter pronouns (which I have chosen to follow) indicates that Jacko was male. At the same time, the article contains no suggestion that he was immature, contrary to tendentious later references to the creature as a 'young sasquatch' (see Sanderson 1961: 47; Krantz 1992: 202). That no mention is made of Jacko's apparent nakedness may be explained either by his evident animality or possible Victorian sensibilities. It may be thought equally odd that the report says nothing about the teeth, especially if, as was later suggested, the captive was a great ape. Consistent with a robust build indicated by the height to weight ratio, Jacko was 'extraordinarily strong'.[4] He was also speechless, uttering only a 'half bark and half growl'.

As interesting as the creature's physical description is what his captors made of him. Seeing Jacko lying motionless on the tracks, the railwaymen 'first thought he was a crazy Indian' (elsewhere in the article expressed as a 'demented Indian'). Apparently after a closer inspection, he was alternatively conceived as an apelike human, or humanlike ape, possibly representing 'a species hitherto unknown in this part of the continent'. The news writer, however, does not entirely discount the possibility that Jacko was an Indian and in a similar vein another British Columbian newspaper later referred to him as 'the wild man' (*The Columbian*, 12th July 1884, quoted in Green 1978: 86). Illuminating this interpretation is a conception, current since the sixteenth century and even into the twentieth, of North American natives as 'hairy', a notion that Dickason (1980: 73) traces to wildman images of late mediaeval Europe (see also Taft 1980: 83–96, regarding a Newfoundland category of hairy hominoids designated as 'wild Indians').[5] Although the news reports do not specify whether Jacko was bipedal or quadrupedal, if he moved mainly on four limbs it is difficult to imagine how he could be mistaken for a human. In addition, the description of his body hair does not correspond, for example, to a person suffering from hypertrichosis, or simply an unusually hairy man. Indeed, the impression generally conveyed is of some sort of ape.

Like most capture tales (including those reviewed in previous chapters), the story of Jacko ends in mystery. While being kept in railway machine shops in Yale, Jacko was cared for by George Tilbury, an otherwise unidentified man who proposed to take him to England to be exhibited. What actually happened to the creature is not known, although speculation has included his dying at sea (since he would have had first to be taken to the British Columbia coast, as the Canadian railway line running east was not complete in 1884) or being acquired by P.T. Barnum and ending up as a sideshow attraction (Napier 1972: 75; Green 1978: 88).[6] Another question concerns Jacko's singularity. According to the article in the *Colonist*, some time prior to his capture three men, including a civil engineer and a shop-keeper, claimed to have seen a 'curious creature' in the area. But their reports were discounted as reflecting 'a bear or stray Indian dog'.[7]

To explain the story of Jacko, two possibilities present themselves: either there was an animal, more specifically some sort of primate, or the entire story was fabricated. Since Jacko was observed by several individuals, some of them evidently educated, it is unlikely to reflect an honest mistake, for example, a misinterpretation of a known local animal. One suggestion is that Jacko was an escaped chimpanzee or 'a juvenile male or adult female gorilla' (Napier 1972: 75; an orang-utan seems ruled out by the colour of the hair.) Yet this seems highly unlikely. For one thing, gorillas and chimps lack the long head hair attributed to Jacko and the weight seems too low for a gorilla of Jacko's height. For another, the area about Yale—in 1884 a small, newly established township—was a rugged frontier and at the time neither circuses nor zoos were a feature of the British Columbian cultural landscape. Following the scientific discovery of the lowland gorilla in 1847 and publicity surrounding Darwin's theory of evolution, interest in apes became widespread in the latter part of the nineteenth century and at least some Canadians (such as the aforementioned Dr Hannington) would probably have been quite knowledgeable about the animals (see Reynolds 1967: 55–57, cited in Green 1978: 34). This suggests that, if the creature were indeed a chimpanzee or gorilla, it would simply have been identified as such, rather than as something combining features of an ape and a human.

As an explanation of the Jacko story, this circumstance would seem to leave only fabrication, a charge actually levelled by the *Mainland Guardian* of New Westminster, British Columbia (9th July 1884, article quoted in Green 1978: 86). The paper claimed that no animal had been caught and that readers of the *Daily Colonist* and the *Columbian*, another New Westminster paper that reprinted the *Colonist*'s story, had been 'duped' (ibid.). However, not only were these claims made by a rival of the *Columbian*, but the *Guardian* does not substantiate its allegation in any way. The article moreover ends with the seemingly contradictory statement that the circumstances of Jacko's discovery and subsequent disposal remain 'a mystery'. What is more certain is that, if the creature were a hoax, its fabrication must have involved a conspiracy of several men—some of them identified by name, whose existences have all been verified (Green 1978: 85) and whose motives remain completely obscure.

The story of Jacko's capture may have spawned at least one later tale. As mentioned previously, in his youth, a Lummi Indian residing in Cowichan heard about a 'little woman', extremely strong, with a voracious appetite and living in a mountain cave, who had been caught in a trap by men repairing railway tracks on Vancouver Island early in the twentieth century. The Lummi man speculated that the creature, after being transported to Victoria, might have been 'sent . . . over to old Queen Mary' (Suttles 1972: 74), an unexplained phrase which may, however, refer to England, the destination also proposed for Jacko. Similarly relevant is a story recorded by Duff in 1949–50, concerning a short, stout 'cannibal woman' living in a 'very smelly cave' above Yale, which was 'blasted away during the construction of the Canadian Pacific Railway' (thus possibly before 1884, when the railway had already passed through Yale).[8] The creature was, according to some reports, captured by 'white people' and placed in 'a big iron cage'. Duff's native informant had heard that a picture of the woman in the cage had appeared in a newspaper 'about fifteen years' previously (Duff 1952: 118).

If Jacko's identity remains uncertain, then so does the connection between the creature so named and the sasquatch. That Jacko's Anglophone captors did not identify him as a 'sasquatch' is simply explained by the fact that, in 1884, the category did not exist—the name, usually attributed to Indian agent J.W. Burns (1929), entering English usage only after the 1920s. Ironically, David Daegling, a physical anthropologist who has forcefully argued that evidence for the sasquatch reflects a series of elaborate and scientifically informed fakes, considers Jacko to have been a real creature, although he rather implausibly interprets it as a chimpanzee or gorilla escaped from Barnum and Bailey's circus (2004: 66–67). By contrast, sasquatch enthusiasts and cryptozoologists (for example, Sanderson 1961; Byrne 1975; Green 1978) place less credence in Jacko than might be expected. The probable reason is that, standing just 1.4 metres tall, the specimen reputedly discovered in the Fraser canyon in 1884 does not live up to the image of the huge, hulking beast associated with the name 'sasquatch' and suggested by twentieth-century reports of sightings and footprints. Yet Jacko was not the only reputed nineteenth-century sighting of a hairy hominoid smaller than an adult human. For example, in 1869, an American hunter in central California reported seeing a 1.5 metre 'gorilla or wild man' which he described as 'disproportionately broad', covered in 'dark brown and cinnamon' hair, whistling and playing with burning sticks from his fire (Green 1970: 14–15; 1973: 5). Moreover, a recent account refers to a creature whose height was estimated by two 1993 observers as respectively 1.2 and 1.5 metres (Hewkin

1993–1996: 73). One is thus reminded of a remark by primatologist John Napier who—after declaring his conviction that 'the Sasquatch exists'—suggests that the animal may nevertheless not be 'all that it is cracked up to be' (1972: 205).

Regardless of his relationship to 'sasquatch', if Jacko were a fake, he might be seen as one of the first in a long North American line of similar fabrications. Indeed, a particular interest of the sasquatch is that it is the one case in which fakery is apparently widespread and well attested. Yet this distinction is hardly surprising. Given financial rewards and the public attention that evidence, whether real or concocted, for the existence of a hirsute human-like creature can command, motives for such fakery are not difficult to discern (see Shackley 1983: 41; Daegling 2004). In addition, an individualistic Anglophone culture, perhaps especially as it has developed in twentieth-century North America, provides a further incentive: a desire, stemming from a pervasive anti-intellectualism and anti-authoritarianism, to 'fool the experts'. No such motivation exists, for example, among rural Sumatran Malays, Chinese peasants, or Florenese highlanders—or at least did not do so until very recently. On the other hand, how or how far this point might apply to the nineteenth-century case of Jacko remains unclear.

Whatever the nature of the creature so named, intriguingly similar indigenous representations quite possibly reflect the influence of new knowledge, introduced by Europeans, of primates and especially anthropoid apes. In fact, McIlwraith's description of the boqs, recorded during fieldwork conducted between 1922 and 1924, sounds almost exactly like an ape and as much like an ape as does Jacko. Pre-existing Amerindian images—such as a feral 'man of the woods' (McIlwraith 1992: 63; and see note 5)—could thus have been reinterpreted and 'simianized' as a result of this new knowledge, conveyed in pictures, verbal descriptions and even live specimens.[9] To facilitate such reinterpretation, however, the indigenous figures would have had to share some—indeed a good many—features with apes. Therefore, one cannot simply dismiss the huge sasquatch that has emerged in Euro-American popular culture as no more than an ironic product of Amerindian exposure to European images of primates. By the same token, it would be difficult to argue that the Euro-American sasquatch owes nothing to the wildman of European folklore and late mediaeval culture. Yet the influence of the European figure probably does not exceed that of Amerindian giants like the heavy-breasted 'zonokwa'. Actually, both possibilities raise similar questions concerning Euro-American familiarity with the European wildman and with native traditions, especially among relatively uneducated members of rural communities who apparently form the majority reporting evidence for the sasquatch. Far more certain is European knowledge of anthropoid apes, which evidently informed representations of the creature called Jacko, albeit to an extent as yet undetermined. At the same time, also playing a part both in interpretations of Jacko and representations of the sasquatch has been a notion of 'hairy Indians', an idea which in its sixteenth- and seventeenth-century origins apparently owed more to the European figure of the wildman.[10]

The Australian 'yahoo' or 'yowie'

Although Australia belongs geographically to Oceania, the images I treat here are a cultural artefact of European (and again, mainly British) settlers as much as is the North American sasquatch. Developed in the late nineteenth and twentieth century, they also pertain to roughly the same historical period. Euro-Australian reports of wildmen cluster in the period from about 1871 to 1920, even though reputed sightings are also reported from the last decades of the twentieth century. Other similarities between Australian and North American hominoids concern the zoogeographical implausibility of a non-human primate native to Australia, the presence in both places of phenotypically distinct native populations and a questionable connection between Australian Aboriginal and Euro-Australian representations.

The antipodean wildman is mostly known from southeastern Australia and is designated by two similar names. One, 'yowie', may be of Aboriginal derivation. The other, 'yahoo', possibly reflects the usage of Jonathan Swift, but in this context may more directly derive from an application of the name to the adult orang-utan (*Pongo pygmaeus*) after a specimen of the ape arrived in England (Joyner 1984: 55–57). Compiled by Joyner (1977; 1980), accounts of the putative creature are extremely varied. Some suggest a hair-covered 'wild man'; others describe an ape-like being, often specified as a 'gorilla'.[11] Several reports claiming sightings of creatures that do not sound very hominoid have been interpreted as reflecting a large wombat or other marsupial (Joyner 1977: 7; Groves 1986: 50–53; Smith 1989: 35–36). However, most descriptions suggest a hirsute, bipedal and generally hominoid form with a simian face and very large feet, sometimes evidenced by associated prints. On the other hand, several refer to creatures that are at least partly quadrupedal (e.g. Joyner 1980: 1–2). Estimates of height, or length, mostly vary between 1.5 and 2.1 metres (see Joyner 1977; Groves 1986, 1988; Smith 1989). Hair colour is usually not mentioned; where it is, the colours are various. Skin colour seems never to be specified, although one report of a creature with a 'red face' (Joyner 1980: 3) might suggest a fair complexion. Among the relatively few fantastical attributes of the yahoo, the one that stands out in a comparative context concerns a reputed eyewitness report of feet 'turned backwards' (Joyner 1977: 5, 1980: 1).

Despite this diversity, Smith (1989) makes a convincing case for the yahoo having a partial basis in European experiences of Australian Aborigines, more particularly isolated, 'degraded and antisocial' male Aborigines of relatively large physical proportions. If this is correct, then of all the hairy hominoids surveyed so far, the Australian wildman would appear to be the image most demonstrably grounded in misidentification of a phenotypically and culturally distinct group of *Homo sapiens*. While Smith mostly locates these experiences during an earlier period of Australian settlement, it is noteworthy that the only verified Euro-Australian sighting of something identified as 'a yowie-like creature' occurred in 1987. Described as a sub-human 'monster' and something 'half-man, half-beast' that 'ran like a gorilla' and had the face of an 'ape', this turned out to be a naked Aboriginal man standing over two metres tall and perhaps weighing between 125 and 155 kilograms. Spotted by a white family, the individual was reported to the police and subsequently arrested (Smith 1989: 31–32, quoting a report from the Brisbane *Sunday Mail*). As regards gender, Smith points out how no reported sightings describe female hominoids; hence there is no reference to female genitalia or breasts, pendulous or otherwise (ibid.: 30, 29). At the same time, the fact that not all reports of yahoo have concerned individuals exceeding Europeans in height would suggest that reputed sightings need not have involved only Aborigines who were relatively large.

Among physical features of Australian Aborigines that might incline Euro-Australians to mistake some for hairy wildmen, Smith notes body hair equal to that of Europeans, as well as 'heavy brow ridges, receding foreheads, broad, flat noses and a marked prognathism' (1989: 31). The author is careful to avoid giving offence, but as he notes, 'all human races possess "ape-like" characteristics which draw the attention of those who do not share them' (ibid.). In support, he cites an experience of the American primatologist Frank Poirier, who in China was mistaken for a (Chinese) wildman (Poirier *et al.* 1983: 37–38). There is a more general point here and one that applies equally to possible connections between North American hominoids and Amerindians and even Asian categories in relation to local minorities such as the Kubu and Vedda. When presented with people of a different phenotype (or 'race') human perception focuses ineluctably and without any necessary intervention of ideas generated by a particular culture, on physical features that are distinctive. Moreover, in these one tends to see resemblances with non-human animals and, for obvious reasons, especially primates. According to Birdsell's detailed study of phenotypical variation among Australian Aborigines (1993: 207–15, 283–91, 308–11), it is precisely in southeastern Australia, the region in which yahoo have mainly been reported, that natives display the greatest amount of body and facial hair, the largest brow ridges and a relatively high degree of forehead slope. It goes without saying that it is these features, all significantly more pronounced than in Europeans, which might invite comparison with apes. Birdsell notes, for example, that Australian Aborigines have 'the largest brow ridges of any living human population' (ibid.: 283). In regard to stature, southeasterners are shorter than some northern Aborigines, but not to a degree that contradicts many accounts of yahoo.[12]

This is not to argue that European experience of Aborigines is the only source of the Australian yahoo. Some of the unusual fauna of the island continent have likely played a part, as in some measure have ideas traceable to the wildman of an earlier European culture and folklore. On the other hand, the evidence for a hairy wildman as a category of Aboriginal languages is equivocal. Some terms (mostly distinct from 'yahoo' or 'yowie') have been identified as possible references to such an image, but among these are words which actually designate birds (Joyner 1977: 12) or refer to renegades, strange tribesmen, or 'wild blackfellows' (ibid.: 26; Groves 1986: 50). Thus, any connection between the European and native Australian representations appears tenuous at best. Furthermore, Joyner (1977: 13) states flatly that an image of wild, hairy hominoids is unknown to Aborigines. If Aborigines do possess a representation of hirsute and morphologically distinct wildmen, then a source might conceivably be found in former or recent 'racial' differences among these populations themselves (see Groves 1986: 48). Palaeontological evidence from Kow Swamp in Victoria State dating from between 9,000 and 12,000 years ago has suggested a demographic component displaying morphological features of *Homo erectus* (Bellwood 1997: 93–94). But the characterization of these finds has been disputed and differences with other human contemporaries may be negligible (Groves 1986: 48).

Given the parallels noted earlier, taken together, the North American sasquatch and the Australian wildman may provide the best argument for hairy hominoids reflecting imaginary categories of a culturally and historically quite specific kind. Especially relevant is a virtual identity in national and cultural backgrounds of the mainly British proponents of these images (or at least of their earlier recorded forms). In both continents, it may be noted, most nineteenth-century settlers derived from either urbanized areas of Europe or agriculturally developed rural settlements far from true wilderness. This, of course, does not mean that all such representations—including the Asian varieties discussed in earlier chapters—should be explicable in the same way, that is, as artefacts of a frontier experience. Also, even if this interpretation fits the North American and Australian cases, it is not yet possible to say what exactly would account for putative encounters, although psychological distress in desolate or unfamiliar settings has been suggested as a factor (Halpin 1980a; Groves 1986: 52; and see Anderson 1990: 240–42)—as, of course, has fabrication. Equally unclear is the possible social and cultural motivation, or 'function', of such hominoidal images. Sometimes, they have been vaguely interpreted as 'symbols' of wilderness, the unknown, or 'liminality' (Halpin 1980a: 23). But questions remain as to why they take the specific forms they do, as well as their similarity to non-western representations that are less susceptible to such interpretations. This is not to say that these explanations are completely invalid, merely that they remain very undeveloped.

East Africa and the 'little furry men'

Besides Southeast Asia, Africa is the only region of the world that is home to great apes, including chimpanzees (*Pan troglodytes*), bonobos (Pygmy chimpanzees, *Pan paniscus*) and Lowland and Highland gorillas (*Gorilla*

gorilla and *Gorilla beringei*). The continent is also the site of possible early or ancient encounters of European and circum-Mediterranean peoples with large primates, perhaps most famously exemplified by Hanno's periplus (Yerkes and Yerkes 1929: 2–3; Reynolds 1967: 29–31). If apes are a significant source of wildman images, therefore, one might expect Africa to provide numerous exemplars. And if surviving or remembered pre-sapiens hominins were their hypothetical origin, then Africa, as the major locus of human evolution, should be prominent for this reason as well.

Recorded both in areas where apes are attested and in regions where they are not, reports of African hominoids have been comprehensively reviewed by Bernard Heuvelmans (1980). One source is a report by Captain William Hichens, an Englishman and former civil servant, who in the 1920s, while hunting lions in the Wembere region of west central Tanzania, observed two creatures emerging from dense forest. Resembling 'little men', the creatures were tailless, covered in 'russet' hair, stood about 1.2 metres tall and walked erect (Hichens 1937: 373). Evidently familiar with local primate life, Hichens remarks that the creatures may have been monkeys, but 'were no ordinary monkeys, nor baboons, nor colobus, nor Sykes, nor any other kind found in Tanganyika' (ibid.). Wembere, it should be noted, lies east of the normal range of chimpanzees and gorillas. Hichens' efforts to follow the hominoids were in vain. Reacting with 'mingled fear and amazement', a native hunter accompanying the Englishman also saw the creatures. He identified them as 'agogwe', rarely encountered beings which, according to what villagers later told Captain Hichens, will weed and hoe people's gardens at night in exchange for food and millet beer.

These particulars suggest that the 'little furry men' are a category recognized by local Tanzanians. Yet available ethnography does not confirm that East Africans maintain a representation labelled 'agogwe' which substantially accords with the physical image reported by Hichens. Nor is it entirely clear from which language the name 'agogwe' derives. Hichens briefly mentioned his sighting in an earlier article, where he writes the name as 'ngogwe'. He appears to gloss this term as 'little men of the trees' and to identify it as a usage of the Iramba (or Nilamba) people (1928: 176). It is also in this article that Hichens specifies the creatures as tailless. Evidently drawing on Iramba lore, he further describes the 'ngogwe' as 'wailing a strange chant' as they travel (ibid.).[13]

Heuvelmans compares Hichens' report to two other accounts of East African hominoids. One refers briefly to 'little red men' inhabiting the eastern Kenyan region of Embu. According to an African who claimed to have seen 'scores' of these beings, they will pelt human intruders with small stones (S.V. Cook 1924: 24); but noticeably missing from this account is any explicit reference to body hair. The same omission characterizes a later report from Mozambique. Writing in response to Hichens' article, another Briton described how, from the deck of a cargo boat in 1937 and with the aid of a telescope, he was able to observe two 'little brown men' between 1.2

and 1.5 metres tall walking on a beach among a troop of baboons (Burgoyne 1938: 51).[14] The baboons appeared undisturbed by their presence. Referring to a similar sighting by an unnamed friend, the author vaguely refers to a native injunction on shooting the little men.

A somewhat more detailed sketch of putative hominoids is offered by a professional big-game hunter, Roger Courtney (1940: 37–49) and concerns apelike creatures called 'mau men'. Found in the vicinity of Mount Longenot in Kenya, the beings were described to Courtney by his guide, a Muslim of mixed Boran and Mkamba descent named Ali (1940: 37–49). Ali had heard the story from his then deceased father, who claimed to have been struck over the head and abducted by a group of mau men while tending sheep on the slopes of the mountain, described as an old volcano full of caves. In this account, the mau are characterized as small, apparently tailless creatures resembling 'monkeys' more than humans. While their skin was 'white', their bodies were covered in long 'black' hair. Hair also hung over the eyes. The reputed eyewitness encountered the mau sitting on ledges inside a cave, around a central fire. Characterized as forest dwellers, the creatures are further described as reaching their caves by way of long underground passages. Apart from the fire and sticks and stones which they employed as weapons, the mau appeared to possess no technology; as Ali points out, they may even have obtained the fire from a natural volcanic source. Chattering like monkeys, they also seem to have lacked an intelligible language. The only indication of their diet is the sheep stolen from Ali's father.

Ali's father explicitly distinguished the mau from less 'wild' pygmies residing in the forests to the west of Lake Victoria, whom he had encountered in his travels. Indeed, in all respects except the use of fire and occupation of caves, the beings sound very much like chimpanzees, which can be similarly light-skinned and with which the shepherd may also have been familiar from his westward journeys. Although caution is always required in translating local colour terms, the long 'black' hair of the mau contrasts with that of the russet-haired agogwe. Even so, since normally dark-haired chimpanzees occasionally possess brown or reddish pelage, wayward chimps, straying some 200 kilometres beyond the eastern limit of their normal range in western Tanzania, could,conceivably explain Hichens' experience as well. Less readily accounted for is the hair hanging over the eyes of the mau. However, it is a point of interest that exactly the same feature is attributed to hominoidal beings reported from Central Africa.

Different from other East African images are hominoids—most of them apparently quite human (see Heuvelmans 1990a)—that numerous Maasai and other Kenyan tribesmen described in the 1970s to the French anthropologist Jacqueline Roumeguère-Eberhardt (1990). Among these is a figure Roumeguère designates anonymously as 'X1' (there are five Xs altogether) and describes as hairy-bodied, possessing long head hair and huge feet, heavy set and extremely strong. The head hair is dark; the body hair is reddish-brown or fawn among younger specimens (who could conceivably be of a size comparable to Hichens' agogwe) but darker or sometimes grey in adults. Standing as tall as a human (1.3–1.85 metres) and occasionally taller, the creatures are further depicted as wielding huge clubs, killing buffalo and carrying away the carcasses, consuming raw flesh and possibly employing a language. Despite their size and strength, they are not aggressive towards humans. These reports, Roumeguère suggests, could reflect a surviving non-sapiens hominin such as *Homo habilis* or *Homo erectus*. In his Preface to Roumerguère's book, Heuvelmans (1990a: 26–34), on the other hand, suggests a robust Australopithecine (presumably a *Paranthropus*).

For a student of Florenese representations, probably the most remarkable feature of Roumeguere's book is a Maasai report of one sort of Kenyan hominoid (distinguished as 'X4') being offered milk in a gourd container and, 'perhaps mistaking it for a fruit', attempting to eat the gourd. It is perhaps curious that, unlike the agogwe and similarly small creatures, nothing like Roumeguère's anonymous hominoid was reported during the colonial period in East Africa. On the other hand, there is arguably some resemblance to the 'Nandi bear', an early twentieth-century British term for a mysterious creature named 'chimoset' by the Nandi of Kenya and sometimes described by European eyewitnesses as a 'big hairy biped' or 'an enormous baboon' (Heuvelmans 1995: 490).[15]

Ape-men of Central Africa

More detailed representations of hairy hominoids come from Central Africa and particularly from the Congo (formerly Zaire, now Democratic Republic of the Congo, or DRC). Reported especially from the Kivu region bounding Lake Kivu in the eastern part of the DRC are a smaller and larger figure distinguished as 'kakundakari' and 'kikomba' in the Konzo (Konjo) language and by other names in other eastern Congo languages. The size difference recalls the bimodalism evident in reports of mainland Asian hominoids, but in this case is rationalized in local interpretations of the pair as males and females of the same species (P. Leloup, cited in Heuvelmans 1980: 589).

Information on kakundakari and kikomba derives mostly from the Swiss zoological collector Charles Cordier (1963; 1973) and Paul Leloup, a Belgian herpetologist long resident in the Congo whose unpublished reports are cited both by Cordier and by Heuvelmans (1980). The creatures are also mentioned by the primatologist George Schaller, who, in one of his books on the Mountain gorilla, states that he can find no reason to deny their existence (1964: 92). Usually estimated as standing about a metre or the height of a young child and sometimes just 60 centimetres, the smaller kakundakari is further described as covered in short, sparse hair (dark in colour but according to one local report, grey), with longer, stiff black hair—or a 'mane' as it is several times called—falling down its back. Long hair also covers the forehead in a 'page-boy' fashion. Kakundakari skin colour is off-white or, according to another account, possibly greyish; an infant creature reputedly discovered lying on a bed of leaves was 'all white'. Other accounts, from Tembo and Lega speakers, however, specify the skin as 'black'. Reminiscent of Oostingh's report of a possible orang pendek encounter in southern Sumatra (Chapter 5), an African described the body of a specimen he encountered carrying an infant animal as remarkably filthy and covered in grime. Consistent with this, two other native eyewitnesses mention an appalling odour.

The kakundakari walks erect and leaves prints (just 12 centimetres long in one instance; Cordier 1973: 188) which sometimes suggest a foot with just four toes; otherwise, the big toe is described as standing out from the others. Lacking prominent canines, the teeth are generally humanlike and contrast with those of African apes; and the creature does not attack adversaries by biting, as do chimpanzees. The hands too are like a human's, but nothing is recorded on the length or proportions of the arms and legs. Like wildmen reported from almost everywhere else, the kakundakari is described as powerfully built and, in spite of its small size, extremely strong.

Africans consider the kakundakari a very rare and possibly extinct creature, shy and elusive and avoiding contact with humans. At the same time, the animal is said to 'wrestle' with people it encounters and invariably to overcome them. Other references to behaviour include the kakundakari's habit of lodging in caves, where it sleeps on beds of leaves, or spending the night in hollow logs. Although the creature has no knowledge of fire, it is given to piling up wood inside caves, as if to build a bonfire. Apparently omnivorous, kakundakari diet consists of freshwater crabs caught by turning over rocks in streams, as well as snails, centipedes, fungi, birds and fish stolen from hoop nets. The animal also takes game from hunters' traps, while fragments of termite nests found in a cave reputedly inhabited by a kakundakari suggest these insects as another food source.

Kakundakari are encountered either singly, in pairs, or in trios. No reports refer to vocalizations and none mentions language. In virtually all respects the creature suggests some sort of ape, although one possessing superior technological ability. Being unable to swim, the kakundakari is said to float across rivers on tree trunks or in stolen dugouts. One man reported a specimen apparently wielding a machete. And there is the further idea that the creatures store foraged food in containers made of leaves, which they carry on the hip. Africans associate kakundakari with herds of Bushpigs (or Red river hogs, *Potamochoerus porcus*), tracks of which frequently occur with the hominoids' footprints. The association obviously parallels the Sumatran belief that beings called 'si bagau', a category closely connected or conterminous with 'orang pendek', herd wild pigs. Similarly fantastic is the attribution of 'white blood' to kakundakari (a reference to a sap-like substance reputedly found on weapons used to wound or kill them), although this is denied by some Africans. According to another extraordinary idea, after defeating humans in wrestling matches the creatures will stop up the victim's orifices with clay.

Not only is the kakundakari more fully described than the East African agogwe or mau, but, as will become clear, it is also depicted far more naturalistically than figures from southern Africa and West Africa. It is, moreover, better attested, being the object of numerous African eyewitness reports, including stories describing the alleged capture and caging of specimens—and even alleged killings, some resulting in putative material evidence in the form of remnant flesh and skins. Most African sightings are apparently reported by male hunters; yet it is a point of interest that a good many derive from women. Female encounters appear partly to relate to experiences associated with collecting wild foods like mushrooms, or visiting streams to catch crustaceans—a women's activity (Heuvelmans 1980: 578) characteristic not only of kakundakari but of chimpanzees and bonobos as well. Since women do not hunt or travel so far into forests as do men, they are very likely less familiar with great apes; thus it is probable that women, observing chimpanzees and gorillas less frequently, will represent these animals differently and more fantastically, than do their male counterparts.

Africans are divided as to how kakundakari should be classified. Some consider them a kind of 'spirit' or 'phantom' ('mushumbi' or 'gitani', in an unspecified African language); but such specifications are sometimes rationalized as references to the creature being rarely or almost never seen or its ability to disappear when cornered (Cordier 1963: 178). Other Africans say the kakundakari is not a spirit as it leaves footprints and must therefore be an animal or variety of human. Some consider it an ape; others suggest it may be a 'degenerate' or feral human. According to yet another indigenous opinion, not necessarily distinct from local identifications of the kakundakari as a spirit, the creature is purely imaginary (Leloup, cited in Heuvelmans 1980: 581–82). Unfortunately, the derivation of names given to this and other African hominoids, including the larger kikomba, is unclear. None is readily susceptible to translation or analysis and there is nothing that could further illuminate their place in local classifications. All one can say with any certainty is that, according to the available lexical evidence, all of the names are distinct from those applied to chimpanzees and gorillas.[16]

Named 'kikomba', the larger Central African hominoid appears to be the subject of fewer native reports than the kakundakari. At the same time, the literature refers to several European eyewitnesses for kikomba, but none for kakundakari. Although much bigger and standing as tall as a human or even taller (1.9 metres, according to one observer), the kikomba is in most physical respects similar to the smaller kakundakari; it is an erect, bipedal, powerfully built and hair-covered hominoid, generally human in appearance but somewhat squatter (Heuvelmans 1980: 588). Possible differences lie in complexion and colour of kikomba body hair, described as 'reddish' in one European eyewitness account and 'brownish' in an African report of an animal reputedly killed in a buffalo trap—descriptions which immediately recall the smaller East African 'agogwe'. Other reports, however, specify the kikomba as possessing black hair and also like the kakundakari, it has a bushy ‘mane’ falling over the eyes; hence it continually shakes its head in order to see.

Kikomba feet are described as humanlike and explicitly different from an ape’s. A broad footprint observed by Cordier measured 20 centimetres; another was 30.5 centimetres long but only 11 centimetres wide (Cordier 1973: 188, 191); and in both instances the second toe was longer than the big toe. Cordier discovered the former in a place where an African claimed to have been chased by a kikomba in January 1960. As there are no bears in Central Africa, humanlike footprints cannot be attributed to these animals, as they can in Southeast Asia and other parts of the world.

According to an account by two native hunters of a specimen they had recently seen, the kikomba—in this instance, described as somewhat resembling a gorilla—has legs proportionally longer than the African ape and arms similarly long. The same men described a smooth, black face, a nose more human than simian, a low forehead, bushy eyebrows and a prognathous jaw. The animal had no crest on the head (such as one finds in male gorillas) and possessed a very short beard and a long ‘moustache’. The zoologist Paul Leloup, who personally conveyed this description to Heuvelmans, specified the moustache as the only divergence from typical descriptions of the kikomba. Further suggesting a difference from known apes is an account by an anonymous European geologist concerning a large primate he saw in 1935, which was neither a chimpanzee nor a gorilla but had a head somewhat like a gibbon’s (Leloup, cited in Heuvelmans 1980: 589–90).

While physically similar to the kakundakari, the kikomba differ rather more from the smaller hominoid in regard to behaviour. Most notably, the kikomba is described as exclusively vegetarian, feeding on honey (which it climbs trees to gather) as well as tubers and fruit of the ginger plant—two foods also exploited by gorillas (Cordier 1963: 180). Another item of diet is larvae obtained by demolishing rotten tree trunks. The kikomba will also take game from hunters’ traps, but does not eat this. The dietary contrast between the kikomba and kakundakari is one of several differences contradicting the previously mentioned local interpretation of the two creatures as male and female of a single species (Heuvelmans 1980: 591–92). Unlike its smaller cousin, the kikomba also seems not to frequent caves. And while no information is given on kakundakari vocalization, the equally speechless kikomba has been associated with howling, roaring or bellowing sounds—although local people alternatively attribute the nocturnal howling to the Water chevrotain (or Fanged deer; Cordier 1963: 180). No odour is described for the kikomba. Moreover, only the kakundakari are reputed to herd Bushpigs.

Kikomba are said often to carry a large piece of wood or a stick (or in one version an old axe handle), implements reminiscent of the club of the large Kenyan mystery hominoid described by Roumeguère—not to mention the weapon commonly attributed to the European wildman and indeed to great apes in early European depictions (Forth 2007; see Plate 8.4). However, while kikomba might use sticks to beat people, like the smaller kakundakari they also and perhaps more typically, assault humans by wrestling with them.[17] Since the creature's strength invariably exceeds the human combatant's, the latter's only defence is to play dead. The kikomba will then go off in search of something with which to cover its victim, thus giving the latter a chance to escape. Interestingly, this behaviour recalls Andrew Battell's early seventeenth-century report of gorillas covering their dead with 'great heaps of boughs and wood' (Ravenstein 1901: 55). Kikomba are also reputed to disturb hunters at night by shaking temporary shelters and pulling leaves off the roofs (Cordier 1973: 187). The resemblance between this behaviour and that attributed by Sumatran forest-collectors to the orang pendek hardly requires mention.[18]

Including eyewitness reports explicitly mentioning a resemblance to known apes and several in which the object was initially taken to be an ape, a good deal of evidence suggests that 'kikomba' and 'kakundakari' label representations grounded in experience of gorillas and chimpanzees. Particularly noteworthy are their generally primate (as opposed to human or hominin) form and behaviour and the attribution to neither of language or knowledge of fire. Their possible use of manufactured tools, if not simply an exaggeration, could reflect a plausible manipulation of discarded or stolen human artefacts not beyond the capacity of apes (Heuvelmans 1980: 593). Specific behaviours pointing to great apes include: wrestling, consumption of termites and honey (for chimpanzees, see McGrew 2004: 115), use of sticks or lengths of wood as tools (chimpanzees; see ibid.: 111–14), feeding on wild ginger, piling up brushwood (which might be linked with gorilla nest-building) and perhaps even covering apparently dead victims. Young chimpanzees in Tanzania have been observed piling up leaves in the dry season and pulling them along the ground, apparently a form of play (ibid.: 122). The bipedalism ascribed to kakundakari and kikomba by no means rules out great apes as their source, since all of these engage in bipedal locomotion, especially when carrying objects or (in the case of gorillas at least) during rainy weather (Schaller 1963: 82). De Waal even characterizes bonobos as 'excellent bipeds' (1997: 47). According to the Dutch zoologist Büttikofer (1893: 337), Liberian hunters not only describe chimpanzees as employing 'cudgels', but also as building 'immense bonfires' of dried wood and then pretending to light these and to warm themselves—a behaviour Central Africans attribute to kakundakari. Reports of kakundakari catching crustaceans similarly echo claims by Ivory Coast peoples that chimpanzees catch 'crabs, fish and crayfish' (Joulian 1995: 278), as well as modern primatological evidence for crustacean fishing by bonobos (De Waal 1997: 66, 80). Finally, the association of kakundakari with Bushpigs (*Potamochoerus porcus*) is conceivably related to the fact that young Bushpigs are regular prey of Tanzanian chimpanzees (Goodall 1986: 275–77).

The Congo region is of course home to both gorillas and chimpanzees (*Pan troglodytes*) and bonobos (*Pan paniscus*) as well. The area around Lake Kivu, where images of the putative hominoids are concentrated, is home to the

Mountain gorilla (*Gorilla beringei*). With regard to the co-presence of these apes, it is further remarkable how several differences between kakundakari and kikomba parallel differences between chimpanzees and gorillas: as respectively smaller and larger primates and as omnivores and vegetarians and in regard to the apparent frequency of claimed sightings. That chimpanzees occur closer to human settlements than do gorillas is also consistent with the incidence of women's encounters specifically with kakundakari. The fact that only kakundakari are described as light-skinned moreover points to chimpanzee complexion, which varies between light and dark. Gorillas, by contrast, are consistently dark-skinned, although so too are bonobos.

Representations of the mystery hominoids might further reflect variations in form, size, or behaviour among local populations of particular ape species, or even individual peculiarities. The physical appearance of the smaller kakundakari somewhat recalls the 'kooloo-kamba', a creature that has been construed as a sub-species or variety of the Common chimpanzee (*Pan troglodytes*) or as a chimpanzee–gorilla hybrid (Shea 1984). First described by Du Chaillu, who shot a specimen and provided a sketch of the living animal (1861: 269–71, 360–61; see Plate 8.5), the ape is more human in appearance than other chimpanzees and accordingly, in African languages, is named separately from these. ('Kooloo-kamba' is an onomatopoeia replicating the creature's call.) This is not to claim that the kakundakari *is* the kooloo-kamba, only that the two terms might well refer to similarly if not identically indeterminate folk categories. An individual chimpanzee whose uncannily human look gave rise to rumours of hybridism was the twentieth-century captive named 'Oliver'. So much did Oliver's appearance and habitual bipedalism point in the direction of *Homo sapiens* that the ape was suspected of having a human parent. The possibility was taken seriously enough for American geneticists to conduct tests, which, however, proved that Oliver, despite his apparent humanity, was pure chimp (Anon. 1996; Ely *et al.* 1998).

Yet another phenomenon conceivably relevant to the derivation of putative hominoids like the kakundakari and kikomba are the 'Bili apes' or 'Bondo apes'. Recently investigated by primatologists, these are a population of chimpanzees that inhabit the Bili forest, some 200 kilometres east of Bondo, in northeastern DRC. The apes have been described as grey-haired, exceptionally big (reputedly up to two metres tall) and leaving large footprints (up to 34 centimetres, thus larger than a gorilla's). They feed on the carcasses of big cats and, unlike other chimpanzees but like gorillas, build nests on the ground. The diet also includes snails. Genetic testing of hair has recently shown the Bili apes to be a peculiar local population of the chimpanzee sub-species *Pan troglodytes schweinfurthii*. Nevertheless, this seems not entirely to have dispelled alternative primatological interpretations of the creatures as a possible new sub-species of chimpanzee or gorilla or—recalling the kooloo-kamba—as chimp–gorilla hybrids (Young 2004; Barone 2007; Randerson 2007). With regard to Congolese folk categories, the size of the Bili apes could suggest the kikomba, while other attributes are reminiscent of the omnivorous, snail-eating and possibly grey-haired kakundakari. The Bili forest appears not too distant from and may even overlap, regions in which people have reported kakundakari and kikomba, or identical creatures designated with different local names.

In regard to Asian comparisons, more noteworthy than correspondences to local apes is the extent to which African representations of both kakundakari and kikomba parallel Sumatran representations of the orang pendek, an ape-like being hypothetically grounded in experiences of a Southeast Asian ape, the orang-utan (or perhaps a large, undiscovered gibbon). Arrestingly specific parallels include hair falling over the forehead or the eyes (a feature further mentioned for the Himalayan yeti); an association with herds of wild pigs; feeding on freshwater crustaceans, larvae from rotten trunks and ginger plants; stealing fish; and disturbing people spending the night in temporary forest shelters. Similarly significant are certain absences. Thus, neither the Sumatran nor the two Central African figures are clearly credited with language. Nor are any of the creatures accused of abducting humans, a charge commonly laid against wildmen elsewhere, although, interestingly enough, also against all of the great apes.[19] And unlike the ebu gogo of Flores and certain mainland Asian exemplars, females of neither the orang pendek nor the kakundakari or kikomba are described as possessing breasts, pendulous or otherwise.[20]

All of the foregoing points to an important conclusion: African and Asian regions inhabited (or in the case of southern Sumatra, recently inhabited) by great apes do indeed produce palpably similar representations of legendary hairy hominoids and the extent to which they do so lends support to known primates as the empirical source of the images. Long hair, sometimes abundant over or about the forehead, may appear contrary to the latter suggestion and moreover to indicate a more human appearance—as it does also in regard to the orang pendek. Yet such hair is not inconsistent with bonobo hair (see photographs by Frans Lanting in De Waal 1997) or even chimpanzee hair, nor indeed with that of the shaggy Sumatran orang-utan. A fringe of hair is also one feature linking the kakundakari, especially, with the similarly sized 'mau men' of Kenya. Another is the light skin attributed to both the mau and kakundakari, which as already remarked, suggests chimpanzees as the most likely source of the Kenyan representation as well. It should be recalled that the description of the mau derives from a single source: a Kenyan whose travels had taken him to the eastern Congo, an area not only within the range of chimpanzees but the very region where kakundakari and kikomba are mostly reported.

The Tanzanian agogwe—or, at any rate, the creatures Hichens saw—could also have been chimpanzees, specifically ones with a variant brown or reddish pelage. Alternatively and perhaps more plausibly (in view of Hichens' knowledge of primates), the sighting, like Burgoyne's observation of 'little brown men', could reflect human pygmies. Supporting this interpretation is the complexion of many pygmies, described as reddish and even as 'white' by some darker and taller Africans (Schweinfurth 1873: II: 137). Some pygmy populations are furthermore distinguished by extensive body hair, manifest partly as a brownish or yellow-brown 'down' (Johnston 1902: 182; Kleiweg 1942: 35; Turnbull 1961: 31; and see Plate 8.6).[21] That the 'little furry men' sighted in the 1930s by Britons in East Africa may have been pygmies is thus by no means impossible. Pygmies have been encountered as far east as the western shores of Lake Victoria and Lake Tanganyika and it is not inconceivable that they could occasionally travel even further east, or have done so in the past. One of the five hominoids described by Roumeguère's Kenyan informants sounds very much like a pygmy, as both the author herself and Heuvelmans (1990a) remark. The silent trade pygmies practise or formerly practised with taller Africans (Johnston 1902: 178–79) recalls the horticultural assistance given in exchange for food and beer which Tanzanian villagers ascribe to agogwe—though, curiously, the pygmy custom is equally comparable to the silent barter attributed to Caucasian wildmen and to certain Asian hunter–gatherers (such as the Vedda, the Sumatran Kubu and the To Ala of Sulawesi). At the same time, the agogwe practice is reminiscent of European folk beliefs concerning nocturnal favours performed by elves, fairies and the like. Also worth noting is the way some Africans classify pygmies with chimpanzees, while positing a human origin for the apes (see Schebesta 1931: 40, who says that black Africans regard pygmies not as humans but as '"man apes", like the chimpanzees'). Taller Africans furthermore ascribe spiritual powers to the smaller folk (Schweinfurth 1873: II: 136, 145). These include the power of invisibility, credited to pygmies and bushmen from Gabon to South Africa (Schadeberg 1999: 32–33).[22] Finally, with regard to the 'appalling odour' sometimes attributed to the kakundakari, it may be relevant that African pygmies are described as possessing a distinctive odour, a 'typical pygmy scent' remarked upon by Kleiweg de Zwaan (1942: 43–44, citing several authors); the odour is found unpleasant by black Africans and Europeans alike.

Southern and West African variants

Notwithstanding the supernatural powers attributed even to attested pygmy populations, hominoid images from southern and West Africa on the whole comprise representations less naturalistic than those from Central and East Africa. A figure widely known from Zimbabwe to South Africa is the 'tokoloshe', a normally invisible creature taking the form of a very small man—knee-high, or just 75 centimetres according to one account. The tokoloshe is generally conceived as a kind of spiritual being and more specifically as something similar to the European 'poltergeist', not least because it is mischievous and most often seen by children. Among the South African Pondo, tokoloshe (here rendered as 'thikoloshe') also serve as familiar spirits of witches (Hunter 1961: 275–82). Indeed, the main link between these beings and hominoids in more northerly parts of Africa is their hairiness. At the same time, this seems mostly to refer to long beards and head hair and only in two accounts (one being a Natal newspaper report of a 'poltergeist') is a hirsute body mentioned specifically (Heuvelmans 1980: 525–26). An account of 'tikoloshe' among the South African Tembu makes no reference to hairiness and moreover describes the creatures as resembling Achondroplastic dwarfs, with short limbs and powerful bodies (Laubscher 1937: 8, 11). Pondo describe tokoloshe as possessing 'squashed up' faces like baboons. There are also stories of specimens being killed or captured. Thus some connection with the hominoids of Central and East Africa is not ruled out. On the other hand, tokoloshe are evidently far more human than these, being able to speak—albeit with a lisp—and wearing animal skins or clothing made from grass; one even carried a tiny brown suitcase.

Noting how tokoloshe are far more fantastically conceived and seemingly more common than the rare Tanzanian agogwe, Heuvelmans ascribes this difference to a universal process of 'mythification'. This entails that the further one moves, in space or time, from places where people offer prosaic descriptions of small, extinct or nearly extinct hairy hominoids, the more they become 'supernaturalized' (1980: 528–29). The author cites a similar contrast between East African hominoids and beings as fantastic as the tokoloshe that are reported from northern Africa. As discussed in Chapter 2, the same variation is discernible in representations of the Florenese ebu gogo, articulated by Nage villagers living close to and more distant from the cave that the extinct wildmen reputedly once inhabited. Such regional patterns raise intriguing questions concerning the possible empirical basis and ontological status of hypothetically pivotal images. Yet, in themselves, they hardly illuminate the actual identity of the possibly natural beings that Heuvelmans characterizes as located 'half-way' between 'supernatural' extremes (ibid.). The tokoloshe may reflect a 'mythified' version of a largely naturalistic representation, but the original subjects of this representation, diffused from a Central or East African source, are more likely to be chimpanzees or pygmies than, for example, surviving Australopithecines—the interpretation Heuvelmans evidently favours. Furthermore, Heuvelmans' initial comparison is specious; for the description of agogwe derives entirely from a putative European sighting, whereas the 'mythified' tokoloshe is an indigenous African image. They relate, therefore, to two quite different cultural contexts.

Figures resembling the Central African kakundakari and kikomba are mostly absent from West Africa. Citing various published and unpublished sources, Heuvelmans (1980: 476–512) provides details of a confusing variety of images of dwarfs and other small hominoidal beings. A few suggest pygmies, whose presence in the region is largely hypothetical, while others suggest chimpanzees, which of course do occur in West Africa. Designated as 'ape-men' by Heuvelmans, one sort of hominoid, recognized by the Guéré of the Ivory Coast, are even called 'mohin-goués', meaning 'like chimpanzees' (1980: 490).

According to Heuvelmans, West African hominoids fall into two main categories. The first comprises 'little red men', an epithet referring less often to hair colour than to a skin lighter than that of black Africans (and thus comparable to the paler complexion of many pygmoid peoples). The second category comprises beings that are 'black' and more specifically black-skinned and are reputed to reside in holes or caves. Although both sorts are commonly depicted as hirsute, only in a minority of instances is there explicit mention of body hair. One example is the tiny and ordinarily invisible 'wokolo', which the Bambara of the Ivory Coast describe as 'soot black', 'hairy like a monkey' and given to shooting people with tiny arrows (Heuvelmans 1980: 483). In another case, also from the Ivory Coast, African colleagues of a French zoologist named Ledoux claimed to have encountered 'pygmies' or 'dwarfs' with long red hair on both the head and the body (ibid.: 500–03; Heuvelmans 1995: 509–13, citing personal correspondence with A. Ledoux). Sometimes combined with non-African physical features like pointed noses, long straight head hair and beards seem to be more distinctive of the darker dwarfs. Africans further credit many of these with spiritual powers, including invisibility, metamorphosis and the 'evil eye', as well as aquatic habits and knowledge of blacksmithing and metallurgy—skills they are sometimes believed to have transferred to their larger human neighbours. Although weapons and clothing are regularly mentioned attributes, some of these 'smiths' (for example, those recognized by Fulani-speaking peoples) are moreover said to possess tails. Bambara describe the aforementioned wokolo as possessing inverted feet, a feature recalling the Sumatran orang pendek and Himalyan yeti but distinguishing these West African creatures from the Central African kakundakari and kikomba.[23]

Suggesting a connection with Central African figures and especially the Congolese kakundakari, the Fang of Gabon credit hominoids called 'séme' with power over Bushpigs. The detail is significant not only in regard to Gabon's position as a western extension of Central Africa, but also in view of other similarities to kakundakari. In a largely naturalistic vein, Heuvelmans thus describes these séme as 'apes' (French 'singes') that resemble humans, or as a 'sort of gorilla with long head hair', standing just half a metre tall and walking erect (1980: 505–06, citing personal communications from H.R. Maudry, August and October 1959). By contrast, other West African images appear less comparable to the kakundakari and kikomba, which in turn are rather more comparable to Southeast Asian hominoids and particularly the Sumatran orang pendek. More than anything encountered in this or in previous chapters, West African dwarfs resemble the elves and brownies of European folklore. By the same token, the images appear more continuous with dwarfs reported from North Africa than with hominoids reported from Central Africa (see Heuvelmans 1980: 402–06, reviewing the work of Robert Grant Haliburton).

Apparently more hirsute than the darker hominoids, West African images of 'little red men' have been interpreted by Heuvelmans as reflecting experience of relict populations of pygmies or a memory of these, revived and transformed by occasional encounters with rare Red colobus monkeys (apparently *Procolobus badius*; 1980: 505, 511). Heuvelmans has considerably more difficulty connecting the altogether more fantastical 'little black men' with either primates or attested humans. Nevertheless, he concludes by suggesting that these may reflect darker-skinned 'proto-pygmies' with straighter hair and noses longer than any present African population (ibid.: 511–12). Noteworthy in his treatment of the 'red' hominoids is an absence of any reference to reddish or brown-haired chimpanzees as a possible source of the images. Not only could this interpretation link West African figures with the East African agogwe and even the kikomba (sometimes described as red or brown-haired); it also evokes a further parallel with the orang pendek of Sumatra. The latter, it will be recalled, is mostly described as 'dark' or 'black' haired, whereas the orang-utan, a plausible source of the Sumatran image, is normally red, although dark-haired specimens, some 'nearly black' (Dubois 1908: 88), also occur. It is therefore no less plausible to propose rare reddish-haired chimpanzees as the source of certain African hominoids than it is to interpret the Sumatran creature as an unusually dark orang-utan.

Whatever the origin of the West African 'red men', it is difficult to see the small, black and long-haired ironworkers as anything other than essentially imaginary. In other words, they appear to be thoroughly spiritual beings. If they are based on anything empirical, then this is most likely present-day humans other than black Africans. Particularly in respect of longer noses, one might think of Arabs, whereas the long red hair of some figures is more readily traced to Europeans, especially as words for 'red' in many languages also apply to browns.

Madagascar: a bridge back to Southeast Asia

Wildman images outside of Asia relate to Southeast Asian exemplars in diverse ways. Insofar as they reveal striking similarities with the Sumatran orang pendek, African hominoids like the kikomba and kakundakari strongly suggest that these and many Asian images reflect human minds responding in similar ways to experiences of similar animals—namely, apes. Further illuminating the parallels is the presence, in both Sumatra and Africa, of food-collecting forest folk who, especially in the African case, are significantly smaller and otherwise physically distinct from sedentary cultivators and whose representation could well have combined with often partial or irregular experiences of anthropoid apes to give the resulting image a more human cast. At the same time, correspondence between the Sumatran and Central African hominoids serves to highlight important differences between the western Indonesian wildmen (and possibly variant images from mainland Southeast Asia) and eastern Indonesian exemplars like the ebu gogo.

Also contrasting with Central African hominoids, the Australian and North American figures—both largely European representations evolving in the course of extensive white settlement of these continents—are not so easily linked with zoological referents, let alone primatological ones. So in this circumstantial respect, they resemble eastern Indonesian hominoids. Although the Australian yahoo may find local support in Euro-Australian experiences of Aboriginal humans, both this category and the sasquatch owe something to an imported figure, the wildman of Europe. Yet there remains a question of how exactly the European figure contributed to these images in the late nineteenth and twentieth centuries, particularly as they relate to what are recorded as actual experiences of the creatures in question. Probably playing a more decisive role in the development of the yahoo and sasquatch was an emerging interest in anthropoid apes and popular versions of scientific constructs of archaic hominids (ape-men, cave men)—categories themselves coloured by older European ideas about wildmen. Of course, the European wildman too may very well have had its ultimate source in ancient European encounters with monkeys and apes. Yet such experience has obviously been highly modified by time and distance and is in other ways quite different from that available to modern Africans in the form of local apes.

Before concluding this chapter, there is another part of Africa to explore, at least if 'Africa' is understood in a purely geographical sense. This is Madagascar, an island settled between 1500 and 2000 years ago by immigrants from Indonesia. If the European wildman and its partial offshoots in Anglophone America and Australia are physically, behaviourally and ecologically quite different from African hominoids, then Malagasy images are equally distinct. Throughout the island, people speak of creatures most often called 'kalanoro', very small hominoids just 60 centimetres to a metre tall or the size of a child but with adult features (Lambek 1981: 187). They are also strong and have long fingernails and red or glowing eyes.[24] Kalanoro are usually described as covered in hair and in one account are further characterized as 'black' (Gaudebout and Molet. 1957: 64). Like the Sumatran orang pendek, the creatures are credited with inverted feet, the prints of which can lead trackers astray (Decary 1950: 207; Sharp 1993: 138–39). Kalanoro live in uninhabited forests and caves and consume their food raw; some accounts suggest they are particularly fond of fish and crayfish. They may also steal scraps of food from isolated huts, which they enter at night to keep warm, but will never enter buildings in which pork is kept (Gaudebout and Molet 1957). According to the Bara of south-central Madagascar, kalanoro roam villages at night in search of food (Lamberton 1937: 169); the same source describes the creatures as good runners and climbers, thereby indicating a partial arborealism. A phrase employed by the Vezo of western Madagascar, 'to eat like a kalanoro', which refers to people who eat what is available without giving thought to future provision (Astuti 1995), may suggest a voracious appetite. There are no reports that kalanoro possess culture or technology. The creatures are, however, capable of speaking, specifically in soft, feminine tones (Gaudebout and Molet 1957; Decary 1950) or in 'quick, choppy and high pitched' voices that are difficult to understand (Sharp 1993).

Other features, mentioned only by Decary (1950: 207) and apparently reflecting less widespread representations of kalanoro current in particular locations, include residence in underwater caves or in reed thickets near lakes and a foot with just three toes. Referring to the central Tanala people, who call kalanoro 'angalampona' (a term adopted from the Betsimisaraka of northeastern Madagascar), Gaudebout and Molet say the creatures live in pairs. If one imitates their distinctive cry, replicated as 'koukou, houlou', they retaliate by shaking huts or throwing stones at roofs and by extinguishing fires and scattering firebrands. One can keep them away by burning a bit of rag and declaiming 'Save yourself, you have no friends here' (ibid.). Speaking of the island of Mayotte (off Madagascar's northwest coast) and apparently basing his account mostly on a reputed local sighting, Lambek (1981: 186–87) describes 'kakanoru' as having short and strong left arms but long and weak right arms, an asymmetrical representation revealing a lateral inversion reminiscent of inverted feet. On Mayotte at least, the creatures are supposed to dislike light, even though the reputed sighting took place at midday.

Recent accounts by western anthropologists identify kalanoro as 'spirits' normally visible only to mediums who can capture and control the creatures and who sometimes present pieces of hair as evidence. Malagasy mediums, however, are not possessed by kalanoro, but only by other kinds of spirits (Lesley Sharp, pers. comm., February 2007). People in eastern Madagascar speak of kalanoro as creatures that used to exist but are now extinct owing to rapid deforestation and survive only in spiritual form (Genese Sodikoff, pers. comm., February 2007). It is interesting, then, how older accounts paint a more naturalistic picture of kalanoro. Referring to stories given 'wide credence' among the Betsimisaraka of Madagascar's east coast, the American explorer (and sometime state governor) Chase Osborn described kalanoro as very short 'wild men of the woods', covered with hair and with long flowing beards. According to a claimed eyewitness account, said to exemplify a commonly told tale, one such 'naked wild man' once came to warm himself by a fire set by a man spending the night alone in a deep forest. To rid himself of the unwanted visitor, the traveller showered the wildman with hot coals, which sent him shrieking into the jungle. Another story concerns a 'naked man' who found some rice in a cooking pot in a foresters' camp. The intruder left temporarily, only to return with a 'nude woman' and the couple, who 'appeared to communicate by grimaces', then 'made love' and 'fed one another with their hands'. Once they became aware of the campers' presence, they too fled (Osborn 1925: 41).

Evidently lacking a fear of fire, the kalanoro depicted in these Betsimisaraka stories seem largely human, thus contrasting with the more spiritual image conveyed by recent writers. Corresponding to Osborn's descriptions is an unnamed 'tribe of nomads' mentioned in an earlier work by Little (1884). These are 'diminutive of stature', naked and covered with hair. Mild and retiring, they inhabit 'tree-dwellings' and communicate in an unknown language. Evidently alluding to something intermediate between humans and apes, Little concludes by suggesting that such creatures 'would delight the heart of a disciple of Darwin, or a philosopher, or student of anthropology in search of the "missing link"' (1884: 51–52).

The same could doubtlessly be said of a creature reputedly captured in 1879 in the densely forested interior of northern Madagascar 130 kilometres west of Maroantsetra. Found sleeping on a tree branch, this 'wild man' was taken by a group of Malagasy in the employ of a trader from the west coastal settlement of Manahar. Some of the captors were severely bitten in the ensuing struggle. Later, at an undisclosed location, a European trader from Mauritius, a Mr. Carmes, observed the captive. According to Carmes, the creature was 'a powerfully built man of about five feet nine inches', with a face and body 'thickly covered in long black hair' and with a peculiar mode of walking: 'he traveled very fast, with his head down, occasionally going on all fours' (Ransome 1889: 301). James Sibree, a naturalist and expert on most things Malagasy, judged Carmes's account 'apparently well authenticated' (1915: 147). Carmes further described the captive's eyes as more like an animal's than a human's and invariably fixed on the ground. Although naked when captured, the hairy man later wore clothes provided by his captors; but he would not eat meat or any cooked food, only manioc and other roots and would sleep only in a squatting, never a recumbent, position. The report goes on to claim that after some weeks the captive learned a few words and partly by this means revealed that he had a father and two brothers living in the same forest. Attempts were made to capture these as well, but 'they easily eluded their pursuers, jumping from tree to tree like monkeys and running on all fours'. 'The captured man died five months after being taken' (Ransome 1889).

This account apparently takes us some distance from late twentieth-century descriptions of kalanoro, particularly in regard to the wildman's relatively large size. Nevertheless, connections are still apparent. Except for the reputed acquisition of speech (possibly an embellishment from local images of kalanoro, invoked by Malagasy when giving an account of the creature's capture), the partly arboreal and partly quadrupedal captive sounds like some sort of ape—particularly a very large chimpanzee. Yet there are neither apes nor monkeys on Madagascar and Carmes explicitly described the hirsute hominoid as a 'man'. Assuming the story has some basis in fact, one wonders indeed whether it might have a connection with extinct large lemuroids, a possibility I return to presently. At the same time, one cannot fail to be struck by parallels with the story of 'Jacko', found reclining not in a tree but by a railway track, just five years later in western Canada. Similarities with capture tales recorded in eastern Indonesia, Mainland Southeast Asia and elsewhere are equally evident.

Like the biting wild man captured in 1879, more recent representations depict kalanoro as capable of aggression towards humans. Not only can they inflict fearful injuries with their long nails, the creatures are also given to abducting people. The Sakalava of northwestern Madagascar claim kalanoro have a particular affinity for young children mistreated by their parents (cf. Decary 1950: 207, who also describes kalanoro 'luring' children away). The parents must then consult a medium to discover what must be given to the kalanoro to retrieve their offspring (Hobbs 2001: 6); on three separate occasions, the most recent in 1998, the reputedly abducted children were successfully recovered inside a particular cave (Hobbs and Salter 2006: 464). An attraction to human children is further indicated by a belief among the Betsileo of south-central Madagscar that female kalanoro will steal babies and leave changelings in their place. The abductors then give the human infants special potions to stunt their growth. Accordingly, the Betsileo call 'ill-favoured' newborns 'children of kalanoro', while in some regions people invoke the name 'kalanoro' as a bogey, to keep children quiet (Decary 1950: 207). Parallels with hominoidal figures elsewhere hardly require specific mention; however, it is worth recalling that substituting their own offspring for human children is a practice attributed to wildmen in eastern Indonesia as well as in China. According to Betsileo, who call the creatures 'angalapono', kalanoro also temporarily abduct adults. A woman claiming such an experience was granted powers of 'divination' (probably meaning mediumship) before her release, while a reputed male captive was taken by a female creature who, apparently like orang-utans in some Southeast Asian folktales (see Chapters 5 and 10), wanted him for a husband (Sibree 1896: 235–36).[25]

Since Madagascar lacks monkeys and apes, the only conceivable primate referent of kalanoro are lemurs and particularly larger, short-tailed species. Referring to a persistent legend from the Ankazoabo region in Bara territory, French palaeontologist Charles Lamberton (1937: 169) suggested that kalanoro could reflect memories of *Hadropithecus stenognathus*, a large, apelike and partly ground-dwelling lemuroid which became extinct within the last one thousand years, thus well within the period of human settlement on Madagascar (Fleagle 1999: 104–05, 108). Morphological parallels between *Hadropithecus* and the Hominidae, including features of the face, have been further detailed by Jolly (1970). However, a more likely referent of *Hadropithecus* may be the 'tratratratra'. This is a legendary simian creature the size of a 'two-year-old calf' (thus much larger than a kalanoro) but with 'a round head and human face', sightings of which were still being reported in the seventeenth century (Decary 1950: 206; Osborn 1925: 429). A few sources suggest an association of kalanoro with the 'vazimba', a putative aboriginal population of Madagascar now represented as spiritual beings (Sibree 1896: 255; Bloch 1986), or—what may sometimes be the same suggestion—with groups related to African pygmies or bushmen (see Sharp 1993: 138; Osborn 1925: 41–42). Yet the evidence for pygmoid peoples on Madagascar is anything but clear and in any case resemblance to the hairy and long-nailed kalanoro would seem to lie mainly in their small stature.

Kalanoro associations with children could suggest some connection with the southern African 'tokoloshe', possibly reflecting Bantu influence in Malagasy culture. Yet the creatures otherwise appear quite different from hominoidal images encountered in continental Africa. By the same token, in a number of respects—including a propensity for abduction, the long nails (see Chapter 3, regarding several Florenese figures), an association with spirit mediums (see Chapter 6, concerning the umang of northern Sumatra) and even the inverted feet—the kalanoro are far more comparable to images from Insular Southeast Asia and, as we shall soon see, from the islands of Oceania. In fact, the name 'kalanoro' (and Mayotte 'kakanoru') bears a striking resemblance to 'kakamora', a category of hairy hominoids reputedly inhabiting the Solomon Islands. It is therefore a matter of considerable significance that the Malagasy people derive in large part from southern Borneo, from where they were transported to Madagascar over 1500 years ago. Like inhabitants of the Solomons and other Pacific islands, they accordingly speak a language belonging to the Malayo-Polynesian branch of the Austronesian language family and in several cultural and sociological respects show obvious affinity to Indonesians. Their Bornean ancestors would of course also have been familiar with orang-utans.

How the same ethno-linguistic connection bears on an understanding of images found among Austronesian-speakers of the southwestern Pacific is the subject of the following chapter. But before moving on, I should register another relevance of Madagascar. Although not concerned with kalanoro (nor particularly with recent lemuroid extinctions), Burney and Ramilisonina (1998) have investigated named categories from the Vezo area of western Madagascar which appear to reflect fauna recorded as extinct during the last millennium. Ever cautious, the authors nevertheless provide compelling evidence for the persistence of seemingly accurate local representations of species that have died out only in the last several centuries and which may even survive in isolated areas to the present. It would therefore not be stretching things too far to suggest that the kalanoro, also, might at least partly reflect one of Madagascar's many extinctions during the same period. Comparative evidence from much further east suggests an alternative explanation, but one that does not necessarily contradict this. For immigrants to Madagascar may well have interpreted newly encountered large lemuroids, currently extinct but then still living, with reference to images imported from their Indonesian homeland.

Plate 8.1 Illustration of an orang-utan in Beeckman.
Source: Beeckman ([1718], 1973: 37).

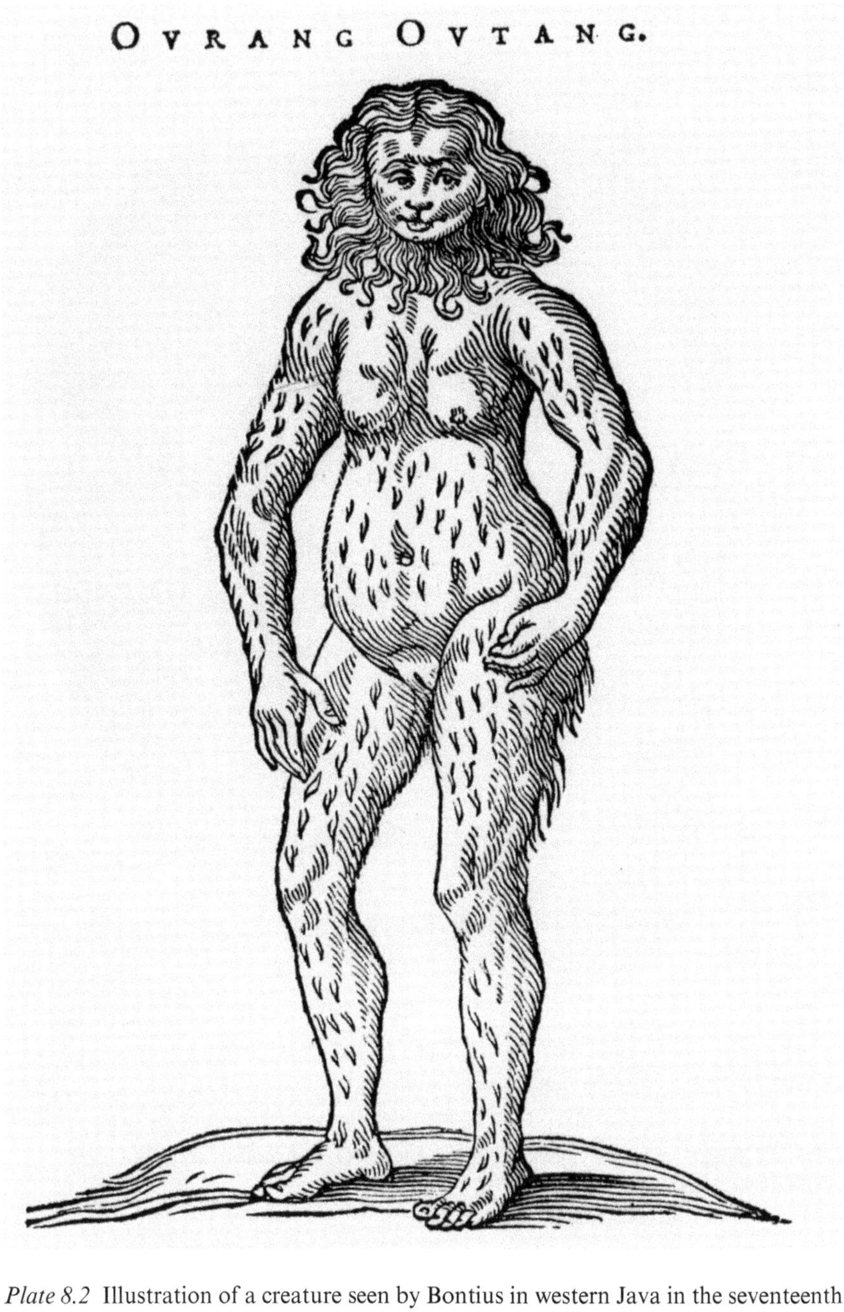

Plate 8.2 Illustration of a creature seen by Bontius in western Java in the seventeenth century.
Source: Bontius (1931: 285).

Plate 8.3 Niska 'monkey' mask collected by Lt. G.T. Emmons in about 1914.
Source: Reproduced courtesy of the Peabody Museum, Harvard.

Plate 8.4 A nineteenth-century painting of an orang-utan holding a staff.
Source: T. Maple (1980), from Mivart (1873).

Plate 8.5 Head of a 'kooloo-kamba' by Paul Du Chaillu.
Source: Du Chaillu (1861: 360).

Plate 8.6 Central African pygmies with Paul Schebesta.
Source: Schebesta and Lebzelter (1933).

9 Pacific images

The preceding discussion of the Malagasy kalanoro brings us back to our point of departure, if not exactly in eastern Indonesia, then in Austronesia. 'Austronesia' denotes the region occupied by speakers of Austronesian languages, a widespread group whose largest division by far is 'Malayo-Polynesian'. Besides Madagascar, Malayo-Polynesian languages are spoken throughout Indonesia, the Philippines, in the Malay Peninsula and other south-easterly parts of Mainland Southeast Asia and in Polynesia, Micronesia and many parts of Melanesia. As an ethno-linguistic category, Austronesia therefore includes peoples discussed in Chapters 2 to 6, as well as the aboriginal population of Taiwan (formerly called Formosa), whose languages compose as many as nine further divisions comparable to Malayo-Polynesian within the Austronesian family.

Turning from Africa (a major focus of attention in the previous chapter) to the islands of Melanesia, Micronesia and Polynesia—that part of the Pacific generally known as 'Oceania'—involves a shift from the continent most closely associated with great apes and human evolution to a region where non-human primates are a zoogeographical impossibility and where human settlement is far more recent. Speakers of Oceanic languages (one branch of Malayo-Polynesian) are thought to have entered the region about 3500 years ago from a point of dispersal somewhere in northeastern Indonesia or the Philippines. However, Melanesian peoples, especially, have much earlier roots in eastern Indonesia. Related historically and linguistically to Insular Southeast Asians, Pacific islanders provide us with an important test. For if Indonesian images of hairy hominoids are primarily cultural in origin, then one might expect to find in Oceania and perhaps also in Taiwan, distinctly similar representations reflecting this common ethno-linguistic heritage.

Melanesian figures

Comprising several island groups extending from New Guinea to New Caledonia, Melanesia is home to several wildman images. Among these are the 'kakamora' of the Solomons, creatures whose name, incorporating 'mora' ('native or original'), suggests a conception of them as the islands' earliest inhabitants (Fox 1924: 354). Solomon Islanders distinguish the kakamora from categories of spiritual beings, yet do not consider them 'quite human'. On San Cristobal Island, the creatures' 'ordinary height' is just over 15 centimetres, but elsewhere they can be as tall as one to 1.2 metres, or between 1.4 and 1.5 metres (Fox ibid.: 138, 355). Specimens reputedly captured by islanders 'are always about the size of very small men, not much smaller than the pigmies of Bugotu of New Guinea' (ibid.: 140). Kakamora are mostly dark-complexioned, although some are 'quite fair' and have long sharp nails and small teeth. They are also exceedingly strong. That kakamora 'bend low when they run' (ibid.: 355) may suggest an imperfectly erect posture, but islanders usually depict them as bipedal and even as given to dancing. While he does not explicitly refer to body hair, Fox describes the kakamora's head hair as long and straight and falling to the shoulders or lower (ibid.: 138). Describing essentially the same image, Codrington (1891: 355), on the other hand, refers to bodies 'covered with long hair'.

Kakamora live in forests and limestone caves, where they subsist on nuts, fruits and possums. One reputed sighting has them catching and consuming raw fish (Fox 1924: 345). Although they possess no fire or other technology, they will snatch firebrands from human fires and 'play with' these—a possible attribute of the similarly named Malagasy 'kalanoro'. Kakamora also speak an 'unintelligible language' (ibid.: 139, 141). While characterized as timid (ibid.: 355), some will kill and eat humans. Recalling the kalanoro, the Sumatran orang pendek and homang and the lolok of northeastern Sulawesi, kakamora also delude people, causing them to lose their way in the forest (ibid.: 139). Solomon Islanders tell stories of the creatures attempting to abduct children, often unsuccessfully. Another story describes how a man captured and married a female kakamora (ibid.: 142–45, 145–47). Indeed, tales of captured kakamora seem as common as those in which kakamora attempt to capture humans. The stories also occur in New Ireland, while natives of neighbouring New Britain tell of 'very strong and active' dwarfs inhabiting the interior of that island, which, despite the best attempts of native hunters, are always able to resist captivity (Brown 1910: 243–44).

Apparently more fanciful attributes of kakamora include their fear of white things (or, in southern Mwala, things coloured red), the special vulnerability of their buttocks (where a strike or stab wound can kill them) and their subjection to a king or queen (Fox 1924: 144, 145, 139). By the early twentieth century, kakamora were considered extinct; yet in 1920, Solomon Islanders started reporting that the creatures were again appearing on the outskirts of villages and stealing from gardens, as they used to do before the introduction of firearms around 1860. Evidently, their reappearance was connected with the British colonial authority calling in all guns; for reported sightings of kakamora resumed just two or three months after the recall (Fox 1924: 139, 354–55).

Impressed by the weight of independent testimony, Fox was open to the possibility of kakamora actually existing in the large forest-covered tracts in the interior of the Solomons. As he noted, around 1920 the islanders were hopeful of capturing kakamora, for which they believed the colonial government would pay twenty English pounds per specimen. Fox alternatively suggested the creatures could reflect an 'earlier race', either in the Solomons or in the islands from whence the Solomon Islanders had come (ibid.: 141). Previously, the missionary R.H. Codrington had argued that figures like kakamora were no more than an exaggerated coastal representation of inland dwellers (1891: 355). Yet, somewhat like Fox, Codrington allowed that the image might nevertheless be 'founded on the memory of large simians in former seats of the Melanesian people' (ibid.), by which he presumably referred to Indonesia or Mainland Southeast Asia.

More convinced of the veracity of local reports of hairy hominoids was Stanley Knibbs, a British government surveyor working in the Solomons (specifically Malaita Island) in the 1920s, who described a 'legend' of rarely encountered 'ape-men' standing about 1.4 metres tall, apparently walking erect but 'with their hands touching the ground' and inhabiting forested areas where they 'lived in trees' (1929: 49–50, 259). Although Knibbs records no local name for the creatures (nor does he mention Fox's book, published just five years previously), his description largely corresponds to Fox's kakamora. Other points of agreement include the ape-men's intense shyness of humans and their stealing from native gardens.

Of particular interest is Knibbs' reference to an eastern Solomons dance, in which dancers imitate the ape-men by assuming 'a habit resembling hair', approach female spectators 'on all fours' and make 'weird noises' (ibid.: 49–50). The women then run helter-skelter and hide in their houses, while the men advance, pretending to attack the pseudo-apes. One curiosity of Knibbs' description is the apparent quadrupedalism of the ape-imitators—although this of course is not inconsistent with real ape locomotion. The mock attack is reminiscent not only of attempts at abduction ascribed to the kakamora, but also charges of assault on human females levelled against apes in both Africa and Southeast Asia. The Solomon Island performance might even be taken for a counterpart of the 'ebu gogo dance' (Chapter 2), were it not for the fact that the Florenese performance only developed in the mid-twentieth century.

While acknowledging the absence of apes in the Solomons, Knibbs argues that these legends of ape-men 'are too exactly true to the habits of the anthropoid apes to be rejected as altogether fictitious and are so universal that one is compelled to take notice of them' (1929: 49). The force of Knibbs's assessment is, however, reduced by the fact that he refers only to his own brief summary and one is given no details of the natives' own accounts. The British surveyor's description of the Solomons ape-men was later taken up by the folklorist Bacil Kirtley (1964: 88–89) who, also noting the absence of apes in the Pacific Islands, concluded that such images could have no basis in zoological reality and must therefore be fictitious. Thus stated, the argument is difficult to fault. Yet as Knibbs correctly observes, there is nevertheless something to explain in regard to the derivation of the images, especially in view of their hypothetical similarity to non-human primates. That explanation, however, may partly lie in a transformation of local images brought about by introduced European knowledge of apes (a possibility mentioned earlier with reference to northwestern North American categories such as the 'boqs'). Relevant here is the perception of Fijians who, on first encountering an imported monkey, immediately remarked that it was 'akin to their woodland sprites', called 'veli' or 'eng-eli' (Brewster 1922: 230). On the other hand, these 'sprites' are characterized neither as hairy nor tailed; so the comparison was apparently based on no more than the possession by both of a generally hominoid form.

In the southern part of Malaita Island, people use 'mumu' to refer to what other Solomon Islanders call 'kakamora' (Fox 1924: 139). At the same time, some creatures thus designated are depicted as large rather than small, while others are 'half-man' figures, possessing just one foot, arm and eye. Other attributes of the category include 'long red hair like horses' tails' and a propensity for killing humans with their long fingernails or by spitting in the eyes (a habit that recalls the aul of western Java). There is, then, a suggestion that 'mumu' can refer to a variety of images, rather more diverse and evidently more fantastic than either the kakamora of San Christobal or Knibbs's ape-men.

According to Ivens, 'mumu' (transcribed as 'muumuu') is an onomatopoeic name referring to the inarticulate babblings of 'wild people' encountered by later immigrants to the Solomons. These he partly identifies with a population the Lau folk call 'gosile', who were reputedly ignorant of fire and ate their food raw (1930: 304). 'Mumu', or the longer term 'mumulou', is also recorded by Codrington (1891), who associates the name with two slightly different images. Florida islanders speak of 'wild men' thus designated as inhabiting part of Guadalcanal (another of the Solomons) where they live in mountain caves and subsist on snakes and lizards. They are about the same size as local humans but have very long head hair, hirsute bodies and long nails. At the same time, these wildmen, one of which had recently been killed by coastal people in Laundari (on Guadalcanal), are said to possess a language and a technology, including a use of spears, nets (which they use to catch humans) and bags filled with obsidian fragments (with which they pelt intruders). Codrington's other image occurs in Sa'a, on Malaita Island, where mumu take the form of very short forest-dwelling humans with long hair and nails. The dwarfs use their nails to kill coastal people, whom they then consume. Neither technology nor language is mentioned for these figures, which are typically seen in trios comprising a male, female and child. Codrington further remarks how Sa'a men who had encountered Australian Aborigines compare them to mumu. This would seem to confirm a conception of the creatures as basically human in form and by the same token lends support to Codrington's interpretation of mumu as an exaggerated coastal representation of interior folk (1891: 354–55). In addition, the implicit evaluation of Australian Aborigines does not contradict earlier discussed interpretations of Aborigines as the main source of the Australian yahoo.

Three tales concerning the mumu of Sa'a depict the creatures as stupid and easily fooled. One relates how a man, stranded in a cave during a flood, was able to kill a group of anthropophagous mumu troglodytes by tying together their head hair while they slept and, by digging, flooding the cave and causing them to drown. Another story describes how two boys, one of whom a mumu had captured in a net, killed the creature by throwing hot stones into its mouth. Prior to this, the younger boy had killed the mumu's grandson and tricked it into cooking and consuming the corpse (Ivens 1927: 415–16, 416–18). How comparable these narratives are to eastern Indonesian traditions is moot. Nevertheless, details such as extermination inside a cave and death caused by swallowing hot stones obviously recall stories told about the Florenese ebu gogo. The first Sa'a story is also comparable to a Malay tale about a pair of cave-dwelling giants that are killed by drowning after being fatally deceived by human adversaries (De Vries 1928: 8–9). And tricking a giant into consuming the flesh of a younger relative, in this instance the giant's daughter, is a theme that further occurs in tales from the Sangir Islands, located between Sulawesi and the Philippines (De Vries 1925: 75–78).

Referring again to Guadalcanal, a brief account by a former colonial land commissioner records 'mumu' as a name for 'undiscovered people' regularly reported from the mountainous interior of the island between 1920 and 1950 (Wilson 1950: 7). Also called 'moka' or 'mola'—terms which might be compared with 'mora', in 'kakamora'—these mumu are described as long-haired, large-eyed and having long fingernails and long sharp teeth (thereby contrasting to the short teeth of Fox's kakamora). They also emit a peculiar wail and move by hopping. The article would be of little interest were it not for one detail: the incredulous commissioner was informed that one of the creatures had been burned to death, apparently by accident, when villagers fired a grass patch. The resemblance may be coincidental, but one might be reminded of the use of fire in deliberate killings of Florenese hominoids (Chapters 2 and 3).

Although located just 300 km southeast of the Solomons, comparable images appear less commonly reported from Vanuatu (formerly the New Hebrides and Banks Islands). A general class of beings named 'vui' includes an unnamed sub-category of 'dwarfs and trolls' taking the form of small humans with skins darker than local people and 'long straight hair'. Body hair is not specifically mentioned. One story specifies a vui inhabiting a cave, while another suggests they possess 'powerful fingernails' like the kakamora (Codrington 1891: 150, 152, 159). But the vui otherwise seem quite different from the Solomons hominoids and, rather more than the latter, suggest a category of spirits. Somewhat nearer the mark are Vanuatan (New Hebridean) creatures, reckoned to be extinct in some places, which locals describe as possessing long hair and long teeth, living in caves, carrying off pigs and killing and eating humans. Recalling the Guadalcanal mumu as described by Wilson, these beings are identified with a nocturnal crying, locally interpreted as the sound of a hominoid infant being washed (Codrington 1891: 355). On a methodological note, it is interesting that, while Codrington discusses the vui and similar Melanesian images in a Chapter entitled 'Spirits' (1891: 150–72), he treats these creatures, as well as the mumu of the Solomons (Guadalcanal), in a separate section labelled 'Wild Men' (ibid.: 354–55).

Before leaving Melanesia, some attention should be given to New Guinea. Hominoidal images comparable to those found in the Solomons are less consistently documented for this large island. This may seem ironic, as distinct populations of 'pygmies' or 'negritos' are better attested for highland New Guinea than for other parts of Oceania (a matter I take up below).[1] Nevertheless, one group of New Guinean hominoids, the light (or 'reddish') complexioned 'kewa cannibals' known to the Mount Hagen tribes of the Western Highlands, recalls both the Solomons mumu and certain Indonesian figures in their reputed possession of reddish hair ('kewa'), sometimes covering the whole body (Dosedla 2006). As we shall later see, on Taiwan some aboriginal peoples similarly describe legendary 'dwarfs' as red-haired. Citing Vicedom and Tischner (1943), Dosedla notes how the kewa are portrayed in legends as an aboriginal population of head-hunters. Individuals of a similar appearance, he adds, 'are still common among the local groups' and are considered descendants of kewa. This 'reddish hair and skin type' further characterize most members of a group in the Ialibu district of the Southern Highlands Province and is also 'typical' of the Tolai people of New Britain (Dosedla 2006: 1, 2). At the same time, the kewa are apparently not described as small.

New Guinea may be the locus of other images, not readily associated with locally known human phenotypes. These include a category of very shy forest 'dwarfs' who avoid contact with people and make apparently nonlinguistic chattering sounds. In 1982, educated New Guineans residing in the interior of Madang Province informed the geneticist Stephen Oppenheimer that, just five minutes prior to his arrival at their house, they had observed a 'small family group' of what they described as 'dwarfs' (Oppenheimer, pers. comm., September 2006). The observers comprised Oppenheimer's New Guinean field assistant, her husband, their child and a child minder. The dwarfs had silently crossed a clearing near their house, located in primary forest and without making eye contact had re-entered the forest. When he arrived, Oppenheimer found his informants in a 'great state of excitement'. They described the dwarfs as 'very short dark people with long curly hair' contrasting with the short frizzy hair common among Papua New Guineans. No information was provided on body hair. Since New Guineans are themselves often short, the dwarfs, Oppenheimer infers, were considerably less than 1.5 metres. Responding to my further questions, he adds that what his informants saw were unlikely to be members of known New Guinean tribes classified as pygmies. According to his assistant, also, they were something else, as they 'didn't look quite human and could not communicate in a normal way'. Although his informants identified the subjects of their sighting with an indigenous category familiar to local people, Oppenheimer did not record the name.[2]

Hein Dosedla (pers. comm., November 2006) has suggested that these dwarfs may have been pygmies of the Ramu River region. Inhabiting the Aiome mountains, these people are described by Moyne and Haddon (1936) as much shorter than their larger neighbours, who fear them for their 'dangerous reputation' and also as lighter skinned and longer haired. If the pygmies were indeed what Oppenheimer's colleagues saw, the case could provide another example of how smaller folk displaying other physical differences may easily give rise to a somewhat more fantastic representation.

Polynesia: wildmen, dwarfs and fairies

Images closely resembling Melanesian wildmen seem rare in Polynesia. The main exception are hominoids the New Zealand Maori call 'mohoao', 'mohowao', 'maero', 'maeroero', or 'mairoero', terms commonly glossed as 'wild man of the woods'. Encountered in mountain forests, these creatures appear naked and weaponless. Their bodies are covered in long, coarse hair and while their foreheads are bare, their head hair trails down to the ground (see e.g. Monro 1898: 262–63).[3]

Like wildmen elsewhere, the maero (the name I hereafter employ as the general designation) also possess great physical strength. According to one account (Gudgeon 1906: 43–4), their arms are very long and—in a way particularly reminiscent of Melanesian figures—they have long, sharp fingernails, which they use to impale birds. Much feared by the Maori, the wildmen are generally aggressive towards humans, whom they will attack, disembowel and consume (Cowan 1977: 64). Further characterized as 'cunning and mischievous' (Monro 1898; Beattie 1919: 212), maero also take human captives, particularly 'young people and damsels', who might later manage to escape (Beattie 1919).[4] In folktales at least, maero laugh and speak (ibid.: 213). In one story, the severed head of a maero, whose body has been cut to pieces by a Maori hero, speaks and with the assistance of its wife, the creature then manages to reassemble its dismembered person. Of greater comparative interest is another particular. The wild couple finally take refuge in a cave on a hill, but are 'smoked out' by a group of Maori pursuers (Gudgeon 1906: 43–44)—a detail comparable to the legendary use of fire in exterminating cave-dwelling hominoids on Flores.[5]

Whereas kakamora and other Melanesian hominoids seem never to have been observed by westerners, at least one European, a man named Sandy, reported encountering a maero and firing his gun at it (Beattie 1919: 213). According to Beattie, the shot should have struck its target but missed, a circumstance he attributes to the creature not being 'mortal'. The New Zealand wildmen might appear to differ from hypothetical Melanesian counterparts in being 'much bigger' than humans (Cowan 1977: 64). Yet it seems to be only

Cowan who describes the creatures as 'giants' and in fact other accounts make no reference at all to their size. At the same time, larger size is one of several features distinguishing the maero from a rather more beneficent class of Maori 'spirits' or 'fairies' designated as 'patu-paiarehe', which are either smaller than humans or no larger.[6]

Whether images of hairy hominoids comparable to the maero and Melanesian figures occur either in the rest of Polynesia or in Micronesia is generally unclear. Among Polynesian possibilities, probably the best known are the mysterious nocturnal 'dwarfs' (or 'pygmies', as they have sometimes been called) that Hawaiians call 'menehune'. These are the subject of a lengthy essay by Luomala (1951). Inhabiting deep forests and valleys in the mountainous interiors of the Hawaiian islands and especially the island of Kauai, menehune reside either in caves, hollow logs, or banana-leaf huts. They are capable of verbal communication, conversing in a language compared to the 'low growl of a dog'. According to one account, the tiny beings also engage in cultivation; otherwise, they are characterized as lacking fire and consuming their food raw (ibid.: 14, 18). Normally invisible, menehune can be seen only by Hawaiians who claim descent from them. They then take the form of small humans, sometimes just 60 centimetres to a metre in height or even shorter, but powerfully built and, accordingly, tremendously strong. Hawaiians further describe the menehune as squat, with distended bellies, ugly faces, large eyes, long eyebrows and low protruding foreheads covered in hair. According to some accounts, their bodies are also hairy. Other descriptions, however, depict them as smooth-skinned (ibid.: 10)—in contrast to beings the Hawaiians call 'mu' which, although otherwise virtually indistinguishable from the menehune, are portrayed as hairy-bodied and bearded (ibid.: 24–27).

Further distinguishing menehune from the more decidedly hirsute maero and kakamora, the Hawaiian dwarfs lack the long fingernails of the New Zealand and Melanesia wildmen. Also unlike the latter, they are generally well disposed towards humans. Hawaiians credit menehune with the nocturnal construction of walls and other structures of stone, tasks which they are obliged to complete in a single night. Luomala (ibid.: 78, 80) makes reference to comparable Oceanic traditions attributing ancient stone structures to the exertions of similarly benevolent beings (see e.g. Rivers 1914: II: 428, referring to stone walls and earth mounds found on Santa Maria, in the Banks Islands).[7] The belief also bears an obvious resemblance to ideas found on Flores and Sumba, where local legends relate how the construction of stone walls and ramparts was crucially assisted by the ana ula and mili mongga (Chapters 3 and 4). At the same time, the eastern Indonesian traditions specify these hominoids, by virtue of their superior strength, as simply hauling and lifting large stones, while the actual construction was directed by human engineers. In addition, the Sumbanese and Florenese wildmen are not specified as working nocturnally, or necessarily completing their labours in a single night.[8]

Notwithstanding their similarities, the Hawaiian menehune differ from Indonesian hominoids to about the same extent that they resemble other Polynesian images like the New Zealand patu-paiarehe, or 'fairies'. The latter are also credited with nocturnal stoneworks, specifically a 'bridge' which is actually a long, black reef formed by an ancient lava flow in what is now Auckland harbour. Indeed, as Luomala concludes (1951: 68–69; see also 50, 83–84), the menehune and patu-paiarehe can be understood as instances of specifically spiritual beings with counterparts in various parts of Oceania. They are thus manifestly different from the New Zealand maero, whom Luomala herself distinguishes as 'wild, long-haired men' occurring not as a 'band' (as do the Polynesian spirits) but as 'scattered individuals' (ibid.: 72). Further distinguishing the largely beneficent patu-paiarehe from the fearful Maori wildmen are their golden or reddish hair, complexions lighter than Polynesians and blue or black eyes—features which recall the nitu spirits, whom the Nage of Flores similarly describe as light-skinned and golden-haired (Forth 1998a: 68). Also pertinent to this comparison is the belief that the mainly nocturnal New Zealand spirits venture abroad only on 'days of heavy clouds and fog' (Cowan 1930: 3; 1977: 64), an attribute suggesting an affinity to the Florenese spirits called 'noa' (Forth 1998a: 99). All the same, the Oceanic contrast of wildmen and spirits is by no means complete. Not only are some of the former not (or not definitely) hairy-bodied, but red hair is a feature of the mumu of Malaita and the 'kewa cannibals' of New Guinea. Also, like the maero and Melanesian wildmen, the generally benevolent Maori fairies will sometimes capture human females and carry them away to the forest (Best 1972: 995–96). On the other hand, only the patu-paiarehe are able to disappear and change shape at will. The Hawaiian menehune are fundamentally similar, as they are normally invisible and able to turn offending humans to stone (Luomala 1951: 16). Noting their 'supernatural qualities', Luomala (ibid.: 68) rules out an ultimately human referent for the menehune. She does so even though the name is cognate with Central Polynesian terms denoting a class of lower-ranking humans (for example, 'manahune' in the Society Islands; ibid.: 57) and despite claims of former intermarriage and the inclusion in a 1824 census of 65 menehune among 2000 inhabitants of Kauai's Wainiha valley (ibid.: 11–12).

Micronesian variants

Images of hairy wildmen are no more evident in Micronesia than in Polynesia (see Christian 1899; Hambruch 1932; 1936; Erdland 1914; all cited in Luomala 1951: 72–75). Instead of hominoids smaller than humans, Micronesian traditions seem mostly to speak of 'giants' or 'ogres', figures that resemble the smaller Melanesian and Polynesian images mainly in their depiction as dull-witted and easily duped anthropophagers often invoked as bogeys (see Lessa 1961; 1980, referring to ogres of the Ulithi atoll). In the Marshall Islands, the giants occur only on the island of Kille, where they sleep for six months of the year (Erdland 1914: 314). While size and physical features of Micronesian giants are rarely specified, some Ulithi exemplars are supposed to possess multiple heads and limbs. In two stories, such fantastically conceived 'evil spirits' are burned to death (Lessa 1961: 62), in one instance with 'coconut cloth' (bark from the coconut palm)—a detail reminiscent of the Nage legend of the ebu gogo, exterminated in a conflagration fuelled by Arenga palm fibre.

Apart from these giants, small Micronesian hominoids have been reported for Ponape in the Caroline Islands, where they are now considered extinct (Christian 1899: 111–112, 113, 137, 382). Islanders describe the beings, called 'chokalai', as dark-skinned and flat-nosed 'dwarf aborigines' who (as suggested by graves attributed to them) stood 1.2 to 1.4 metres tall, whose only weapon was the bow and whose vocalizations resembled the gibbering of bats. According to present-day Ponapeans—the descendants of Malayo-Polynesian-speaking settlers who, Christian suggests, gradually exterminated them (ibid.: 382)—some chokalai may still survive in remote forests and may furthermore have descendants among a darker, flat-nosed and goggle-eyed segment of the population living near a particular estuary. Apparently comparable to the Caroline dwarfs are good-natured 'forest spirits' inhabiting bush and thick grass in the Marshall Islands and sometimes mating with and fathering offspring by human females. While described as 'little people', however, the physical form of these beings is not specified (Erdland 1914: 314). And neither they nor the chokalai of the Carolines are characterized as hairy.[9]

The extinct dwarfs of Taiwan

Generally comparable to these Micronesian exemplars and equally pertinent to an understanding of Indonesian and Southeast Asian images, are small humanlike figures reported from Taiwan. As noted, Taiwanese aboriginals, inhabitants of Taiwan before large-scale immigration of Chinese from the mainland in the seventeenth century, speak Austronesian languages related to the Malayo-Polynesian languages spoken in Oceania and Insular South-east Asia. Indeed, owing to the marked diversity of Taiwanese branches of the Austronesian language family, Taiwan has long been regarded as the Austronesian homeland. The special relevance of Taiwanese images to the ontological status of Florenese and other eastern Indonesian hominoids should thus be apparent.

Images of small people, sometimes called 'dwarfs' and mostly considered long extinct, occur in the legends of several Taiwanese aboriginal tribes. Although the literature sometimes refers to the little people as 'negritos', this usage reflects an anthropological interpretation of the representations as possibly reflecting a former population comparable, for example, to the Agta negritos of Luzon (northern Philippines). One source of this interpretation may be a Taiwanese Chinese notion of 'black ghosts', partly based in a memory of African slaves introduced by the Dutch in the seventeenth century (Ferrell 1968: 67). However, as Ferrell points out, aboriginal Taiwanese do not specify the little people as dark-complexioned. Nor, it seems, do they describe them as in any sense hairy.

In fact, the traditions contain few details of their physical appearance at all. The Saaroa people of southern Taiwan describe the dwarfs' height as one to 1.2 metres, while the Bunun (among the shortest aboriginals on Taiwan, with an average height of less than 1.59 metres) say 60 to 90 centimetres (Ferrell 1969a: 61). One group of Paiwan, too, say 60 centimetres; the Saisiat estimate the dwarfs' height as about one metre (Pan 1999: 27). Saaroa also describe the hair of the little people as curly and 'red' in colour—thus evidently contrasting to the dark and straighter hair of all current Taiwanese populations. Bunun say the little people had tails, but lacked an anus, for which reason they could not eat solid food and only 'sniffed the vapor'.[10] Other aboriginals (Tsou, Rukai and Paiwan) who subscribe to images of 'underground-dwelling food-sniffing people' do not characterize these as dwarfs (ibid.: 66, citing Mabuchi 1964); hence the tailed creatures recognized by the Bunun may represent a conflation of two distinct sorts of figures. On the other hand, it is interesting that a myth from Sulawesi describes the lolok as consuming only the steam of cooked rice (see Chapter 4).

Like wildmen virtually everywhere, the Taiwanese dwarfs were physically strong. They inhabited mountainous regions, including in some instances specific named locations (for example, Mount Maibalai, according to the Saisiat) where they resided in caves or holes in the ground. They also formed relatively large groups, with as many as sixty occupying a single location. Inhabitants of Taiwan's southeastern coast, on the other hand, describe the small people as living in stone houses and as the builders of the stone coffins and other megalithic remains found in this region—a detail reminiscent of the Karo umang of northern Sumatra (see Chapter 5). According to the Bunun, the little people's diet included frogs, for which they had a particular liking. However, everywhere they are supposed to have cultivated rice, millet, or tubers and—again recalling the umang—in several traditions they are credited with teaching people agricultural techniques. Bunun, who, as just noted, attribute tails to the little people, describe them as climbing trees with the facility of monkeys (presumably the Formosan macaque, *Macaca cyclopis*, Taiwan's only known non-human primate) but, at the same time, possessing manufactured weapons, specifically bows and arrows. Other aboriginals credit the little people with pottery, spears and knives. Also suggesting a partly arboreal habit is a variant of a Saisiat story which describes the dwarfs as regularly taking their rest in the branches of a particular tree (Pan 1999: 27) and an Atayal tale that has them crossing a creek by way of the overhanging branch of a 'giant tree' (Ferrell 1968: 64).

Language seems not to be explicitly attributed to the little people; however, linguistic ability is implicit in their reputed liking for singing and their teaching certain songs as well as agricultural techniques to Taiwanese.

Although Ferrell appears occasionally to contrast them with 'humans', the dwarfs are mostly described as a kind of 'people'; indeed, barring the tails mentioned by the Bunun, in most respects they seem no more than very short human beings. In addition to advanced agricultural knowledge, the dwarfs are said to have possessed 'powerful sorcery', for which they were feared (Ferrell 1968: 63). But this too is something regularly ascribed to other humans, including of course ethnically distinct neighbours.

The Taiwanese case would be of no special interest were it not for traditions according to which the aboriginals, some time in the distant past, rendered the dwarfs extinct. Probably the best known of these is a Saisiat legend. The Saisiat are the least numerous of the aboriginal groups (just 1,300 people in 1931) and inhabit a foothill region in northwestern Taiwan (Mabuchi 1974: 184). The legend is connected with a biennial agricultural ritual performed by the Saisiat partly to honour the spirits of the murdered dwarfs. Saisiat first came across the dwarfs in a cave, having been drawn there by their singing (Pache 1964: 39). The story further relates how Saisiat used to invite the little people to attend harvest ceremonies, but on one such occasion the dwarfs seduced their women and the men decided to exact revenge. In order to return home, the little people had to cross a river by way of a log bridge, so the Saisiat men cut the log and when the dwarfs began to cross they fell into the river and drowned. According to a variant of the myth recorded by Pan (1999: 27), the men cut the branches of a tree in which the little people used to take their rest, disguising the cut marks with mud. The version recorded by Pache (1964: 41) appears to combine these two motifs: the 'bridge' was formed naturally from branches of trees that grew over a stream, from either side. Because of its coolness, the dwarfs liked to repair to this place for mutual delousing. Saisiat then killed them by cutting down the trees.[11] All variants of the story, however, relate how a male and a female dwarf escaped destruction. Before departing Saisiat territory and moving to the east, they taught the Saisiat the 'pas-ta'ai', an agricultural ceremony now performed every two years (Ferrell 1968: 63–64, citing Lin 1956: 51).

Other Taiwanese peoples have similar myths of dwarf extermination. The Atayal say their ancestors cut the branch of a large tree, once used by the small people to enter Atayal territory and rape their women (Ferrell 1968: 64). Ancestors of the Bunun cut down a bridge; however, their motive was not the dwarfs' erotic tendency but their habit of firing arrows at Bunun. In further contrast, the Saaroa claim the little people simply removed to the east after Saaroa ancestors had engaged them in 'bitter fighting' (ibid.). None of these accounts links the disappearance of the dwarfs with any modern ceremonial remembrance. Therefore, on this and other grounds, Ferrell's interpretation of the Saisiat legend of the extinct dwarfs as something that has been grafted onto what was 'originally some sort of fertility-deity/ancestral-spirit ritual' (1969a: 50) appears well supported.[12]

The Saisiat tradition, especially, reveals several elements obviously comparable to the Nage legend of the ebu gogo. Both concern hominoids shorter than local people residing in a mountain cave, initially living in relative amity with villagers and even attending village rituals. Eventually, a negative act by the troglodytes breaches this peace. Local people then undertake an act of extermination involving trickery, which results in the annihilation of the cave-dwellers, except for a male and female who escape and move to another territory. There are, to be sure, differences between the two legends. For example, while the Nage wildmen were cultureless, the Taiwanese dwarfs possessed technology, indeed superior agricultural and ritual abilities. They were also apparently glabrous and had a fully human physical form. And the misdemeanours that led to their destruction were quite different from those that caused the Nage to take action against the ebu gogo. Even so, similarities between the Taiwanese and Flores traditions are a sufficient indication that both may be rooted in a common cultural heritage. Pointing in the same direction is the story told by the To Lage of central Sulawesi, concerning the dwarfish To Ligowi who once regularly attended To Lage feasts but, owing to a misunderstanding, ceased to do so and left the territory for good (see Chapter 4).

If Ferrell's interpretation of the Saisiat ritual is correct and in view of the lack of connection between other aboriginal representations of extinct dwarfs and current ceremonial usages, then the little people obviously cannot be a product of the ritual, or an ideological refraction of a rite (as Edmund Leach might have supposed). Rather, the legend of the little cave-dwellers must have preceded its synthesis with the Saisiat agricultural ceremony. But this of course sheds no light on the derivation of the image of the dwarfs or the story of their extermination by ancestors of modern Taiwanese. Indeed, all Ferrell is able to do in this respect is point to parallels in other Pacific cultures and particularly the Hawaiian menehune. Especially relevant to this comparison are an aboriginal (Paiwan) report that the exceptionally strong dwarfs could 'carry huge stones' and their association with stone-works found in currently uninhabited sites (Ferrell 1968: 65)—ideas which, again, reveal parallels with representations found in eastern Indonesia.

Views from the Philippines

Although Taiwanese aboriginals do not speak of their legendary 'dwarfs' as dark-skinned, the image could yet have roots in historical contact with the Philippines (see Bellwood 1997: 211–17) and particularly knowledge, direct or indirect, of the short-statured negritos of Luzon. In contrast to Oceania, legends of seemingly mythical little people appear to be lacking in the Philippines.[13] Nevertheless, these islands, whose inhabitants all speak Western Malayo-Polynesian languages, are relevant to our topic insofar as outsiders' representations of the Agta negritos of Luzon are reminiscent, if not so much of Oceanic or Taiwanese images, then more certainly of ones from eastern Indonesia. I should stress that I refer to depictions of fully attested human beings, hunter–gatherers comparable to the Kubu of Sumatra and historical Vedda of Sri Lanka and not to legendary (and perhaps not fully human) hominoids. Emphasizing their dark complexions and relatively small size, taller Philippine lowlanders thus describe Agta as displaying long, pendulous breasts (Minter, pers. comm., February 2006), going naked and constructing houses in trees (Worcester 1912: 838, 849).[14] Whether lowlanders characterize Agta as hairy-bodied is unclear. However, western observers have so described them, with reference to the 'peppercorn' hair 'distributed very abundantly over their bodies' (see ibid.: 838, who also mentions their 'disproportionately long arms'; regarding hair, see also Bean 1913).[15] It has more recently been claimed that 'perceptions' of Philippines foragers like the Agta are 'partially derived from Western European conceptions of the Noble Savage and Wild Man' (Rosaldo 1982: 322). Be that as it may, it seems rather more certain that another source has been similar representations independently developed and maintained by neighbouring communities of taller and lighter-skinned Philippine cultivators (called 'pute', meaning 'white, light', by the Agta; Estioko and Griffin 1975).

Although it also does not refer to hairy hominoids, a myth known to the Buid people of the Philippine island of Mindoro is of similar comparative interest. This concerns what their ethnographer describes as 'predatory human spirits' named 'fangablang' ('those who are encountered') which take the form of ugly, anthropophagous 'giants'. Formerly these beings were visible to everyone but can now be seen only by spirit familiars of mediums. Owing to their man-eating proclivity, a Buid hero once exterminated a group of these giants in a field of spear grass. This he accomplished by surrounding the group and setting fire to the grass. All were then killed except for 'two small children', from whom all present fangablang descend (Gibson 1986: 137). It is obvious how this story resembles the apparently historical report from Guadalcanal of a 'mumu' killed early in the twentieth century when villagers fired a grass patch. There is an equally noticeable resemblance with Indonesian tales concerning the extermination of pestilential monkeys (Chapter 3) and, of course, Florenese legends in which fire is used to destroy hairy hominoids.

Local differences and Asian origins

Just as Philippine lowlanders emphasize or exaggerate physical differences between themselves and neighbouring negritos, so Oceanic (Melanesian, Micronesian, Polynesian) images of often hirsute hominoids may ultimately owe something to a similar meeting of distinct phenotypes. People of distinctively small stature—variously designated as 'pygmies', 'negritos', or 'negritoids' (see, for example, Christian 1899; Rivers 1914; Dixon 1920: 264; Brewster 1922; Ivens 1930; Hambruch 1932; Birdsell 1975: 515–17)—have long been proposed as a demographic component of the chain of islands running from New Guinea eastward, from New Britain through Bougainville in the northern Solomons and southeast as far as Espiritu Santo and Malekula, in Vanuatu (Bellwood 1979: 27). Writing on highland regions of Espiritu Santo,

Speiser reported encountering ‘pygmies’ with an average male height of 1.52 metres who, although everywhere mixed with taller Melanesians, in some places approached 70 per cent of the local population (1913: 165–68). At the same time, Speiser judged these Santo pygmies to be less hairy than their larger neighbours and also lighter skinned, a specification that would partly contradict their identification as ‘negritos’. By the same token, if these small Vanuatans are to be taken as a model for Oceanic wildmen, their physical appearance seems at odds with such instances as the dark-skinned Melanesian kakamora and Micronesian chokalai and with hairy exemplars like the kakamora, mumu and maero. Tressol briefly alludes to ‘*hairy* pygmies living unclothed in the mountains’ of Vanuatu (1961: 51, emphasis provided), but how these may be connected with Speiser’s subjects is unclear.

Since several legendary hominoids are described as straight-haired, a factor further weighing against ‘negritos’ or ‘pygmies’ as their ultimate referent is the usual frizzy or woolly head hair of the human populations so designated. Profuse body hair has been reported for generally dark-skinned New Guinean negritos (Kleiweg 1942: 35), who in regard to height are probably the best attested Oceanic instances of the category. For example, Rappaport (1974) gives average male and female heights among the Tsembaga Maring of Papua New Guinea as respectively 1.48 and 1.38 metres, while for the Tapiro pygmies of West Papua (Irian Jaya), Rawling (1913: 264) recorded a comparable average stature of just over 1.44 metres. Yet as we have seen, New Guinean images, including the long, curly-haired dwarfs reported by educated Papua New Guineans, seem quite different from these. Although Oceanic pygmoid or negritoid peoples have sometimes been conceived as an aboriginal population, moreover, the case of Vanuatu, which may not have been occupied before the arrival of Malayo-Polynesian speakers some 3000 years ago, suggests that Speiser’s ‘pygmies’ were as much descended from these immigrants as are taller coastal folk. It is yet possible that images of dark-skinned Oceanic hominoids like the chokolai and kakamora reflect an encounter between larger and lighter-skinned Malayo-Polynesians and darker and hairier ‘Australoids’, who entered the region from Indonesia (eastern Sundaland and adjacent islands) about 40,000 years ago. Nevertheless, it seems still to be a majority view in anthropology that pygmoid peoples, rather than reflecting earlier waves of migration, represent local adaptations, sometimes to scarce food resources on relatively small islands.

Even if encounters between larger and smaller and sometimes lighter and darker, people have contributed to the Oceanic images, it is evident that these have been substantially modified in a partly fantastical direction. Indeed, given their similarity to Indonesian figures, it is more likely that the images themselves, or significant parts of them, were imported into the Pacific and thus reflect a cultural heritage shared with other Malayo-Polynesian speakers. Looking westwards, this applies with even more force to the kalanoro of Madagascar, an island peopled by Indonesians less than two thousand years ago. Indications are, then, that Indonesia has been the centre of a radiation of images extending at present from Madagascar in the west to the Pacific islands in the east.

This thesis is supported by a series of quite specific resemblances. Apart from similar heights (between one and 1.5 metres) and in most instances hairy bodies, prominent among Oceanic, Malagasy and Indonesian images are long fingernails, a feature of comparable figures in Flores, northern Sumatra, the Mentawai Islands and the Malay peninsula, as well as of the Melanesian kakamora, Malagasy kalanoro and Maori maero. (Although partly distinguished from more prosaically described hairy hominoids, the Florenese beings named 'ség̈e re'e' or 'fége re'e' similarly possess long nails, as does Rangda, the mythical character of Balinese dramatic performance and so did the creature named 'sindai', reputedly captured in Sumatra in the 1950s.) Among wildmen reported outside of Oceania and the Malayo-Polynesian-speaking world, by contrast, long nails or claws are regularly mentioned only for the Sri Lankan 'nittaewo'.

Although globally more widespread, several common behavioural attributes include: cave-dwelling, stealing from gardens or occasionally from dwellings, abduction or attempted abduction, susceptibility to capture and a tendency to deceive people and cause them to lose their way. Also shared by Southeast Asian and Oceanic figures, at least, are anthropophagy and intermarriage with humans. Irrational fear, for example of white or red things among Solomon Islands wildmen, is perhaps comparable to Florenese hominoids' aversion to combs. Rather less symptomatic of a shared cultural heritage but suggestive nonetheless, are stories of wildmen consumed by fire or suffering extermination in a cave (recorded for Flores, the Solomons and New Zealand —and, with qualifications, Micronesia and perhaps even the Philippines) as well as traditions of hominoid participation in the construction of large stone-works (Flores, Sumba, Taiwan and, in a qualified sense, Oceania). Extermination of cave-dwellers, of course, also links Florenese with Taiwanese traditions, while several attributes of the Taiwanese dwarfs directly recall hominoids reported from northern Sumatra and Sulawesi.

It also is noteworthy how some Oceanic traditions depict hairy hominoids as stupid and easily deceived, features equally characteristic of the Nage ebu gogo and the Sumbanese mili mongga, although not, interestingly enough, of other eastern Indonesian wildmen. Furthermore, in Oceanic narratives it is giants more often than smaller hominoids that prove fatally unintelligent, a circumstance which in fact accords with a more widespread and possibly worldwide, pattern.

Several differences might appear to weigh against Southeast Asia as a source of Oceanic and Malagasy images. Some of the latter concern hominoids that move quadrupedally and are partly arboreal, while certain Oceanic figures possess artefacts and technological ability. None of these traits is typical of Indonesian representations (except, perhaps, for the occasionally quadrupedal lae ho'a of eastern Flores). Like the Malagasy kalanoro, some Oceanic wildmen (for example, the kakamora) are also more aggressive than Indonesian exemplars, most notably the Sumatran orang pendek and more naturalistic representations of the Florenese ebu gogo. Finally, one could by now hardly fail to note that, whereas several eastern Indonesian figures possess pendulous breasts, these appendages are not mentioned for the hominoids of Oceania or Madagascar.

The Taiwanese traditions might also call into question an ultimate Southeast Asian derivation of Oceanic and Malagasy representations. For if Taiwan is the homeland of Austronesian speakers, then Taiwan could also be the source of images of small, physically distinct cave-dwellers ultimately exterminated by larger humans. A recent interpretation of genetic evidence has suggested that Austronesian languages may ultimately have originated in Insular Southeast Asia, especially the Philippines or Borneo (Hill *et al.* 2007). Yet this is a controversial view. Also, even if, contrary to the bulk of linguistic and archaeological evidence, Taiwan were not the Austronesian homeland, other evidence reveals contact between Taiwan and coastal locations in the southwestern Pacific—including Oceania, the Philippines, and Indonesia—over a long period following a hypothetical Austronesian (or proto-Malayo-Polynesian) departure from Taiwan between 4000 and 5000 years ago (Ferrell 1969a: 13–15; see also Hung *et al.* 2006). As suggested earlier, the Taiwanese dwarfs might even have a source in representations of Philippine negritos. On more than one ground, therefore, it is entirely possible, indeed I would say probable, that Taiwanese traditions of dwarf extermination derive from Insular Southeast Asia.

On balance, manifest similarities among images shared by Austronesian and especially Malayo-Polynesian-speaking peoples are best explained by common cultural heritage and moreover a common origin in Southeast Asia. It is a point of some importance that the most noteworthy resemblances between Southeast Asian and Oceanic figures pertain specifically to hominoids from eastern Indonesia and wildmen of Melanesia, especially the Solomons. In view of the much longer human contact between Indonesia and Melanesia and the relatively recent occupation of Polynesia effected by the expansion of Malayo-Polynesian speakers, these resemblances could point to an introduction of images from eastern Indonesia, including the Flores region, well before the appearance of Austronesians in Oceania and perhaps tens of thousand of years ago. Noteworthy in this connection is archaeological evidence that humans, passing through eastern Indonesia, had moved to the Solomon and Bismarck Islands by 40,000 years ago (Bellwood *et al.* 1998: 233, 260). Alternatively, the Melanesian representations could have been brought from Sulawesi or the Philippines, regions where a proto-Malayo-Polynesian language hypothetically split into Western and Central (or Central-Eastern) branches, the latter then eventually giving rise to the Oceanic languages (ibid.: 233; Bellwood 1997: 119–24).

An association of wildman images particularly with Malayo-Polynesian speakers appears consistent with the fact that New Guinea, where such images are less well attested, is inhabited mostly by speakers of Papuan (and thus non-Austronesian) languages. By contrast, in the Solomons, where images comparable to Southeast Asian exemplars are especially prominent, the majority of languages (56 of 62) are Malayo-Polynesian (Taki and Tryon 1997: 364). Further support for a Malayo-Polynesian connection may be found in the particular resemblance between the kakamora of Melanesia and the maero, the wildmen of the Polynesian Maori. Yet this interpretation fails to explain the relative absence of comparable hairy wildmen in other parts of Polynesia and in Micronesia. Interestingly, the resemblance between maero and kakamora calls to mind older theories regarding a Maori derivation from Melanesia; however, these have since been rejected and Central Polynesia is now generally accepted as the origin of New Zealand's first human inhabitants (Bellwood 1987: 133–34; Spriggs and Anderson 1993).

Some evidence therefore favours a pre-Malayo-Polynesian date for a spread of Southeast Asian wildman images into the Pacific, while some does not. But whatever the timeline, the cultural-historical connection does not disclose their ultimate ontological status. The figures might be purely imaginary, having no basis in any empirical reality, human or non-human. On the other hand, especially the Melanesian instances could reflect representations which in a Southeast Asian homeland were at least partly informed by some local experience. A Philippine connection might suggest inspiration from taller and lighter Malayo-Polynesian speakers' perceptions of negritos. Yet one is also reminded of Codrington's suggestion that wildmen of the Solomons may ultimately reflect memories of 'large simians'. If these were orang-utans or gibbons, then for Melanesia, one would probably have to posit the long survival of a memory passed on by pre-Malayo-Polynesian-speaking ancestors deriving from Sundaland, more specifically areas coinciding with the present islands of Borneo and Sumatra. A source in some Malayo-Polynesian homeland is less probable, unless this were Borneo (see Hill *et al.* 2007) or unless one is to go back as far as late Pleistocene times, when orang-utans still survived in southern China—the place of origin of the ancestors of Austronesian-speakers before leaving the Asian mainland for Taiwan, and is still the home of gibbons.[16]

Yet, whatever its historical or geographical locus, an origin of the images in experience of apes remains questionable. For apart from body hair, great physical strength and possibly long fingernails, details pointing to decidedly simian features are largely absent from Pacific images. Indeed, with the questionable exception of Knibbs's ill-described 'ape-men', such features are arguably more apparent in the distended bellies, 'long eyebrows' and 'low protruding foreheads' of the Hawaiian menehune. It may therefore be premature to rule out, as an alternative source of both Insular Southeast Asian and Pacific images, a sapient memory of pre-sapiens hominins such as *Homo erectus*, a species that may have survived on Java until as recently as 27,000 years ago (Curtis *et al.* 2000: 264). And one might also not want to exclude prehistoric (or even more recent?) contact with hominins corresponding to the eastern Indonesian remains recently interpreted as *Homo floresiensis*.

10 Conclusion

What were the ebu gogo?

This chapter completes our return to the eastern Indonesian figure of ebu gogo. If a shared culture history and long memories of Asian apes can account for Oceanic images, then they could equally account for the Nage wildmen. But even if the first interpretation were secure, to leave things thus would be premature. For one thing, the thesis does not explain why ebu gogo and comparable representations among other Malayo-Polynesian speakers have for so long been sustained and maintained. To address this question, we need to reconsider the possible symbolism of ebu gogo and the social and cultural value such images might possess. Additionally, we should consider how far hypothetically durable images can be explained as particular realizations of an 'archetype' or, otherwise expressed, some pan-human cognitive disposition that can find expression in the most diverse cultural, geographical and historical settings. Finally, in the light of comparative evidence presented in the last six chapters, more attention must be given to possible connections between eastern Indonesian images like ebu gogo and empirical creatures, present or formerly present in the local environment.

As shown in Chapter 2, the ebu gogo reveal few signs of a social construct or any obvious connection with Nage social institutions or cultural values. The extinct hominoids possess no religious or ritual significance and Nage insist they were empirical beings, radically different from supernaturally powerful spiritual beings. At the same time, some features of the legendary wildmen are evidently imaginary (such as the over-the-shoulder breasts and the remarkable swallowing ability). As I have demonstrated, these can be ascribed to conflation with originally separate images of malevolent female spirits or an implicitly fictitious bogey figure (also called 'ebu gogo', but more often 'gogo meo'). Nevertheless, similar stories told elsewhere in central Flores cast doubt on the historicity of the Nage legend of the wildmen's extermination—or, at any rate, the manner in which it was executed—and several features of the narrative point to a connection with widespread Indonesian tales concerning the extirpation of pestilential monkeys and the destruction by fire of other harmful beings.

This sort of 'deconstruction' might be taken further. Various authors have treated the figure of the European wildman as an allegorical device, one that has served to confirm, or occasionally to criticize, such antithetical conditions as 'civilization', 'sanity' and 'orthodoxy' (White 1978: 151). Bernheimer, it will be recalled, interprets the wildman of late mediaeval Europe as a figure illustrating the consequences of a fall from Christian grace and removal from Christian society. By the same token, evaluations of the European figure have been shown to differ from one historical period to another in accordance with changing views of both humanity and society. Even though small-scale non-western societies obviously differ from European society, such an approach might yet have validity for the Nage wildman. Ebu gogo, it will be recalled, were unintelligent, naked and voracious hominoids leading a crude and cultureless existence, reduced to stealing from villagers' fields. In these respects, they could be construed as the very model of what human beings should not be. What is more, it was their thieving ways (linked, arguably, with their lack of culture and technology, implying an inability to provide for themselves) and in some accounts their proclivity to child abduction and perhaps even anthropophagy, which brought about their destruction. As a moral tale, therefore, the legend of ebu gogo's extermination might be seen as promoting the value of human culture, a settled existence involving agriculture and respect for property.

However, quite apart from the risk of fabricating a tropical version of the Protestant Ethic, there are several problems with this analysis. Besides indications that Florenese hominoids are culturally unimportant images (as mentioned in Chapter 1), the interpretation accounts only for the persistence of the representations and not for their origin. More particularly, it does not explain the physical features of ebu gogo. As shown by the Bornean myth summarized in Chapter 3, thieving monkeys could do just as well and indeed, the allegorical function could be better served by straightforwardly human antagonists. As for abduction and anthropophagy (or cannibalism), Nage associate these offences against culture and social order much more closely with quite different figures, including witches, malevolent spirits and human 'head-stealers'—all of which are far more prominent than ebu gogo in the general scheme of Nage life. If the ebu gogo legend does serve a general allegorical purpose, one must further ask why it is so localized, being associated with a single Nage village. Also, while the extinction of the creatures accords with an interpretation of destruction as a punishment, one must wonder why the wildmen are, as it were, not kept alive as a model of the sub-human condition. After all, elsewhere in the Malayo-Polynesian region where wildmen are similarly depicted as abducting humans and stealing from fields, they are not the subjects of extermination tales. And if the ebu gogo were constructed as a means of demonstrating the negative consequences of removal from human society as interpretations of the European wildman might suggest, why are neither they nor other non-western hairy hominoids represented as feral or degraded humans, that is, creatures that were once human but have since *become* wild?

One can just about see how the legend might serve a moral purpose, but this is speculative and difficult to substantiate, especially in the absence of ritual applications of the tale (it is not told as part of initiation ceremonies, for example) or religious activities informed by it.[1] There is, moreover, nothing positive about the ebu gogo. Hence unlike some European images of the wildman, the Nage figure cannot be understood as a 'noble savage', positively free of the ills of established society (see White 1978: 168). This is hardly surprising, since eastern Indonesian culture (like non-western culture generally) entails a quite different view of the relationship between society and the individual. Apart from possible allegorical functions, however, we might consider that the figure of ebu gogo is somehow intellectually compelling—or 'good to think', to employ a famous anthropological phrase. It might, in other words, be thought to evoke a boundary, defining what is essentially human: again, culture and technology and articulate speech. This could explain why the form of the figure is almost human yet physically distinguished by the possession of certain features of non-human animals, a combination contributing to local ambivalence over how such creatures should properly be classified. The most notable of these features is body hair, a defining quality of wildmen virtually everywhere.

Some anthropologists have construed hair as a universal social symbol and though the focus has mostly been on head hair, there is no reason to suppose the same analysis could not also apply to body hair. Leach (1958) has interpreted long, unshorn hair as a symbol of sexuality, either socially sanctioned or excessive; by contrast, Hallpike (1969) construes long hair in a more general way, as expressing a relative absence of social control. In regard to wildmen, however, such interpretations leave much to be desired. Not just ebu gogo, but most of the world's hairy hominoids are not represented as particularly sexual. Male sexual organs are rarely mentioned and sometimes gender is not clearly distinguishable. Even the pendulous breasts of certain exemplars are not considered sexually attractive and in fact are sometimes described as contributing to a generally ugly appearance. To be sure, some people do depict hairy hominoids—including the wildman of Europe, the yeti and the north Sumatran umang—as occasionally engaging sexually with humans. Yet such depictions pertain only to a minority of figures and, in the case of the yeti at least, are largely confined to narrative contexts separate from more prosaic representations of the same creatures.

Construing the hairy bodies of wildmen as symbolic of absent social control may therefore seem nearer the mark. Yet this does little more than restate the hypothetical function of such figures as allegorical images of essentially human creatures separated from society and is thus suspect on the same grounds. As wild animals exist independently of human society, a hirsute condition could indeed be taken as symbolizing a lack of social control. But hairiness might just as well be understood as expressing undomesticated animality *per se* (an assessment in which Nage, in regard to ebu gogo, would certainly concur) and rather than an imaginary symbol, it could of course be a real symptom (though one that is perhaps exaggerated) of a real creature. In addition, if the hair were a symbol of asocial animality, why, it must be asked, is relatively long or profuse hair the main physical feature distinguishing wildmen from normal humans? That is, why are they not also provided with horns or, indeed, a tail?[2]

This question invites two others. First, speaking just of the ebu gogo, it may be asked why, rather than a hairy, anti-social and over-sexed monster, Nage as it were content themselves with a figure which, palaeoanthropologically speaking, seems plausible? Second, if the Nage did conceive the ebu gogo as an undesirable combination of human and animal figures, one must wonder why they do not construe them as human–animal hybrids, or—in view of evidently absent social functions—more specifically as the products of degraded or anti-social humans mating with animals.[3]

Wildmen and spirits

In some measure, to argue that wildmen are purely imaginary connects them with spiritual beings and suggests that they may be understood in the same way. By the same token, to show that they are quite different from spirits casts further doubt on their non-empirical character. Like wildmen elsewhere, ebu gogo do indeed share certain features with spirits, many of which similarly haunt desolate and uninhabited places. Some spirits share with wildmen such specific behavioural tendencies as abducting humans and although the idea does not pertain to ebu gogo, some wildmen, like many spirits, are considered a source of mystical powers (Indonesian examples include the Florenese lae ho'a, the lolok of Sulawesi and the northern Sumatran umang). Arguably, it is this sort of resemblance—possibly implying a 'metaclass' of strange beings that would subsume spirits, wildmen and perhaps some rare or unusual animals—which makes possible the sort of conflation identified in Chapter 3, whereby features more typical of malevolent and mostly female spirits are transferred to otherwise independently conceived hairy hominoids.[4]

Yet, in most respects, wildmen are radically different from spirits. As noted, Nage deny that their local wildmen were a variety of spirits, as evidently do people in many other parts of the world (see Chapters 5 to 9). Analytically as well, attributes of ebu gogo and comparable images do not fit the category of spiritual beings to which ethnographic research has repeatedly attested. Spirits can be defined as a culturally universal construct whose most decisive characteristics include a fundamentally human psyche, invisibility or incorporeality, an ability to transform and assume animal or human guise and various powers manifestly superior to those of human beings. Such powers include the ability to disappear and otherwise defy laws of physical nature, or more simply extraordinary powers of vision or hearing (cf. Atran 1990: 219–20; Boyer 1994: 113–14, 118–19; 2001). In mysterious ways, spirits are also typically able to affect humans and human affairs—for example, by bestowing fertility or wealth (sometimes by transforming worthless items into objects of value) or causing illness. It is, of course, particularly by virtue of these powers that spirits are focal to religious ideology and practice and commonly figure as the express objects of ritual action.

None of this applies to the ebu gogo, whose lack of mental agility, relative rarity (before their extinction, as an apparently remnant population), vulnerability and mortality tend further to distinguish the creatures from spirits. In contrast to extraordinary perceptual abilities of spirits, representations of hairy hominoids in general rarely mention their eyesight or hearing, from which one may reasonably infer that these senses are not considered superior to those of humans. Also important are the detailed and often disparate and uneven, descriptions local people give of the physical appearance and behaviour of hairy hominoids, exemplified especially by the eastern Indonesian materials reviewed in Chapters 2 and 3. By contrast, when referring to their assumed physical forms, people typically describe spirits in more perfunctory and stereotyped ways—for example, as appearing as 'beautiful women' or, simply, as taking the form of snakes. Like similar hominoids elsewhere, ebu gogo are not considered a physical guise that spirits temporarily assume, as spirits are frequently claimed to do in respect of various animals (including not only snakes, but fish, birds and even monkeys in the case of Nage spirits). Of course, wildmen elsewhere are not so strictly contrasted with spirits as are the ebu gogo, particularly when conceived as a source of mystical powers. Yet to the extent that some wildmen are thus 'spiritualized', this, as I have often had occasion to point out, does not compromise their empirical status since, worldwide, numerous attested animal kinds are identically credited with supernatural ability.

As demonstrated by cognitive anthropologists like Atran and Boyer, spirit categories are memorable and continuously represented owing to 'counterintuitive' attributes such as disembodiment and an ability to walk through walls, combined with intuitive features spontaneously inferred from experience of real persons. One might well consider whether empirical animals, especially ones rarely seen or currently extinct, might become equally memorable—and perhaps comparably 'spiritual'—because of extraordinary yet *real* features that may seem equally counterintuitive. A creature incorporating an uncanny combination of human and animal features would be a good example. Yet, unlike such human–animal combinations (or what may be construed as wildmen) spiritual beings can only be fully understood as imaginary refractions of social experience, particularly of social statuses and relationships—a position supported by a wealth of ethnographic evidence and anthropological analysis. And it is for this reason that spirits, typically represented as persons despite their fantastic powers, are imagined as possessing essentially human, in addition to super-human, features.[5]

The case of ebu gogo shows how sub-human hominoids do not so readily admit, let alone require, a sociological interpretation. Even so, wildmen might, like spirits, be construed as 'symbolic humans' in a more psychological way. Specifically, the images could be considered as reflections of panhuman childhood experiences of powerful parents or other senior kinsfolk and perhaps especially of maternal parents.[6] As shown, the names of several Flores hominoids include terms denoting mothers or grandparents (for example, 'ine' in 'ine ngiu' and 'ebu' in 'ebu gogo'). There are of course also the large breasts, attributed to female wildmen on Flores and elsewhere. Arguing against this thesis, however, is the relatively small size of the ebu gogo and many other comparable hominoids. In fact, the interpretation only begins to work where wildmen are giants. On Flores and elsewhere, one finds a representation of early human ancestors as giants (Forth 1998a: 234–35); but ebu gogo were neither ancestors (Forth 2006) nor giants. A view of the Nage wildmen as reflecting powerful parents also hardly fits the former's reputation as unintelligent and generally vulnerable.[7] Finally, there is the fact that neither ebu gogo males, nor male wildmen generally, are credited with large genitalia, as one should expect were their psychological origin to lie in powerful parental (and not just maternal) figures.

In view of the feminine character of much supporting evidence, this sort of psychoanalytical interpretation has more force when applied to the partly independently conceived Nage bogey named 'ebu gogo' (but more often 'gogo meo') and also the possibly related malevolent female spirits construed by earlier writers as Florenese variants of the Malay 'pontianak' (Chapter 3). Bogeys everywhere are ultimately reducible to threatening parents, who invoke them for disciplinary ends. And rather than fathers, it is more often mothers, or other senior female kin, who do so. But as shown, these representations probably have a separate origin from the more prosaically conceived population of currently extinct wildmen Nage also designate as 'ebu gogo'. In fact, a similar consideration counts against explanations of hairy hominoids anywhere as nothing more than imaginary bogeys. For this function is more often served, in a range of cultures, by a variety of fully attested human and non-human creatures, including animals, ethnic others and socially marginal individuals, as well as witches and malevolent spirits.

The wildman as 'archetype'

A different sort of psychological explanation may be found in an interpretation of wildmen as a universal 'archetype', a mental construct that is non-cultural in the sense that it occurs independently of any particular cultural tradition or social experience. A concept most closely associated with the work of Carl Jung and briefly deployed in social anthropology by Rodney Needham (1978), 'archetypes' may be compared to the more recently proposed 'cognitive dispositions' which result from 'the evolutionary history of the [human] species' and which 'account for certain general and recurrent properties of cultural representations' (Boyer 2001: 57). Whereas 'archetype' denotes quite specific and invariant images, the 'dispositions' of modern cognitive anthropology mostly refer to mental constraints inclining people to produce, retain and propagate representations with certain relational properties (a conjunction of intuitive and counterintuitive features in the case of spiritual figures). This difference aside, strikingly similar representations of wildmen found in very diverse cultural and geographical settings do provide *prima facie* support for some interpretation of these as a product of a pan-human cognitive proclivity. Still, we need to be cautious.

As a figure discernible in the traditions of all continents, 'the wildman' is a rather sparse image, consisting fundamentally of a bipedal creature with a generally humanlike form and hairy body. The image is also not particularly distinctive, for it almost corresponds to known animals, particularly the anthropoid apes. In this light, the wildman's hypothetical universality may find more support in a series of specific features which, by virtue of their representation by culturally unrelated peoples, present themselves as arresting similarities. Documented in previous chapters, these features include: long or pendulous breasts, inverted feet, long or sharp fingernails, an often foul odour, red or reddish hair, extreme strength, bimodalism in regard to size, a reputation as abductors and susceptibility to capture by humans. The distribution of each of these attributes deserves closer attention.

Often described as so long they can be tossed over the shoulders, the breasts are mentioned for wildmen of China and Central Asia, the Himalayan yeti and the European wildman, as well as for the ebu gogo and other Florenese figures and the mili mongga of Sumba.[8] As features of hairy hominoids, they do not occur in other parts of Indonesia or Southeast Asia (including Sumatra and parts of central Flores), nor in Australia, Africa, Madagascar, or Oceania. (The long-breasted 'wewe' of Java and the Balinese figure of Rangda, it may be recalled, are glabrous.) Among Malayo-Polynesian-speaking peoples, therefore, pendulous breasts are specific to Flores and Sumba and even on Flores they are not encountered everywhere. In Himalayan traditions, furthermore, the appendages are prominent only or mostly in folk narratives and are absent from reputed eyewitness accounts. On the other hand, exceptionally long and pendulous breasts have further been attributed the world over to attested human populations, including the Agta negritos of the Philippines and, in early European travellers' tales, Hottentots and Patagonians (Pigafetta 1975: 43), as well as to a variety of spiritual beings (Thompson 1946: 248). Further attesting to the image's widespread incidence are modern western appearances of large and unusually manipulable breasts in ribald jokes and pornographic imagery.

Such appendages may appear to be a product of pure fantasy. Yet even over-the-shoulder breasts have a possible empirical source, as an exaggerated representation of normally sagging or low-slung human (or hominin) breasts; accordingly, experience of this physical feature can motivate their inclusion as a component of more fantastic or at least scientifically unattested figures. At the same time, breasts draped over the shoulders might be construed, like back-to-front feet, as a symbolic inversion, since the breasts would then point backwards rather than forwards. Either way, pendulous breasts are actually attributed to a minority of hairy hominoids. They are, moreover, discontinuously distributed; are not exclusive to wildmen; and are largely associated with spiritual representations that further admit more decidedly counterintuitive features (such as metamorphosis or powers of invisibility).[9]

The association of pendulous breasts with both the wildman of Europe and the ebu gogo of eastern Indonesia, for example, still leaves something to explain. Nevertheless, rather than a defining feature of a universal archetype, the breasts are better understood as an optional elaboration, not essential to the core image of the wildman. In fact, given their more general occurrence, as a property equally associated with spirits and ethnic others, a figure identifiable as 'female with over-the-shoulder breasts' is more widespread and perhaps also more 'archetypal', than is the wildman *per se*—and it is therefore arguable that the cultural representation of some (but by no means all) putative hominoids has been influenced by this other, more ubiquitous image. As occasional attributes of wildmen, prominent breasts can hardly be a symbol or sign of animality, if only because non-human primates lack these exclusively human appendages. And they are not obviously symbolic of fertility or sexuality, nor of infertility, even though the pendulous breasts of certain figures are expressly linked with female ageing and ugliness (Forth 2007).

Much the same assessment applies to inverted feet, another attribute of a variety of spiritual beings. Taken literally, such feet may be the only implausible or truly fantastic physical component of wildman images. Yet, like extraordinarily long breasts, inverted feet can sometimes be explained with reference to empirical phenomena such as peculiarities of tracks left by primates and bears, or as a hyperbolic depiction of human feet or footprints. (As noted, Central Asian wildmen, for example, are described as having only inward-turning feet, rather than feet turned completely backwards.) Obviously, the attribute can have an independent origin, particularly when it is not linked with footprints or occurs in places lacking bears and apes. Even so, as a characteristic of hairy hominoids, its distribution is highly restricted. Inverted feet are not ascribed to ebu gogo, nor indeed to Oceanic images. Otherwise, the feature is mostly confined to Malayo-Polynesian-speaking regions: Sumatra and western Java in respect of the orang pendek and aul, Madagascar and Flores in regard to figures like the ine weu of Manggarai and the eastern Ngadha ebu ngiu. Outside of Malayo-Polynesia, inverted feet are attributed to the yeti and Central Asian wildmen, but not to the European wildman or the sasquatch.[10]

Once again, then, we have a feature which, although widely encountered and by virtue of that fact reasonably treated as an instance of a transcultural and pan-human imagery, is neither exclusive nor essential to a universal wildman archetype. The same applies to another and empirically more plausible, attribute of some wildmen: long or sharp fingernails. However, except for their occurrence as an attribute of the Sri Lankan nittaewo, this feature is reported only for images recognized by Malayo-Polynesian speakers (in Oceania, Flores, Sumatra, the Malay Peninsula and Madagascar); hence even more than the inverted feet, the nails appear symptomatic of a specific ethno-linguistic tradition. Long nails invite an observation of a different sort. While sometimes ascribed to primates, notably orang-utans, they can hardly reflect an accurate observation of humans (*Homo sapiens*), whose nails, unless deliberately grown long through avoidance of labour, typically become short and blunt through constant manual activity. And the same would presumably apply to other, less modern species of the genus *Homo*. Like profuse body hair, long nails might therefore be construed as a symbolic expression of a more general animality. Still, a question remains as to why, as a property of wildman images, they are not more widespread and occur mostly in places where people either have experience of orang-utans or recognize local hominoids whose representation could reflect memories of such creatures.

More common than long nails, or even back-to-front feet, is the attribution to wildmen of a strong and generally unpleasant odour, often resembling the smell of a known animal. This is mentioned for hominoid categories in Sumba and Flores including ebu gogo, but not elsewhere in Indonesia or Malayo-Polynesia and thus not for wildmen of Madagascar or Oceania. In Southeast Asia, the feature otherwise receives mention only for the enigmatic creatures reported from Trolak in the Malay Peninsula in 1953–54 (Chapter 6). A strong smell is associated with the yeti, Central Asian wildmen, the Australian yahoo (or yowie) and with some qualification the African kakundakari and North American sasquatch. But odour forms no part of descriptions of putative hominoids from Sri Lanka and China, nor of the European wildman.

Whatever is to be made of this relatively common component of wildman images, it is hardly exclusive to these. Besides certain humans (as represented by others, at least), not only do certain animals give off a strong odour, but some spiritual beings are represented as doing so as well (Classen 1992: 159). Since many wild primates seem not to exude pronounced or distinctive natural smells, a strong body odour, rather like prominent breasts, suggests a particularly human, or hominin, trait. It is, moreover, a quality people often ascribe, accurately or inaccurately, to ethnic or 'racial' others (Moncrieff 1966: 209; Classen 1992: 133). As noted, Sumatrans and Europeans alike attribute an unpleasant odour to hunter–gatherers like the Kubu and Lubu (though, ironically, not to the orang pendek) and taller Africans and Europeans ascribe a distinctive smell to pygmies.[11]

Among the most widespread attributes of wildmen everywhere is red, or reddish, hair. In fact, it occurs in wildman images reported from all continents except, interestingly enough, Europe. The exception may seem curious in view of the popular association of red-haired Europeans with personality traits linked with some concept of wildness, notably a violent temper. And there is an equally notional connection of red-heads with body odour. At the same time, not all hominoids in regions where red-haired exemplars are reported have hair this colour; on Flores, for example, the ebu gogo are described as dark-haired, as for a large part are the Sumatran orang pendek. In any case, the apparent ubiquity of red hair is less remarkable than it seems. As words for 'red' in many languages can apply to a variety of shades (including browns), it is relevant, first of all, that the pelage of many mammals, including primates, can be described as reddish. In fact, broadly construed, 'red' and 'black' (or 'dark') describe the hair colour of a great many mammalian species. Most humans, of course, have 'black' hair, or what speakers of most languages would classify as such; yet red or reddish hair occurs not only in Europe but occasionally among Asian and African populations as well. Furthermore, a reddish complexion is ascribed to some African pygmies and to other groups living in proximity to more populous darker-skinned neighbours. As an association of wildmen, 'red' could of course carry a symbolic load, as it does universally in a variety of contexts. However, its natural occurrence provides alternative explanations. It may be of more than incidental interest that Dubois, the discoverer of *Homo erectus*, had his sculpture of the hominin (now kept in the Natural History Museum in Leiden) 'painted in lifelike colors, with brown skin and orang-utan red hair' (Shipman 2001: 320). Red hair supports an interpretation of Asian and Oceanic hominoids as contemporary reflections or cultural memories of orang-utans. The colour attribution occurs well beyond current or known prehistoric ranges of these apes; but other cases (from Africa, for example) may be explicable as references to reddish-haired monkeys and chimpanzees. In addition, where reddish hair occurs as an unusual variant in normally dark-haired mammals—such as some bears, primates and even humans—a relatively rare red-haired specimen might well be taken for a creature of a different and unfamiliar, kind.

Similarly mundane and at least equally ubiquitous as mention of reddish hair, is the specification of wildmen as extremely strong and where small hominoids are concerned, strong in marked disproportion to their size. Rarely is this strength described in a way that appears completely implausible. Greater-than-human strength could be understood as a symbolic expression of the wildman's animal nature. Yet the attribute would of course be perfectly factual if the referent were an ape and probably also an archaic hominin.

A global resemblance among wildman images initially revealed as a variation within particular regions concerns the creatures' size. As shown by the yeti, Chinese wildmen and the African kakundakari and kikomba, hairy hominoids sometimes comprise two varieties, one smaller and one larger than local humans. Similarly, while eastern Indonesian wildmen are generally small, the Sumbanese mili mongga is big. And coexisting with general images of larger than human creatures are occasional or isolated reports of small individuals otherwise resembling the larger ones—for example, 'Jacko', sometimes interpreted as a young sasquatch and the Sumbanese 'Apu Kalita'. Such bimodalism is not of course distinctive of wildmen. Imaginary creatures of all sorts, including the giants and fairies of European tradition, can simply be imagined as smaller or larger than humans. But these are often very much smaller or larger and given that size is no object for human fantasy, it is surprising that so many wildmen are depicted as small rather than large and for the most part as not very much smaller than humans—which is to say, not minute (as spirits can be). Such representations, therefore, by no means rule out a basis in empirical creatures. Moreover, the subjects might include ones whose other features (general robusticity, long arms) can, by way of a form of synaesthesia, convey the impression of large size or be reflected as such in verbal portrayals, or whose overall threatening or imposing appearance can encourage exaggeration. As noted, people occasionally describe the normally small ebu gogo and orang pendek as large and minority reports of large size are sometimes traceable to inferences drawn from other attributes. Where attested species are concerned, it is further noteworthy how, just a few decades after its European discovery, the gorilla had assumed the proportions of King Kong.

A more widespread attribute of wildmen is a reputation as abductors and, less often, as anthropophagers. As demonstrated by Nage denial of these behaviours as a realistic capacity of ebu gogo, the notion may sometimes be fanciful, even for people who otherwise subscribe to the empirical existence of the creatures so accused. In addition, such aggressive tendencies appear contradicted by a more common depiction of hairy hominoids as timid, retiring and largely vegetarian. Somewhat illuminating this ambiguity, accusations of kidnapping and man-eating are obviously connected with a frequent deployment of hairy hominoids as bogeys. Like the bogey function generally, a reputed habit of abduction—most often of children—is of course yet another trait which is not exclusive to the wildman, being shared with a variety of other categories, including ethnically distinct neighbours, disparaged social groups and non-human primates. The point applies also to stories of changelings, the substitution of wildman infants for abducted human babies. None of this, then, sets off the wildman as a distinct figure; nor indeed, insofar as rumoured abduction provides an instrument of parental discipline, is it clearly an article of adult *belief*.

In some places, adults have occasionally claimed experience of abduction by wildmen (for example, in Madagascar and North America). Yet, on a global scale, noticeably more common than tales of particular abductions are stories of hominoids themselves being captured, often accidentally in traps set for other animals. In fact, stories of this sort, which usually take the form of rumours rather than detailed narratives, have been reported from most parts of the world. (As a theme in graphic art, literature and folk drama, the hunting and capture of wildmen in European traditions is somewhat different, but comparable nonetheless.) Two features of these capture tales are especially noteworthy. First, the victims—or their bodies where death follows capture—typically disappear without trace. Second, the captives are frequently anonymous, which is to say not identified as instances of locally named categories; examples include the Sumbanese Apu Kalita; creatures allegedly caught in recent decades on Sumbawa and Flores; and hairy hominoids reputedly captured in Vietnam, western Canada and Madagascar.[12]

Myths sometimes recount the capture of spirits, including female spirits who might subsequently marry their human captors.[13] But unlike these, tales of captured wildmen are typically told in a prosaic idiom, devoid of the fantastic details that characterize stories about spirits. At the same time, accounts of wildman capture from distant parts of the world reveal remarkable coincidences. To that extent they suggest a culturally underdetermined and distinctive feature of a distinctive category. But how might these similarities be explained? So arresting are the stories, they arguably appeal to a pan-human interest. This in turn suggests they may be a product of diffusion, passing rapidly between distant locations (as can rumours in general) and assisted in recent times, in many parts of the world, by print media. One could of course argue that the stories have the currency they do precisely because they evoke a human-wide disposition to conceive of creatures of a particular form. Yet, as the widespread incidence of such capture tales would suggest, what may be more basic than the figure itself, or the core image of a bipedal hairy hominoid, are human attitudes towards creatures regarded as sub-human, even if only in a marginal sense.

Capture stories from Flores, Sumba and Sumatra portray the victim as crying or weeping. Insofar as weeping and shedding tears appear to be exclusively human traits, the captives, if not completely fictitious, would presumably have to be human—or hominin. On the other hand, behaviours of monkeys and apes have sometimes been taken for weeping or sobbing (see Darwin 1998: 134–35, 136, 159). Writing in 1718, Beeckman described his purchased Bornean orang-utan as sobbing and crying (1973: 38) and Bontius and Noble gave similar accounts of creatures they observed on Java (see Chapter 6). Nage, too, say that injured macaques will weep from pain. Hence, non-human primates would seem to be another possibility. Attributing weeping to wildmen may seem to credit them with a semblance of human feeling. Yet it is more likely a narrative device lending, if not a note of tragedy, then a certain dramatic appeal to tales of their capture.

Apart from the image's simple, generic core—a hairy, bipedal human form—it is still not entirely clear what should be meant by calling the wildman an 'archetype'. In view of the non-exclusive character and variable combination of most attributes, one might consider the wildman a widespread 'synthetic image' (Needham 1978).[14] But this largely restates rather than resolves the question. On the other hand, it can be concluded that representations of hairy hominoids manifest a cognitive predisposition, realized in many different cultural times and places—though by no means all—to form images of a certain kind. Basic to this proclivity, indeed I would say its precondition, is a pan-human recognition of a fundamental contrast between humans and animals (Atran 1990: 218; Astuti *et al.* 2004) coexisting with an apprehension of a certain ontological continuity between the two. The extension of these categories can vary culturally, as for example when certain human beings are categorized as 'animals' (which is to say, sub-human). But this is a function of ideology and symbolic thought and does not signal radical differences of perception. By the same token, only contextually can human cognition elide the human-animal opposition, either by way of a cultural or religious belief that animals have some sort of qualified existence as human beings (see Viveiros de Castro 1998), or humans as animals (Forth 1998a: 149–51), or in the scientific notion, equally contrary to ordinary intuition (Wolpert 1992), that 'humans are animals'.

The wildman is thus a readily conceivable human figure with limited features of an animal (most notably profuse hair) who lives like an animal.[15] Yet for such a diffuse image to find expression as an extant, or once extant, creature requires a catalyst or trigger. The trigger can be empirical phenomena, as when seventeenth-century Europeans represented newly discovered apes, if not explicitly as instances of the wildman portrayed in mediaeval or Renaissance art and literature (Forth 2007), then as some sort of hairy humans. A signal outcome of comparing African images like the kikomba and kakundakari with Sumatran hominoids (Chapter 8) was the demonstration of how, in these otherwise quite different parts of the world, similar triggers, namely, different species of great apes, can catalyze remarkably similar representations.

With qualifications I introduce later, abnormal hirsutism in human individuals may be another phenomenon that can call up an image of some distinct kind of hair-covered hominoid. In regard to morphological features distinguishing the wildman from full humanity, yet another is palaeontological materials, interpreted both in ancient times and more recently in fundamentally similar ways (see Mayor 2000, who shows that European images of giants have a plausible source in the exposed bones of large prehistoric mammals). Conversely, culturally constituted images of wildmen can shape conceptions of empirical creatures—as when Europeans, after encountering great apes, initially and for some time considered them as more humanlike and in the case of the gorilla at any rate more violent, than in actuality (Heuvelmans 1990b: 9). Of course, insofar as apes have been subject to representations molded by pre-existing wildman images and have simultaneously provided empirical substance to representations of more humanlike hairy hominoids, the relation can obviously be reciprocal. It is arguably due to this mutual reinforcement of experience and pre-existing image that descriptions of wildmen often bear an uncanny resemblance to scientific images, notably of anthropoid apes and ancient hominins as reconstructed from palaeontological and archaeological evidence. At the same time, the European wildman, the figure specifically implicated in the development of primatological and palaeontological images (Stoczkowski 2002), may itself very well have an origin in ancient, albeit irregular and discontinuous, experience of apes.

Zoological catalysts may not always be necessary for the emergence of an imaginary wildman. Rather more certain, however, is the insufficiency of social and ideological factors alone to produce such a figure, as the example of the ebu gogo attests. Equally obvious are ways in which cultural factors can act against acceptance of wildman images as empirical realities, even in the presence of perceptual stimuli. If intellectual orthodoxy is taken as an instance, then several examples come to mind, including the regular dismissal of sasquatch sightings as bears or tree stumps (which is not to say that what people see are actually sasquatch).[16] In the same connection, it is also worth noting how images of violent chimpanzees suggested by local African representations—and reminiscent, moreover, of the European wildman—have only recently been borne out by primatological research. This has overturned an older view of the creatures as peaceable, which was consistent with a surviving western ideology that views animality, or nature, in a basically positive, optimistic way, in contrast to humanity and human society—a variant, indeed, of the earlier 'noble savage'.

Bases in experience: non-human animals

If either empirical or social factors can give rise to wildman images, it follows that different instances need not be, epistemologically or ontologically, of a piece. Partly on zoogeographical grounds, I consider the North American sasquatch and Australian yahoo to be empirically quite unfounded and largely rooted in western representations, including popularized knowledge and images of apes. At the same time, especially in the second instance, an empirical catalyst is available in European experiences, perhaps ideologically tempered, of indigenous humans. On the other hand, Sumatran and Central African hominoids (including orang pendek, kikomba and kakundakari) are explicable as distorted representations of anthropoid apes, possibly influenced as well by exaggerated images of ethnic others (hunter–gatherers like the Kubu and African pygmies). There is, furthermore, reason to suspect that, if not recent experience, then memory of orang-utans plays a part in mainland Asian representations. Malagasy and Oceanic images I am similarly inclined to interpret as reflecting a cultural inheritance from Southeast Asia, based originally in experience of apes—or of something else.

But what of eastern Indonesian figures and especially hominoids like ebu gogo? One could interpret these as distant memories of northwestern Indonesian or even Mainland Asian apes. One might by the same token consider a diffusion of fantastically elaborated images, although in view of the several physical and other attributes recurring in wildman figures worldwide—for example, inverted feet and pendulous breasts—a centre of diffusion somewhere in Central Asia might appear more likely than somewhere recently inhabited by orang-utans. There are, however, several more local possibilities to consider. On Flores, hypothetical empirical sources for ebu gogo and similar hominoid images include non-human primates, a formerly separate group of human beings (*Homo sapiens*), or some other humanlike creature.[17]

At present, the only known non-human primate on Flores, or on most islands east of Wallace's Line, is a monkey, the Long-tailed macaque (*Macaca fascicularis*). Like several other larger mammals, these were brought to Flores some 3,000 years ago and while modern humans first arrived on the island perhaps 15,000 years ago, at that time there were no anthropoid apes, if indeed there ever had been.[18] As demonstrated previously, Florenese legends concerning the extermination of ebu gogo and comparable creatures may well owe something to folktales concerning the killing of devious or pestilential monkeys. I now consider the different question of whether monkeys, specifically macaques, could have provided an experiential basis for the general representation of the Nage wildmen.

In several respects, the ebu gogo obviously resembled monkeys. They had hairy bodies, simian faces and like macaques stole from cultivated fields. Yet, here, the resemblance largely ends. Unlike the wildmen, the macaques have long tails. They are also arboreal, rarely venture near caves and, despite a local notion that monkeys will assume a bipedal posture when humans are not looking, are basically quadrupeds. Nage are thoroughly familiar with Long-tailed macaques—as agricultural pests, animals regularly killed and sometimes consumed and occasionally as pets. To be sure, Nage do represent monkeys in a generally anthropomorphous way that might seem to bring them closer to the image of the wildmen. Thus Nage say that monkeys bury and perform funerary rites for their dead, engage in traditional pugilistic competitions like humans (see Forth 1998a), blow on wounded companions in an attempt to revive them and produce medicines from the disinterred bones of other monkeys. In folktales, moreover, monkeys speak and attempt to engage in sex with human females; they even speak in a way similar to ebu gogo (see Chapter 2). However, these themes apply exclusively to narrative contexts and otherwise Nage deny linguistic ability to simians (even though they sometimes impute to them a sexual interest in women). What is more, while ebu gogo were reputedly able to talk, erotic inclinations form no part of representations of the Nage wildmen, nor those of other Florenese hominoids. And while attribution of cultural practices such as burial and ritual pugilism portray monkeys as humanlike, Nage describe the ebu gogo, ironically, as lacking in all culture and technology.

There are other noteworthy differences between the wildmen and monkeys. In myths and also as a more general conviction, monkeys are for Nage a physical form assumed by spirits ('nitu'; see Forth 1998a: 40–41). One story tells how a man once unknowingly became acquainted with a community of spirits and fell in love with a spirit maiden whom he would visit at night. In the daytime, their residence became a large tree which they inhabited in the guise of monkeys. No such ideas pertain to ebu gogo. It is also noteworthy how, not just on Flores but throughout Indonesia, monkeys are depicted in folktales as tricksters, characters who, while themselves often ultimately deceived, are cunning deceivers.[19] The guileless and generally unintelligent ebu gogo, of course, are portrayed only as victims of human deceit. The differences can be put down to a use of monkeys, like other animals, as metaphors for human beings; Nage generally describe macaques as intelligent, often uncannily so, but intelligent only by comparison with other animals. Ebu gogo, by contrast, were unintelligent by a human standard.[20] As shown in Chapter 3, putative eyewitness accounts of the supposedly extant Lionese hominoids called lae ho'a seem often to reflect experience of macaques, not least when they are described as possessing tails. Yet even the lae ho'a and therefore much less their Nage cousins, are unlikely to originate simply in experience of monkeys. The roots of ebu gogo thus appear to lie elsewhere.

Earlier, I discussed another way in which monkeys and apes could have shaped non-western representations of wildmen: through images and knowledge of simians, or even live specimens, introduced by Europeans in the late nineteenth or twentieth century. Fijians, it may be recalled, identified an imported monkey with a local category of spiritual beings. There is, however, nothing to suggest that this identification was motivated by anything other than the small size and generally hominoid appearance of the creature, features also attributed to the anthropomorphous spirits. More generally, explaining features of indigenous categories with reference to European illustrations of anthropoid apes shown to people previously unfamiliar with the creatures is tenable only insofar as indigenous and exogenous images share a good deal in common in the first place and therefore fails to illuminate their pre-existing similarity.

Other humans, or human others

Connections explored in earlier chapters, for example between the largely Malay representation of orang pendek and forest-dwelling Kubu people, raise the question of whether comparable Florenese representations may also be grounded in experience, ancient or recent, of relatively small, economically and politically marginal groups which are, or were, ethnically distinct, technologically simpler and otherwise different from larger, more settled populations of cultivators. Lending plausibility to this sort of explanation are attributions of hairiness and darker complexions to aboriginal populations in Manggarai origin mythology (Chapter 3) and indeed current depictions, by more recently arrived western Florenese, of surviving indigenes (or people who are thus regarded) as similarly dark and hairy. Such hypotheses actually implicate two sorts of motivations for the wildman image. The first is empirical: experience of physically different human others. The second is social or ideological, since representing other populations as hairy hominoids, or in some other way as less than fully human, can reflect cultural values and interests and can reinforce social boundaries.

As hypothetical sources of wildman imagery, ethnic others may be phenotypically distinct or they may simply be culturally different. In regard to the first possibility, it should immediately be noted that Flores at present provides very little evidence of physically distinct localized populations living in close proximity. While usually characterized as displaying a generally 'Australoid' phenotype (see Plates 1.1, 2.1 and 2.3), Florenese do vary noticeably in body size, skin colour and hair texture. However, these differences exist within single communities and, certainly in the Nage region, are observable even within single clans and villages. From human remains excavated from rock shelters, the missionary-archaeologist Theodor Verhoeven identified a negrito, or 'proto-negrito', component of the population of Flores dating to 3000 or 4000 years ago. Features of the remains included hyper-prognathism (protruding jaws), large teeth, small chins and robust mandibles. The shortest skeleton, that of a mature female excavated at Liang Toge, a rock shelter in west central Flores, was 1.46 metres tall (Verhoeven 1958: 230; Jacob 1967: 77–114). Yet, if only on the basis of height, it is questionable whether these are correctly classified as 'negrito', or 'pygmoid' as they have also been described (Jacob 1967: 94).[21] Moreover, even if the populations so identified were ever distinct from larger 'non-negrito' Florenese, they appear to have been for a long time but one component of a considerable mix. Keers' identification of a negrito element, detectible to some degree in all Flores populations, may reflect such admixture (1948: 147), but it could alternatively be interpreted as evidence of a clinal variation long present throughout eastern Indonesia.[22]

Attempting to link the diminutive skeleton interpreted as *Homo floresiensis* with local humans, the Indonesian palaeoanthropologist Teuku Jacob has recently spoken of 'Rampasasa pygmies' inhabiting the Manggarai village of Waemulu (Jacob *et al.* 2006). It is unclear, however, whether these comprise anything other than a few families exhibiting relatively small stature (Forth 2006: 344). The 'Rampasasa pygmies' studied by Jacob during a visit to Flores in April 2005 included 76 individuals, 41 females and 35 males (Jacob *et al.* 2006: 13423, 13426). Their average stature is given as 1.46 metres (thus the same height as the Liang Toge female). But this is not exceptionally short for Flores generally and, although not common, one can encounter adult Nage men standing less than 1.5 metres. Considering the amount of ethnographic research conducted in Manggarai during the twentieth century, not least by J.A.J. Verheijen and other missionary scholars, it is quite extraordinary that no one has ever reported these pygmies before. Manggarai villagers I spoke to in 2005 were also doubtful about the existence of very short people composing distinct local groups (Forth 2006: 344).

There is thus precious little reason to believe that negritos, or any other population physically distinct from modern Nage, inhabited the vicinity of Lia Ula cave some two to three hundred years ago, or when the ancestors of the 'Ua people arrived in the Nage region. The hypothesis could gain credibility if the legend of ebu gogo reflected a much older encounter (perhaps 1000 to 2000 years ago, say); we might even consider an earlier meeting of lighter-skinned 'Mongoloid' peoples with darker and perhaps hairier—but not necessarily shorter—'Australoids'.[23] But this is equally speculative and in any case it cannot simply be assumed that lighter-skinned immigrants will invariably represent a darker-complexioned indigenous group as lacking fire and tools and possessing hirsute bodies. The attribution, in most accounts, of long head hair to the ebu gogo also does not match the short and usually frizzy or woolly hair of the typically glabrous negritos of the Andamans, Malaysia and the Philippines. In an earlier book (Forth 1998a: 105), I suggested that ebu gogo 'may well have some empirical basis in a former component of the human population of Flores that is no longer present'. When I wrote this, I had no definite opinion on what this component might have been and although I mentioned 'negritos', in the same statement I observed that negrito features, as usually conceived, are more plausibly reflected in representations of the black-skinned spiritual beings Nage call 'noa'.

Interpreting the Lia Ula wildmen as former negritos also does not illuminate the pendulous breasts of female ebu gogo. Still, such breasts could reflect experience of some locally associated phenotypical peculiarity, perhaps highlighted by some cultural practice. It may be recalled that lighter-skinned Philippine lowlanders describe the Agta negritos as long-breasted. Writing on the Donggo people of highland eastern Sumbawa in the mid-nineteenth century, Zollinger (1850: 127) stated that even the young women had 'extraordinarily long hanging breasts', so 'deformed' that, 'in this degree and the generality of their occurrence', they surpassed anything he had ever encountered in the Indonesian archipelago. The author prefaces his statement with the observation that Donggo women in general are 'hideously ugly'. In a similar vein, Winter referred to nursing women among the Kubu of Sumatra as possessing 'withered breasts, stretched long and reaching almost to the belly' (1901: 219–20); and he also described the Kubu as generally ugly.[24] Since both assessments evidently reflected the authors' direct witness, they undoubtedly betray a western prejudice. But Nage too sometimes offer comparable, although less negatively evaluative, accounts of local female bosoms. As discussed in Chapter 2, Nage men relate tendentious stories of women with long, pendulous breasts encountered in Namu and Tana Wolo, northwesterly regions they associate with possibly surviving ebu gogo. A few men further claimed that, in pre-colonial times (or 'until about three generations ago'), such breasts were widespread among Nage women as well, but are nowadays very rare. According to one of my regular male informants, the breasts descended to the waist and were indeed so long they could be tossed over the shoulders.

The fact that Nage characterize their own women in this way suggests a commentary on modernity, connected especially with twentieth-century changes in female clothing practices. Among the Nage of Bo'a Wae, these included a general adoption of female upper garments, including brassieres, beginning in the 1950s and 1960s. Women in regions to the northwest went bare-breasted until rather more recently. Yet the long-breasted ebu gogo are unlikely to reflect an evaluation of 'backward' neighbouring populations, since this difference would have become apparent only in the last few decades. Tracing the representation to an unconscious conception of their own forebears might seem somewhat more plausible, were it not for the fact that Nage definitely do not identify the ebu gogo with human ancestors. And in any case, the extraordinary breasts of the female wildmen are, as demonstrated in previous chapters, more readily explained as the result of a conflation of hairy hominoids with images of long-breasted female spirits, maintained independently of wildman figures in other parts of Flores and elsewhere in Indonesia.

No more likely than pendulous female breasts as a local phenotypical basis for ebu gogo imagery is abnormal and excessive hair growth, including the extremely rare condition known as hypertrichosis lanuginosa. Some central Florenese are familiar with such conditions. Keo people described a local boy born in the 1940s as being covered entirely in hair; he was also relatively short. Taunted by schoolmates with the biblical name 'Esau' (after the hairy son of Isaac), he did not finish primary school and it is not clear what later became of him. Although one man described the boy's face as simian and claimed he was the product of a mother who was raped by a monkey, 'Esau' was recognized as otherwise normal and was not identified as an 'ebu gogo' or anything similar. This is not surprising. Apart from the fact that Keo do not recognize a local category of hairy hominoids, the case evokes the well-established principle according to which individual abnormalities cannot account for completely general beliefs or practices—in this instance, representations of whole populations, in other ways non-human or sub-human, living separately from modern humans. Hypertrichosis lanuginosa, it should further be remarked, entails the growth of long hair on virtually every part of the head and face and produces an appearance quite different from what is typically described for wildmen, in eastern Indonesian or anywhere else (see Chapter 6, Plates 6.2 and 6.3).[25]

All of the foregoing suggests that, if the ebu gogo could be traced to a thoroughly human group of ethnic others, these would have differed from Nage more in regard to culture than physical appearance. They might, for example, have been a remnant group of hunter–gatherers who kept themselves separate from neighbouring cultivators to the extent that there is no longer any memory of trade or any other positive relations with them. This hypothesis, too, finds little support in the available evidence; nevertheless, it invites several comments. Although Nage profess no knowledge of their forebears having employed stone tools, villagers in eastern Ngadha, immediately to the west of Nage territory, claim occasionally to find stone artefacts, buried or on the surface, in ancestral settlements. During the 1990s, their ethnographer was shown a stone spear point by an elder 'whose family had a reputation . . . for making fine weapons and tools'. While no age is mentioned for the artefacts, eastern Ngadha people regard this former lithic technology as something distinguishing them culturally from both the Nage to the east and other Ngadha to the west (Molnar 2000: 30). There is, then, a suggestion that people using stone tools may have been living close to Nage during a period—perhaps within the last several hundred years—when Nage had for some time been in possession of iron tools. If so, it is also possible that Nage overlooked or disparagingly discounted this lithic technology, thus giving rise to a myth of people completely lacking a material culture.[26]

If this is to explain the ebu gogo, however, it fails to account not only for the hairy hominoids' physical peculiarities, but also the fact that eastern

Ngadha maintain their own representations of such figures (see Chapter 3). There is, moreover, no association of reputedly lithic eastern Ngadha folk with caves. A better model might be found, for example, in the To Ala, nineteenth-century cave-dwellers of Sulawesi (Chapter 4). Or, considering that the group in question could have derived from cultivators who for some reason turned to foraging, the Tasaday of Mindanao in the Philippines, who according to one interpretation descend from nineteenth century Monobo villagers who became separated from a larger population and took to food collecting and living in caves (Headland 1992: 218). Yet this too is entirely speculative, and there is simply no comparable ethnological evidence for recent cave-dwellers on Flores. Recalling the 'gibar bohot' of the Moluccan island of Buru (Chapter 4), one might with equal plausibility trace the ebu gogo legend to a band of vagabonds, perhaps refugees from local wars holed up in a cave. Further arguing against the image of ebu gogo reflecting former vilified ethnic others is the circumstance that Nage do not ascribe animalistic or primitive physical features to disliked contemporaries. The group to which they express most antipathy are the Ngadha, more particularly the western Ngadha. But while Nage characterize Ngadha as dirty, ill-mannered drunkards given to various disapproved practices, they do not describe them as physically different from themselves.

Rather than a distorted representation of another ethnic group long disappeared from the Nage region, a somewhat more credible explanation for the ebu gogo is suggested by a provisional interpretation of the nittaewo of Sri Lanka as a complex reflection of Singhalese views of the creatures' Vedda exterminators (Chapter 7). Although other Nage denied it, several regular informants, individuals who at the time I had known for some fifteen years, described the people of 'Ua—the exterminators of the ebu gogo—as physically different from other Nage and as generally ugly. According to a variant idea, 'Ua folk were distinct in former times, when they mostly married endogamously, but nowadays, through regular marriage with other Nage, they have come to look much like everyone else. Details of this distinct physical appearance included short stature (or specifically, very short lower legs), large ears, noses or faces like monkeys, thick eyebrows, thin head hair, protruding lips, 'red' (light brown) patches in otherwise black head hair, coarse skin, genital hair infested with lice of a kind normally found on deer and an unpleasant odour (like a goat).

The last several features of this stereotype are evidently linked with a better founded reputation for dirtiness, owing to a shortage of water in the 'Ua region that has only recently been alleviated. Other Nage do not describe 'Ua people as hairy-bodied or having pendulous breasts. Nor, it should be remembered, do they claim that 'Ua folk ever mated with the wildmen.[27] Nevertheless, it is conceivable that the ebu gogo developed from a derogatory representation of the 'Ua folk maintained by outsiders, which 'Ua themselves then adopted, elaborated and used to construct an imaginary sub-human population residing further upslope. In other words, ebu gogo might be understood as a deflected image, a negative stereotype which the 'Ua countered by transferring it to an even more primitive category of beings whom they furthermore claimed to have annihilated.[28]

On the other hand, the interpretation does not explain similar representations of small hairy hominoids in several other parts of Flores (such as Poma and Lio). The 'Ua might have drawn on these in fabricating their own legend. But not least because of the apparently greater development of the legend in the Nage region, this is unlikely on other grounds (see Chapter 3). Also, derivation of the image from elsewhere does not account for such fantastic features as the long breasts and extraordinary swallowing capacity of the ebu gogo (which are *not* ascribed to the Lio and Poma hominoids). A derogatory stereotype of 'Ua villagers articulated by other Nage could conceivably be the origin of the ebu gogo. Yet things could equally be explained the other way around, by viewing the stereotype as a reflection of the 'Ua people's close association with the legendary wildmen.

Or something not quite human?

Local representations of wildmen encountered in Asia, Africa and Oceania may be understood as the equivalent in small-scale non-western societies of European representations of 'primitive' societies—once, indeed, described as 'wild' people and 'wild tribes'. They might, in other words, be considered as 'the primitives' primitives'. Yet whether, like the non-western objects of western stereotypes, figures like ebu gogo ultimately refer to former or existing populations of *Homo sapiens* remains to be seen. Clearly, if they do not and if the Nage wildmen are not purely imaginary, or fantastic distortions of Long-tailed macaques, then this leaves just one other empirical possibility: an undiscovered species, still surviving or that has disappeared from Flores during the last millennium or perhaps earlier.

No evidence indicates that any ape or tailless monkey, nor any other vaguely hominoid creature, survived on Flores until geologically recent times. In view of recent palaeoanthropological discoveries, therefore, a more obvious candidate is the small hominin whose remains have been interpreted as a new species, *Homo floresiensis* (Brown *et al.* 2004; Morwood *et al.* 2004, 2005). Comprising the holotype (or type skeleton) as well as remains of eight other individuals excavated in Liang Bua,[29] a cave in the Manggarai region, this population was apparently extinguished by volcanic activity some 12,000 years ago. Nevertheless, other populations may have survived in more easterly parts of Flores. Moreover, even at the terminal date for Liang Bua, *floresiensis* could very well have been sharing the island with *Homo sapiens*. Some anthropologists have challenged the thesis of a new species, construing the holotype—an individual with a cranial capacity of just 380 or 430 cubic centimetres according to different estimates—as a microcephalic human dwarf (e.g. Jacob *et al.* 2006). The evidence for a new hominin species and one that has evolved quite independently of modern humans, is by now quite considerable (see Morwood *et al.* 2005; Falk *et al.* 2005; Argue *et al.* 2006; Gibbons 2007). But even if the hypothesis of a new species is confirmed, the notion that the ebu gogo straightforwardly reflect a memory of *Homo floresiensis* is open to question on several grounds.

The first concerns agreement between physical features of ebu gogo and *floresiensis*. From Nage descriptions, features of the former that arguably fit the latter include a generally primitive facial morphology. The very similar mandibles (lower jaws) of the *floresiensis* holotype and another individual and the absence of a chin, for example, appear to match more detailed accounts of the lower face of ebu gogo. On the other hand, evidence relating to canines and other teeth (which for ebu gogo is in any case less than consistent) is more equivocal. The height of nine hominins excavated at Liang Bua did not exceed 1.10 metres (1.06 in the case of the holotype). The ebu gogo were also small, but not definitely as small as this. (Informants, it may be recalled, indicated stature between 1.0 and 1.5 metres, while explicit estimates ranged from 1.0 to 1.25 metres.) Whether or not island dwarfing is the cause of *floresiensis*'s small size, there is no reason to believe the species would have increased in size had it survived into the last millennium. On the other hand, it is conceivable that robust features could give an impression of greater size.[30]

Among several features of the newly discovered hominin not reflected in descriptions of ebu gogo are proportionally long arms and sloping shoulders (bound up with limited turning capacity of the upper arm; Culotta 2006: 983–84).[31] Another is the relatively long feet of *floresiensis* (Morwood and Van Oosterzee 2007: 105, 176–77). These features may not have been so noticeable as to register in the cultural memory, or may at any rate have been less memorable than, for example, hairy bodies. I leave open whether, in this regard, any significance can be attached to the 'bent or crooked lower arms' of the enigmatic human or hominoid creature observed in the Nage region in the 1970s, or the very long arms and limbs that swung inward (as it walked 'like a cripple') of another mysterious being encountered a decade earlier (see Chapter 2). Equally doubtful is the possibility of linking these apparently idiosyncratic sightings with the sloping shoulders and restricted humeral torsion of *Homo floresiensis*, or indications that the small hominins, though equally bipedal, walked differently from modern humans.[32] The evident strength of *floresiensis* forearms suggests they may have been 'very adept at tree climbing' (Morwood and Van Oosterzee 2007: 103); as noted, ebu gogo are depicted as strong climbers, though not specifically of trees.

Skeletal material of course provides no indication of the prominence of female breasts, so the pendulous paps of the ebu gogo are, for this comparison, completely neutral. For the same reason, the hairy bodies of the Nage wildmen are also irrelevant. Yet given the close relation between the genus *Homo* and tropical apes (especially chimpanzees), it is more than likely that non-sapiens *Homo* were substantially hairier than modern humans, and that our present glabrous condition is relatively recent and probably a product of advanced cultural factors. One should also not lose sight of the fact that some modern humans are notably hirsute, circum-Mediterranean populations and the Japanese Ainu being two cases in point.

Archaeological evidence suggests an association of *floresiensis* with stone tools, use of fire and hunting (including animals as large as juvenile Stegodons; Morwood *et al.* 2004: 1089). Especially in view of the small brain size, doubts have been raised about whether this cultural complex actually belonged to the hominins. If the doubts prove correct, then a relatively cultureless *floresiensis* would accord with ebu gogo's lack of technology and fire—and also with the putative hominoid's largely vegetarian diet. A recent interpretation of the wrist bones of *Homo floresiensis* also suggests the hominin may have possessed limited technological ability (Gibbons 2007; Tocheri *et al.* 2007). However, simple stone artefacts excavated in 2004–05 in central Flores and dating to at least 700,000 years ago have been construed as evidence of technological continuity with tools associated with *Homo floresiensis* at Liang Bua; the technology is, moreover, distinguishable from later lithic artefacts associated with *Homo sapiens* on Flores (Brumm *et al.* 2006). Language is a different, although related, matter. All one can say in this connection is that hypothetical cultural associations of *floresiensis*, especially as these involve lithic tools and hunting, are consistent with the ebu gogo's reputed ability to speak. But language does not accord with the Nage wildman's complete lack of material culture.

A fascinating speculation concerns how *Homo floresiensis* and *Homo sapiens*, considered as separate species co-existing on Flores possibly for thousands of years, would have related to one another. Their interaction 'may have involved little or no direct contact, symbiosis, competition or predation' (Morwood *et al.* 2004: 1090–91). In his recent book, Morwood argues that vulcanism cannot have caused the complete demise of the tiny hominin species and that modern humans were responsible for their extinction (2007: 178–79). This view clearly provides an intriguing parallel to the story of the ebu gogo's extermination by the 'Ua folk. If *Homo floresiensis* were indeed a separate species, perhaps occupying a different ecological niche from Florenese *sapiens*, it is possible that contact between the two populations would have been irregular and would have occurred only in situations of *sapiens* demographic expansion and local migration of the sort described for the 'Ua people. Similarly, Morwood has suggested that the first modern humans on Flores may have occupied coastal regions and only later moved inland to areas inhabited by *floresiensis* (2007: 238). In any case, rather than trade or some other symbiotic relation (such as one would expect had the ebu gogo been a group of human foragers), it is likely that, if *floresiensis* survived until the development of cultivation on Flores, the hominins would, like ebu gogo, indeed have stolen from cultivators' fields and otherwise proved a threat or nuisance. For the same reason, it is likely that modern humans would have adopted violent countermeasures.

At best, images of ebu gogo and *Homo floresiensis* reveal an approximate concordance. As representations, nothing decisively contradicts their identification; the palaeoanthropological reconstruction and the Nage image could refer to the same thing, yet there is nothing to show conclusively that they do. The same applies to other Florenese hominoids. Nevertheless, it is of some interest that the ana ula of Poma and the Lionese lae ho'a, usually depicted as standing about one metre tall and sometimes even less, suggest a closer agreement, particularly in regard to size. In fact, in the context of this comparison, one of the most astonishing features of the remains unearthed in Liang Bua is the small size of the nine individuals, whose average height appears to have been not much over one metre (Morwood *et al.* 2005). Clearly this is very small, even by comparison with known populations of human pygmies and negritos.

Further weighing against an identification of locally recognized hominoids and *Homo floresiensis* is the likelihood of the chrono-species surviving so recently as to be retained in local memory. Given that relations with humans could well have been competitive, even hostile, it seems highly improbable that *floresiensis* would survive to the present. Survival to 500 or even 200 years ago is marginally more likely and by the same token, more likely still is survival until one or more thousands of years ago. Yet this of course raises a question of the probability of a reasonably exact cultural memory of the creatures surviving for thousands, or even hundreds of years. This, it should be stressed, is a question the answer to which remains indeterminate. Since oral traditions can evidently register geological events that took place thousands of years ago, such long term cultural memory is by no means impossible. Australian Aboriginal mythology, for example, appears to reflect volcanic activity and sea level rises occurring 5,000 to 10,000 or more years ago, while Aboriginal classifications include categories identifiable with sub-fossil marsupials which became extinct up to 400 years ago (Sharpe and Tunbridge 1997; see also Burney and Ramilisonina on Malagasy fauna, discussed in Chapter 8).[33] We should recall, then, that Nage oral history locates the extermination of the ebu gogo not long before or after the 1830 eruption of the volcano Ebu Lobo, thus possibly not more than 200 years ago.

Other evidence for cultural traditions retained over long periods concerns connections among narrative traditions themselves. To cite just one Malayo-Polynesian example of particular relevance for the present study: virtually identical stories concerning humans abducted by and forced to mate with primates are told in northern Sumatra and Borneo, where they concern orangutans and on the eastern Indonesian island of Roti (see Map 4), where the primate is a monkey. Shared details include: confinement in a tree; birth to the female partner of a human-primate hybrid; bringing of food, including coconuts, to the human captive; the human escaping by way of a rope made from coconut fibre; and the primate's violent killing of the hybrid offspring on discovering the human's escape.[34] This is not to say that narrators necessarily believe these stories pertain to actual events. However, since it is highly improbable that either the north Sumatran or Bornean tale has diffused to Roti and even more improbable that numerous details of the narratives have been independently invented in places as far apart as Roti and Sumatra, they evidently represent nearly identical variants of a tradition dating to the time of the split between Western-Malayo-Polynesian languages (like those of Borneo and northern Sumatra) and Central-Malayo-Polynesian languages (like Rotinese). On current evidence, this would have taken place somewhere in northeastern Indonesia around 3,000 years ago. So it would appear that people have been telling—and remembering—essentially the same story for this length of time. In regard to wildmen, it should also be remembered that hominoid images are commonly transmitted, if not in standard myths or folk-tales, then in what can generally be called 'stories'.

Colleagues have suggested to me that if *Homo floresiensis* or something similar had survived into the twentieth century, Europeans and especially European naturalists concerned with documenting zoological species present on Flores, would have been unlikely to miss such a creature. I agree, but only to a point. First, there were no European naturalists on Flores 500 or even 200 years ago (roughly the date which Nage legend suggests their ancestors exterminated the ebu gogo). Second, before the full establishment of a Dutch colonial administration on the island in the early years of the twentieth century, very few Europeans visited Flores. Since then, their numbers have not been much greater and their activities have generally been confined to coastal and lowland areas. Finally, Europeans working on Flores, including Catholic priests engaged in ethnographic, linguistic and zoological research, appear to have been largely unaware even of the names of local categories that might have revealed unfamiliar zoological kinds, as I noted with regard to ebu gogo in Chapter 1. Those who were familiar with these categories, like Paul Arndt who recorded a version of the name 'lae ho'a' in his Lionese dictionary (1933), interpreted them as kinds of spiritual beings ('böse Geist', to cite Arndt's favourite German gloss). Such scholar-priests would no doubt have been among the first Flores residents consulted by visiting naturalists. But since the priests understood local hominoid categories as references to spirits, it is hardly likely that they would have mentioned them to zoologists.

Even if one is able to overcome a general incredulity concerning surviving non-sapiens hominins, the case for ebu gogo or some other Florenese hominoid reflecting *Homo floresiensis* is obviously difficult to make. Yet the thesis may have one saving grace. A relatively small, morphologically primitive hominin like the one identified as *floresiensis* provides a better empirical reference for ebu gogo than do either Long-tailed macaques, hypothetical demographic isolates phenotypically distinct from other Florenese, or equally hypothetical and necessarily even more exaggerated elaborations of other human groups displaying only cultural differences. Although the evaluation is obviously more subjective, the hypothesis of surviving non-sapiens hominins is also preferable, all other things being equal, to the argument that figures like ebu gogo are completely imaginary symbolic constructs of enduring, but indefinite, social value. Given the shortcomings of mostly functionalist sociological explanations, one might place more value in the idea that closely comparable but geographically widespread representations of hairy hominoids directly express innate features of human thought. Yet this approach (which in any case does not preclude and may even require empirical inputs) raises as many questions as it answers and so is not obviously superior to the thesis that a non-sapiens hominin like *Homo floresiensis* might have survived until recent times or at least so recently as to leave a reasonably accurate imprint on local collective memory. That the hominins are still to be found on Flores seems, intuitively, to border on the impossible. Yet it is rather more conceivable that they survived well into the period when modern humans occupied Flores, for example, until one or two thousand years ago. How likely it is that they hung on until just several centuries ago is another question.

It is a common maxim that, by showing one instance of a reputed phenomenon to be impossible or highly unlikely, one casts serious doubt on the class as a whole. How this might apply to wildmen, however, depends on what are counted as instances. Orang-utans and chimpanzees, for example, are not imaginary. Nor were *sapiens*-contemporary Neanderthals. According to analyses advanced earlier, most reports of wildmen—including the orang pendek, yeti and African categories, not to mention the sasquatch and yahoo—are better explained by factors other than surviving non-sapiens hominins or even undiscovered species of apes. In contrast to these, eastern Indonesia and particularly Flores, seem rather singular—and not just because of palaeontological evidence for what might be called a 'cryptozoological body'. Accordingly, Florenese hominoids (or some of them) might be considered as 'signals' disguised by 'noise' deriving from other parts of the world (cf. Groves 1987). As we have seen, another challenge to eastern Indonesian hominoids reflecting memories of local non-*sapiens* hominins is the occurrence of variants of a fairly distinct wildman image in several parts of the Malayo-Polynesian-speaking world. Yet these variants could reflect an empirically derived image emanating from the same source (somewhere in eastern Indonesia if not in Flores) or, indeed, the same creature once surviving in several parts of insular Southeast Asia. In this connection, a particular significance may attach to arguments for a derivation of *floresiensis* from Sulawesi (Morwood and Oosterzee 2007), also a possible centre, some three thousand years ago, of the dispersal of Malayo-Polynesian speakers and the evolution of the language family into western and central and eventually Oceanic, branches (see Bellwood 1997: 119–24). On the other hand, as an empirical source for the Florenese images, the little hominin may still have a competitor in orang-utans, especially if, as noted in the last chapter, Malayo-Polynesian (or Austronesian) dispersal can be traced to Borneo.

Regardless of the eventual verdict on the remains interpreted as *Homo floresiensis*, one of several developments of their unexpected discovery is the way it has highlighted arresting similarities between palaeanthropological reconstructions of archaic hominins and representations of hominoidal creatures maintained by non-western peoples like the Nage. The resemblances may not seem so remarkable when one considers the widely accepted view of pre-sapiens hominins as in some degree ape-like humans and at the same time various cultural interpretations, including recent as well as older western ones (Forth 2007), of non-human primates and especially the larger apes, as virtual humans. However that may be, there is a further point. A consideration of the evidence, including the cultural evidence, argues against a derivation of representations like the ebu gogo from more fantastic images, such as a fictitious bogey or the malevolent female 'pontianak' spirits separately represented in other parts of Flores and elsewhere in Indonesia. But if ebu gogo were in fact purely imaginary and particularly if the representation were somehow derivative of more classically conceived spiritual beings, then prosaic descriptions of the Nage wildmen, especially in regard to their putative morphology and behaviour, would imply an independent development, in a small-scale non-western society like that of the Nage, towards naturalism. This would then parallel the emergence of scientific thought, with a concomitant process of 'disenchantment' (as Max Weber called it) usually regarded as specific to Europe in relatively recent times. Surely such possible convergence should by itself be sufficient to stimulate anthropological interest in images of wildmen.

Notes

1 Introduction

1 The exclusion of tailed humans, a widespread image, is justified on at least two grounds. Being more humanlike than apes, wildmen are not normally credited with tails. Conversely, tailed humans are not usually described as hairy or otherwise apelike. As elsewhere, individuals with tails—that is, an elongated coccyx—are known to occur on Flores; but people do not identify them as instances of local wildmen.

2 In this newer taxonomy, 'hominid' (or the family Hominidae) includes, in addition to hominins, the African apes (grouped with humans as members of the subfamily Homininae) and orang-utans (subfamily Ponginae). In scientific usage, 'hominoid' can refer to all species of the superfamily Hominoidea, which besides the foregoing comprises the Hylobatidae, the gibbons and siamangs. I depart from this usage insofar as I reserve 'hominoid' for scientifically unattested and possibly imaginary, creatures. Some categories I call 'wildmen', or 'hairy hominoids', have previously been designated as 'parahominids', meaning 'almost hominid'. Because of its relative unfamiliarity, however and also because the newer sense of 'hominid' includes the great apes, I have found no advantage in its use. To refer to 'unidentified tropical Asian hominoids', McNeely and Wachtel (1988) have proposed the acronym 'untrahom'. But even if it possessed other merit, this neologism would cover only part of my topic.

3 Anthropologists generally define 'witchcraft' as a power, possessed by specific individuals, to cause harm to others by purely psychic, or non-physical, means. A 'witch', therefore, is a superficially ordinary person thought to possess this power.

4 Exemplifying this distinction for western culture are, on the one hand, representations of rodents in mundane practice and scientific discourse and, on the other, images like Mickey Mouse. British readers may prefer to compare what they know of bears from school textbooks, or nature documentaries, with Rupert Bear. It might be objected that no westerners (except perhaps children) 'really believe in' storybook mice or bears. But this largely misses the point. Even if it could be shown that non-westerners 'really believed', for example, that animals appearing in myth can speak or possess an intellect comparable to humans—a claim my field research concerning Indonesian mythology has led me to doubt—it still does not follow that this sort of representation is entirely of a piece with ordinary knowledge of the same animals.

5 Field research in 2005 further took me to parts of Flores where I had not previously conducted extensive ethnography, including eastern Ngadha, Poma, Rawe, Lio and Manggarai.

6 The fact that members of the *Homo floresiensis* discovery team were told about the ebu gogo by Nage villagers in 2004 (Morwood and Van Oosterzee 2007: 154) might seem to contradict the suggestion that such categories are not prominent in

local traditions. Indeed, many Nage are familiar with the representation. However, if this particular outside interest was not inspired by mention of ebu gogo in my 1998 book, it was nevertheless raised in conversation with local people years after I had initiated enquiries about the category in 1984. Nage quickly and correctly, realized that I was very interested in the ebu gogo legend; and they very likely assumed that other westerners would be as well. As I noted in the Preface, my discovery of the Florenese category derives from a report by a Nage informant prompted by my mention of the Sumbanese mili mongga. And it was only through further investigation of the ebu gogo that I came across the ana ula and lae ho'a described in Chapter 3. How likely it is that I would otherwise have heard about any of these categories is difficult to say.

2 The story of ebu gogo

1 'Ua now comprises a series of villages. In the strictest sense, 'Ua denotes a territorial and cultural entity partly distinct from Nage, but I henceforth employ 'Nage' in a broader sense, which includes 'Ua and several other districts to the west.
2 Of six men who provided descriptions in 2005, four said ebu gogo were short, standing between one and 1.5 metres, while the other two said they were as tall as local humans. These results do not differ significantly from information obtained before media reports appeared in 2004 and 2005 of the discovery of the diminutive *Homo floresiensis.*
3 While it is difficult to generalize, many Nage, being of an 'Australoid' phenotype, possess quite profuse body hair. Thin and sometimes thick, male beards, moustaches and side-burns are also common.
4 As the names of children abducted by ebu gogo, Sedho and Lilo are known only to a minority of Nage and to none outside of 'Ua. Despite a general knowledge of various ancestors supposedly alive at the time of ebu gogo's extermination, moreover, no one could say whose children these were. When I asked him in 2005, Julius Poi (who had not previously mentioned Sedho and Lilo) said the names belonged to the two ebu gogo that escaped the conflagration at Lia Ula—a curious idea insofar as this is the sole indication that the wildmen might have had personal names. Another 'Ua elder, describing the names as those of two characters in a variant of a widespread creation myth concerning the separation of the sky and earth, insisted they had nothing to do with the legend of ebu gogo.
5 Discussing a sound recording attributed to the North American 'sasquatch' which includes /g/ sounds, Kirlin and Hertel (1980: 276) note that implosive consonants like /k/ and /g/ are unique to human vocal tracts and cannot be produced by known monkeys and apes. Therefore, it would seem that, if 'ebu gogo' referred to real creatures that were capable of producing such sounds, then they must have been either *Homo sapiens* or a closely related species of hominin.
6 In stories, monkeys alternatively use 'ja'o' for the first person singular, thus further paralleling ebu gogo speech.
7 An example of an attributed chant lyric is: 'dala sasa ila méze, ala mm mm mm'. Another is 'kai kébe ara (or 'bara') mére gasa sara mm mm mm'. 'Ila mére' and 'ara mére' are both references to a fire (around which circle dances are performed) burning high and bright, while 'kai kébe' seems to mean 'to open and close'. I am unable to make any definite sense of 'dala sasa' and 'gasa sara' and nor were Nage I questioned. The /r/, it is worth noting, no longer occurs in most Nage dialects and in 'mére', for example, is replaced by /z/. The sound, however, is found in all neighbouring languages and is occasionally employed in song by Nage themselves. According to another idea, while circle dancing ebu gogo would also utter what my field notes describe as 'rhythmic hissing sounds' (tss-tss-tss-tsssss).

8 Comparison with churches and chapels reflects the fact that most Nage have long been converts to Catholicism.

9 My informant gave his account in the presence of a large group of 'Ua villagers, including an elderly man who, serving as a guide to officials from the Department of Education and Culture wishing to document local traditions, had visited the site in 1976 and found the entrance—or what he thought was the entrance—sealed. The older man's statement corresponded with information provided by Department officials when I spoke to them in 1988 about the attempt to locate Lia Ula.

10 I raise these points since reputedly extant creatures, in various ways comparable to the Nage wildmen, are associated with shamans and mediumship in northern Sumatra and Madagascar (see Chapters 6 and 9). A parallel connection does not exist in the case of the ebu gogo.

11 Nage, it should be stressed, do not believe that the ebu gogo made any use of palm fibre, either as clothing or bedding. Employed in making rope and twine, the very coarse fibre is also not used as clothing or matting by any human group, or at least none known to Nage. Contrary to what is related in a popular book concerning the discovery of *Homo floresiensis*, no one I questioned had ever heard that the fire that killed ebu gogo was started with 'clothing' soaked in cooking oil (Goldenberg 2007: 47). The fuel is always specified as palm fibre.

12 In another Nage myth, numerous maggots emerging from the carcass of a huge snake become a source of wealth for the killer as they transform into water buffalo (Forth 1998a: 83–84). Similarly, a Rotinese story (De Vries 1925: 5–14; 1928: 11–26) relates how an orphaned hero kills an anthropophagous old woman and buries her in a cave; after three days the body becomes infested with maggots, which then all turn into sheep. Although the theme of mysterious acquisition of wealth is absent from the ebu gogo legend, other narrative features shared by the Rotinese tale and the Nage legend of the wildman holocaust nevertheless suggest a common origin.

13 That neither Wolo Pogo nor anyone else claims precedence in 'Ua territory is surprising in view of the general tendency in central Flores to claim territorial precedence in respect of land occupied by late-comers (Forth 1998a; 2001).

14 Dates and details derive from the website of the Global Volcanism Program of the Smithsonian Institution, National Museum of Natural History, http://www.volcano.si.edu/index.cfm, 30 August 2004. Other eruptions of Ebu Lobo were recorded in 1888, 1910 and 1924. Apparently far less catastrophic than the eruption in 1830, these more recent events seem not to be remembered by Nage. In any case, even the earliest of the three dates, 1888, is much too recent in relation to genealogical and other evidence, as I show in a moment.

15 In 1996, Julius Poi thought that the extermination had preceded the eruption by three or four generations. Yet, in 2005, he stated that the eruption had come first, although both events had taken place in the generation of his ancestors named Meo and Dike. This was one of surprisingly few inconsistencies in Poi's two accounts.

16 The genealogy producing the most recent date of 1885, in which the ancestor concerned was the informant's great-grandfather, is in all likelihood contracted or otherwise incorrect. Other genealogies record descent from an ancestor living four, five, or (in one case) six generations before the informant.

17 The average of dates where a generation is calculated as 30 years is 1823; where the calculation is based on 25 years, it is 1845. The average of all dates is 1834.

18 A myth from the Sangir (also Sangihe) Islands, off the northeastern arm of Sulawesi, relates how a giant in his death throes swears to his human killers that he will 'become' a volcano and take revenge by causing an eruption and killing their

children and grandchildren (De Vries 1925: 75–78). Despite the similar names of the wildmen (ebu gogo) and the Nage volcano (Ebu Lobo), the fact that Nage do not consistently locate the eruption after the extermination counts against the Nage legend reflecting this theme, as does the representation of the ebu gogo not as giants, but as hominoids smaller than local humans. The resemblance of the names of the wildmen and the volcano consists solely in the common, apparently honorary epithet 'ebu', as I explain below.

19 Since the names are very similar, it is important to note that Wolo Ua (Mount Ua) is not to be confused with Wolo 'Ua (or, locally, Wolo Rua), the place of origin of the 'Ua people. 'Ua' means 'rattan', while "Ua' (with an initial glottal stop indicated by the apostrophe and corresponding to /r/ in certain western and northeastern Nage dialects) means 'two', or 'twin' (as in Wolo 'Ua, 'Two Hills'), although nowadays Nage often understand it as the term for 'honey bee'.

20 I first recorded this tale in 1999. In 2005, I asked the informant to repeat it. He kindly did so and although the second version was more detailed than the first, it did not differ in any substantial respect. The only particular appearing in the 1999 version that was not repeated in 2005 concerned the bent or crooked appearance of the subject's lower arms. Also a native of Ngadha, the friend who accompanied Lulukumi was last known to be resident in Kupang, on the island of Timor. Unfortunately, I was unable to contact him.

21 Lulukumi used the Indonesian term 'makluk', 'creature' or 'being', which in this context meant something other than a human being, but not necessarily an animal.

22 The idea that Soda Ule had seen a 'transforming witch', an interpretation provided by village elders to whom he related the incident, is supported not only by the individual's apparent nakedness, but also by the fact that the weather at the time was turning stormy. Transformation into a witch, according to Nage, is always accompanied by a wind and rainstorm (Forth 1989).

23 It took some effort to explain to him, after our 2005 interview, why I was interested in what he had seen and he definitely did not connect it with recent reports of *Homo floresiensis* (for example, as possible evidence that such a creature still existed). Rather, he was inclined to explain his subject simply as a disturbed local person and as a manifestation of what he described as backward social and economic conditions during the 1970s.

24 In 1999, a young female school teacher mentioned how she had learned in high school about the giant hominin *Meganthropus*—something that could account for her opinion and that of some other relatively young informants, that the ebu gogo were large and tall.

25 Another referent of 'gogo' is a large bamboo container for liquids, as in 'gogo ae', a water container. Although not obviously germane in this context, the image of a large hollow vessel is perhaps suggestive in relation to the child-devouring monster. 'Gogo' also occurs as a man's proper name.

26 In regard to 'ebu' in the sense of 'grandparent', it is worth noting that terms for senior kinsmen are applied to bogey characters in a variety of cultures (see, for example, Ter Haar 2006, regarding the Chinese 'Auntie Old Tiger'). In the previously mentioned tale from the Sangir Islands (see note 18), children address an anthropophagous and apparently hairy giant who emerges from the ground as 'grandfather'; he in turn addresses them as 'grandchild' (De Vries 1925: 75).

27 The only context in which the wildmen might be designated simply as 'gogo' is the metaphorical expression 'gogo lau lia mona apa bhia', 'gogo down in the cave, there is nothing he will refuse'. This alludes to the reputed voracity of the creature so named, which in this poetic idiom is compared to that of the crocodile; thus the paired expression is 'ngebu lau ma'u mona apa bau', 'crocodile down at the coast, there is nothing it does not want'. Insofar as 'ngebu' is a variant of 'ebu',

the usage suggests another context in which 'ebu' and 'gogo' are associated. Yet the only indication that the referent in this case is the wildman, rather than a similarly voracious bogey, is 'lia', 'cave', a term otherwise motivated by the spatial contrast it effects with 'ma'u' ('coast, seashore'). Moreover, while the parallel phrases denote a person with an excessive sexual appetite, Nage ideas concerning the wildmen of Lia Ula do not associate them with any such proclivity. The fact that the 'cave', like the coastal habitat of the crocodile, is specified as 'seaward, near the sea' ('lau') can be explained by a metaphorical association of this direction with the genital region (Forth 1998b: 280).

28 Nowadays, rather than 'gogo meo' or 'ebu gogo', Nage are more likely to threaten children with the arrival of 'witches', 'crazy people', 'cats', or 'mo gele', a term identified especially with white foreigners who are supposed to abduct children and steal their heads (Forth 1991).

29 The last appearance of an 'ebu gogo dancer' was in early 2005, when a man performed for a film crew from the American television network CBS which had come to 'Ua to investigate the legend of the wildmen as part of a documentary focused on *Homo floresiensis*. 'Ua people I questioned subsequently about this performance critically described it as a rushed affair in which an inexpert dancer was not properly attired, unlike the figure shown in Plate 2.5.

30 Like English colloquial usage, Nage expressions translating as 'kind of' do not consistently distinguish resemblance from class inclusion (Forth 1995).

31 I too may be accused of this in my treatment of the Sumbanese 'makatoba', or mili mongga (Forth 1981), reconsidered below in Chapter 4.

32 Numeral classifiers are not relevant to this distinction. There is no special classifier for spirits and Nage use the human or animal classifier according to whether spirit referents are manifested in one or the other form.

33 Since the incorporation of western New Guinea as part of Indonesia, Indonesians generally have been exposed to government propaganda representing Papuans as wild, uncultivated people, much in need of guidance and governance by the Indonesian state. In a similar vein, the Mula Koli man mentioned earlier wished to compare the language attributed to ebu gogo to the Tetum language of East Timor.

34 A period of two or three centuries falls within the period estimated by Van Gennep (1910) for the survival of oral traditions as accurate accounts of past events. Other anthropologists and folklorists, however, have maintained widely different opinions on this issue (Vansina 1965: 8–18) and it has long been recognized that the speed with which accounts of actual events are transformed in oral transmission can vary according to several factors, including subject matter, the identity of narrators and the extent to which reports are incorporated into established narrative genres.

3 Other Florenese hominoids

1 In fact, 'ana ula' is locally interpreted as meaning 'small child', but this is etymologically suspect. 'Ana' can indeed mean 'child', but it also has the sense of 'person, people', 'member (of a group)'. The meaning of 'ula' in this context remains unclear (but see Forth 1998a: 101–02). Another Poma name I recorded for ana ula is 'dho dhodho', a term on which I was unable to obtain further information.

2 Translating colour terms is always problematic. The Indonesian/Malay 'abu-abu' (meaning 'like ash') denotes shades classified in English as brown or beige, though a usual dictionary gloss is 'grey'. The Nage equivalent is 'foa', conceivably derived from 'fu 'o'a', 'monkey hair, fur'. Malay 'pirang', which two Poma men applied to ana ula body hair, usually refers to a reddish brown or auburn colour; it is not ruled out that this attribution reflects recent exposure to pictures of orang-utans.

3 Nage people tend to associate this tradition with Rawe (Forth 1998a: 103). This probably reflects the fact that Rawe was partly settled by Poma people in the early twentieth century and also the greater familiarity of Nage with Rawe folk.

4 According to information recorded in Nage (Forth 1998a: 103), ana ula were also given puppy dogs, for which they had a general liking (ibid.: 103). I never heard this from Poma people, who described the ana ula as fearful of dogs. Still, the two attributes are not necessarily contradictory. Equally wary of dogs, the ebu gogo are also described as consuming dogs.

5 As elsewhere in central Flores, 'toro' means 'red'. What its significance may be in this context I was unable to discover, although comparative evidence suggests the term may denote a variety of frightening images. In Ngadha, 'toro gogo' refers to a spider (Arndt 1961).

6 So'a naming coventions and the surname of the murdered girl (Bepe Ruma) suggest she may have been the daughter of Ruma Réwo. However, since she was beaten by a 'grandparent', the girl's subsequent death cannot definitely be linked with Ruma's killing of the toro gogo infant—for example, as an act of revenge.

7 Named Mala Pare, the site, which I visited in 2005, lies no more than 200 metres east of the present village of Libu Nio. In the days of the toro gogo, however, this area was enclosed in thick forest and the Libu Nio ancestors then inhabited Wolo Pora, a village about a kilometre to the south. Over time, I was told, the cavity once occupied by the toro gogo has become filled with earth and mud; hence it is now no more than a large depression. Some 25 metres to the west, thus in the direction of the present village, are several long and narrow vents, said to be very deep and connected by underground chambers to what was the main entrance to the cavity.

8 In the eastern Ngadha village of Wolo Rowa, the complete name 'ebu ngiu' apparently refers to a bird. More specifically, it denotes an unidentified species whose call presages death, but also signals the presence of wild pigs near cultivated fields (see Forth 2004a: 72, 210, note 10, regarding Nage and Sumbanese ideas concerning owls).

9 The main sources are a personal communication from Andrea Molnar (pers. comm., 1996) who conducted ethnographic research in Sara Sedu in the 1990s and several references in Molnar (2000). In 2005, I was able myself to discuss the ebu ngiu with Sara Sedu villagers.

10 At least when conceived as malevolent shape-shifters, kedho bear a general resemblance to Nage witches ('polo'). In both cases the supernatural abductor is said to demand that the victim lick his or her anus, in which regard there is also an arresting parallel with European notions of witch behaviour (see Sanders 1995: 151). As with victims of Nage witch abduction, to recover a victim of capture by a kedho, relatives must yell and beat on gongs while conducting the search; besides possibly scaring away the abductor, the percussion brings victims out of the semiconscious state in which, in both Nage and Sara Sedu, they are typically found.

11 'Ibu' is not clearly related to 'ebu', as in the Nage 'ebu gogo', nor is it to be confused with the Malay term for 'mother' ('ibu'). In his dictionary, Arndt (1961, s.v. 'cibu') glosses the term as 'small, treelike plant', 'showing few sprouts' and 'strike, beat; striking sound', but these senses are not obviously relevant to 'ibu ngiu'.

12 Although the primary sense of 'ré'e'—and the Nage variant "é'e'—is 'ugly', the word can also mean 'hateful, malevolent'. In the second sense, the referent need not be simultaneously physically ugly; however, as was once pointed out to me, these two qualities often coincide.

13 The feet were mentioned by a Manggarai man with reference to the north central Manggarai district of Lamba Léda, where he said the creatures were more usually called 'ngiung'. While these creatures also possess long humanlike

breasts, they can further assume the general form of a pig and thus move quadrapedally as well as bipedally. In all probability, these human and porcine traits refer to two separate guises which the beings are thought equally able to assume.

14 I was unable to discover a relevant meaning for 'wolo'. It is not, however, to be confused with Ngadha and Nage 'wolo', meaning 'hill, mountain'.

15 Referring to the penis, 'La'e', I was told, alludes to the ancestor's original nakedness. According to Van Bekkum (1946: 71), the entire name Paju La'e means 'penis' and relates to the ancestor's act of introducing rice to the Cibal district by secreting a stolen grain under his foreskin.

16 Being subject to ritual restrictions, I myself was never able to inspect the hair. I was however told that, with the appropriate ceremony, Empo Paju's descendants may touch it as well as view it.

17 Further attesting to his human character, Empo Paju is described as originally deriving, with other Manggarai ancestors, from Poso Mando Sawu, a high and thickly forested mountain in central Manggarai. Van Bekkum (1946: 71) describes Empo Paju as the descendant of Rendong Mata Leso, the apical ancestor of Cibal, who also resided on Poso Mando Sawu.

18 The Desu originally inhabited a village or district of the same name, where some of their descendants are still to be found, as well as a place called Lale, near hot springs at Ulu Mbu. They also once lived at Todo Koe. Although Todo informants mentioned a particular man, described as extremely hirsute and an 'original Desu' type, owing largely to a shortage of time I was never able myself to observe any such phenotype.

19 It is worth noting that the name of this population is almost certainly unrelated to 'Rua' (or 'Ua), the name of the human group, originally from Wolo Rua, credited with exterminating the ebu gogo. Not only does Nage 'rua' ("ua') have a different meaning, but the 'Ua folk have no tradition of deriving from Manggarai.

20 Another apparent exception to my general characterization of western Florenese images may be suggested by a story concerning a figure named 'Reba Ruek' and other reputedly hairy and dark-skinned hominoids living in Liang Bua, which found its way into the Indonesian media in late 2004. As explained elsewhere, however, the story—originally published on the internet by a Manggarai civil servant—is very likely a recent artefact, possibly based on traditional narratives (such as the legend of Empo Paju) but in any case more definitely inspired by the palaeoanthropological discovery, in Liang Bua, of the remains identified as *Homo floresiensis* (Forth 2006: 343).

21 Likewise, many Lio people are bilingual in Nita (or Sikka). The original Nita is an interior region, located about 8 kilometres from the north coastal town of Maumere. Apart from the eastern section of the modern Ende Regency, Lio people are found especially in western and southern parts of Sikka Regency, located immediately east of Ende Regency. For convenience, I use 'Lio' to refer to people familiar with the Lio category 'lae ho'a', even though some of these identify themselves ethnically as Nita.

22 It is no coincidence that the reporter, Richard Shears, spoke to some of the same informants I had interviewed in 2003; they were introduced to him by Mr. Endy Paji, a Nita man and the curatorial assistant at the museum of the Catholic Seminary in Ledalero, who assisted me in 2003 and again in 2005. In 2005, I spoke to all the individuals mentioned in Shears' article, which appeared in the *Daily Mail* on 6 November 2004. What they told me, however, was for a large part at variance with what appeared in the newspaper article.

23 Answering my question, the man said he later suffered no ill effect from this encounter. This, he suggested, was because he was accompanied by his dog as well as by a human companion. I was not subsequently able to contact the latter.

24 Originally, I thought this description might reflect familiarity with Schouten's painting of *Homo floresiensis*. However, the man claimed never to have seen any such illustration and circumstances suggest that he was truthful. Certainly at the time of my enquiries no one locally seemed to be in possession of this or any other illustration depicting an archaic hominin. During my visit, another villager produced a photograph of an orang-utan, but my informant said he did not think the ape shown in the picture particularly resembled what his father had observed.

25 Other people familiar with the Ma'u Lo'o man's story of the female corpse said they had never before heard him say that the thing lacked arms. In response to my question, the man himself was unable to say whether the creature appeared never to have had arms or whether these appeared to have been severed. Curiously, both in regard to the absent arms and the three toes, the creature, which he otherwise described as a hominoid, sounded rather birdlike. As I note below, the man further described vocalizations attributed to lae ho'a as resembling those of a raptorial bird.

26 Curiously, this ability has been attributed to the Nage ebu gogo by members of the *floresiensis* discovery team (Morwood and Van Oosterzee 2007: 155). I myself have never heard Nage mention it.

27 A Lio migrant to Nage, now long deceased, had told people how he had once observed monkeys burying a monkey carcass near the Nage village of Nage Mi.

28 Although Leko Sétan means 'Devils' Bay' and, like the adjoining seas, is considered a haunt of dangerous spirits, I was able to affirm that these spirits are something separate from the local lae ho'a.

29 Rites periodically performed by the owner of the objects, to invoke the protection of lae ho'a, require an offering of a raw egg and human menstrual blood, or alternatively the blood of an entirely black fowl. Performed outside his village in a spot said often to be frequented by lae ho'a, this is supposed to bring about the appearance of a number of the creatures, whom the informant in one context described as his 'relatives'.

30 I am grateful to Professor Owen Beattie, a forensic anthropologist and Mr. Wayne Roberts, Assistant Curator of Vertebrates, both of the University of Alberta, for inspecting the fragment and comparing it with samples of various animal skeletons. Much the same interpretation was obtained by Dr Raymond Corbey of Leiden University, who showed the fragment to a number of archaeological and palaeontological colleagues.

31 Only in 2005 did I identify the particular village—Nabe—from which most of the share-croppers, five or six in all, apparently derive; and owing partly to its remote location, I was unable to visit it. As my field assistant, Endy Paji, suggested, some of the people I did speak to in Feo Ndari may indeed have heard the story but may have been reluctant to say so, particularly to a westerner and possibly for fear I might report it to the authorities.

32 Suggesting a functional equivalence between the back hole and the considerable cleavage associated with long breasts, the ibu ngiu of western Ngadha similarly employ their breasts in capturing and confining abductees (Arndt 1930). From a structuralist perspective, by contrast, the two attributes may be seen as constituting an inversion, comprising a deficient back and an excessive front. A remarkable parallel to certain Florenese images is found in the Swedish 'skogsnufra', a female being with 'hanging breasts' which can also take the form of a 'quite desirable maiden' whose back is 'hollow like a dead tree-trunk' (Mazur 1980: 9).

33 Among some Malays, 'langsuir' and 'pontianak' are contrasted as deriving respectively from women who die in childbirth and from stillborn children. However, other Malays and also the Sundanese of western Java, ascribe the first derivation to pontianak, so the distinction is by no means absolute (Osman

1989: 86; Hidding 1935: 53; see also Echols and Shadily 1989, s.v. 'kuntianak', a variant of 'pontianak').

34 As noted in the previous chapter, these authors also specify 'Ebu Gogo', a female character who appears in a So'anese story, as a witch. Translated as 'worn-out rice-winnow', Ségе Re'e occurs as the name of a character in a Nage myth, identified with the moon. Nage, however, do not recognize categories of beings named 'ségе re'e', or 'fégе re'e'. Interestingly, on quite independent grounds, Arndt (1932: 151) interpreted the figure of the pontianak as an instance of moon symbolism.

35 In western Java, a fishnet is one of the things used to protect newborn infants against pontianak (Hidding 1935: 39).

36 Another feature shared by both sorts of Florenese images and which might be explained in the same way, is long fingernails, an attribute, for example, of both the lae ho'a and the Sikkanese and Lionese ségе re'e (or fégе re'e), as well as the Malay pontianak or langsuir (Skeat 1900: 326). According to a Nita man, ségе re'e appear only during cloudy weather, also an association of lae ho'a, ana ula and noa.

37 Among Nage, the story of the killing of the hominoid infant (by the 'Ua ancestor Huma Léli) is told separately from the extermination legend. Nevertheless, the changeling theme could have been original to this Nage tale, especially as it is otherwise unclear why the female ebu gogo would leave her own infant in a human dwelling.

38 Frequently characterized as dull-witted, giants or ogres, it may be noted, appear to be far more common victims of killing or injury in world folklore than are generally more intelligent dwarfs (Arne 1973: 360ff). Apparently even rarer are folktales in which wildmen, as I have defined this term, are killed (see ibid.; De Vries 1925, 1928; Thompson 1946, 1975). Hence the ebu gogo legend, for example, does not unequivocally exemplify universal folklore themes or motifs.

39 Although usually concerning a single animal, other stories of foolish monkeys deliberately killed by burning occur in Timor (De Vries 1925: 221), Sumba and elsewhere. In accordance with the status of monkeys as ambiguous 'tricksters', sometimes a monkey is the deceitful killer, while in some such cases, including tales from western Java and the Sangir Islands, the victim of burning is a giant (De Vries 1925: 19–22; 1928: 5–6, 106–10, 246–50). In a Sumbanese variant of the Bornean tale I recorded in 2005, a frog, in revenge for a particular monkey's thieving and deceit, tricks a group of monkeys into climbing a tree full of ripe mangoes. He first convinces them to stack piles of soft dry grass beneath the tree, so they will not be injured if they fall. When all are in the tree, the frog fires the grass, killing all except one monkey which manages to escape. The only Florenese variant of this tale I heard has the frog fatally deceiving a single monkey by causing it to jump on sharpened stakes or palm ribs. Nage described to me several actual methods of catching and killing pestilential monkeys, but none involved fire.

4 Other eastern islands

1 Other names include 'mila mungga' and 'mila katoba'. Possibly the source of 'mili', the word 'mila' means 'poor, impoverished'. 'Makatoba' is a common name in southeastern Sumba, where 'mili mongga' is commonly rendered as 'mini mongga' (cf. 'mini', 'man, male person'). Further west on Sumba, comparable images are labelled 'maramungga' in Mamboru and Lewa and 'muga' in Wanukaka (Kapita 1982: 161, s.v. mongga). In the westernmost district of Kodi, the creature is known as 'maghu rumba' (Janet Hoskins, pers. comm., 1996), a term partly cognate with eastern Sumbanese 'meu rumba', although 'maghu' is the word for 'shadow'. In view of the creature's somewhat simian appearance, an interesting association with

monkeys ('buti') is suggested by the standard eastern Sumbanese parallelism 'buti-meu rumba', even though in this context 'meu rumba' can alternatively be understood as a reference to the wild cat.

2 Information from Mahu, more particularly the vicinity named Meu Rumba, derives largely from enquiries made by Bernardus Retangu, a Sumbanese currently teaching English in Java, who has a special interest in eastern Sumbanese traditional narratives and who generously assisted me throughout my 2005 researches. Retangu himself remarked how Mahu people generally describe mili mongga as no taller than humans, whereas in his home village of Lambanapu, not far from Waingapu, people say they are much bigger. As this would suggest, the contrast may pertain less to educated townspeople versus less educated villagers, than to a distinction of coastal Sumbanese and people living in interior regions, close to forested areas reputedly inhabited by mili mongga.

3 As genitalia are not shown, the creature's sex is not otherwise evident. The statue, as well as a modern replica constructed on a newer grave, show a lateral bulge on the lower back, about the waist. What this might represent is not known. Apart from the replica, the Rindi statue appears to be one of a kind, as I have never found anything similar in other parts of Sumba.

4 An adjective applied to mili mongga speech is 'hayamingu', which also refers to the hesitating or imperfect speech of children. Another phrase denoting imperfect speech is 'pakareuku apu kamingu', 'to talk like Apu Kami (Grandmother Kami)', a character commonly occurring in narratives, who is sometimes identified as a mili mongga (Forth 1981: 112).

5 Whether this sound is onomatopoeically replicated in the term 'mongga' is uncertain. The comparison with a tiger's roar is evidently a modern idea connected with the fact that meu rumba ('wild cat'), one name for the creatures, nowadays refers not just to small wild (or feral) felines, but to big cats as well. The only known feline on Sumba is *Felis cattus*.

6 I am aware that, in my monograph on Rindi (Forth 1981), I described the mili mongga—or 'makatoba'—in effect as 'earth spirits'. This reflects an attempt—ill-considered and undiscerning as it would now appear—to classify them in terms of the conventional analytical categories of cultural anthropology. What is more, the evidence on which this was based (which was rather less than I now possess) consisted largely of references to the creatures in myth. To be sure, the myth in which a mili mongga is transformed into a normal human ascribes the creature's initial wild condition to its possession of 'wàndi', a spiritual property that departs when the creature's hair is burned off. Nevertheless, this wàndi is something otherwise possessed only by witches (Forth 1981: 112); hence its attribution in myth to a mili mongga does not compromise the empirical status of the latter any more than it does that of human beings whom Sumbanese conceive as witches.

7 According to Wielenga, humans can gain good fortune by obtaining women's skirts which 'meu rumba' (i.e., mili mongga) weave from their long head hair, but in return they must forfeit the souls of their parents (1910: 311; 1913: 264). This, however, is another idea that appears to find expression only in myth.

8 The similarity between this figure and the Sumbanese mili mongga is especially marked in a So'anese myth, where 'Sége re'e' steals the identity of the eldest of a pair of sisters by entering her house and donning her clothes (Mommersteeg and Dirkzwager 1999: 133–42). As noted, such 'identity theft' by female mili mongga also occurs in Sumbanese narratives.

9 'Kalita' denotes material from the boughs of the gewang palm, used in making snares of the sort in which she was caught.

10 I am most grateful to Dr Hitchcock, currently a professor at London Metropolitan University, for allowing me to summarize extracts from his field notes and information he provided by electronic correspondence in 2004 and 2005.

11 I again discussed this matter with Professor Hitchcock when I met him in London in October 2005. What he told me then did not differ from his previous accounts. Our correspondence in November 2004 occurred shortly after media announcements of the discovery of *Homo floresiensis*. But while he was certainly aware of these, I have no reason to believe that his account was affected by news of the discovery. The incident was, after all, recorded in his 1982 field notes.

12 I searched editions of *Kompas* dating to 10th October, thus past the end of September, when Herman was planning to travel to Taliwang. A search of *Jakarta Post* from 14th to 20th September suggested that the story was not carried in the English language press. A letter I wrote to Drs. Herman, a few months after the report appeared, never received a reply.

13 I am grateful to Hans Hägerdal, a Swedish historian and Indonesianist, for bringing this source to my attention and kindly translating relevant sections. Although Blomberg facetiously refers to the prospect of catching a 'small forest troll, hairy and with a tail', Dr Hägerdal interprets this as probably reflecting a Swedish representation of trolls as tailed creatures, rather than descriptions provided by Blomberg's Indonesian companions.

14 I make this claim following discussions with anthropologists and other scholars specializing in these regions. In this connection, I am especially grateful to Dr Andrew McWilliams of the Australian National University, an anthropologist who has conducted research in several parts of Timor and to Dr Aone van Engelenhoven, a linguist and ethnographer now working in Leiden University who hails from the southern Moluccan island of Leti. Although absence of evidence cannot be taken as evidence of absence, when experienced investigators such as these report unfamiliarity with categories like the ebu gogo, I take it as a strong indication that, even if such categories do exist elsewhere, they are not nearly so prevalent as on Flores.

15 This story is repeated in the newspaper *De Sumatra Post*, 15 June, 1932. Although the source is evidently Kopstein (1930), curiously this version refers to just one foreign fugitive. It is moreover reported in the context of the reputed killing, in 1932, of an infant 'orang pendek' in Sumatra (a specimen subsequently revealed to be a fake; see Chapter 5) and the item purports to show how orang pendek, or similar images, also occur on other Indonesian islands. Kopstein himself equates 'gibar bohot' with the 'orang pendek' or 'sedapa' of southern Sumatra (Chapter 5) but this, evidently, is his inference alone. There is no evidence that the Buru Regent, or Deninger for that matter, was aware of the Sumatran category.

16 According to anthropologist Barbara Dixon Grimes (pers. comm., December 2005), 'gebar bohot' (the plural of 'geba bohot') is still regularly heard on Buru. It refers primarily to people who, owing to some misdeed, supposedly flee into the jungle or to a distant settlement for fear of reprisal. Like wildmen elsewhere, the putative desperados are invoked on Buru as bogeys, to dissuade children from wandering too far from home. In accordance with its very general sense, the category can include legendary head-hunters called 'kuwihil', which apparently correspond to the head-stealers Nage designate as 'mo gele' (Forth 1991).

17 Coomans remarks how footprints locally attributed to the lolok resemble tracks associated with the hirsute 'orang pendek' of Sumatra. Since Dammerman (1924), probably correctly, identified the Sumatran prints as belonging to the Malayan sunbear, it is important to note that, whereas bears occur throughout Sumatra, there are none in Sulawesi.

18 In regard to the small size of the To Uta, it is noteworthy that Riedel estimated the average stature of human inhabitants of northwestern Sulawesi at just 1.5 metres.

5 The 'short man' (orang pendek) of Sumatra

1 'Orang pendek' was recorded by Raffles in 1817 (Volume 2, p. lxxvi), but simply as a Malay equivalent of English 'dwarf'.
2 Suggestive of Malay 'labuan' ('harbour'), which occurs in many place names, 'Labun' is unidentified, but was apparently located somewhere near Bencoolen. Following standard practice, I use 'Malay' to refer to the majority population of village cultivators and fishermen of Sumatra.
3 De Santy does not say what became of the body. Apparently referring to the same case, however, Van Herwaarden (1924: 104) describes how a villager of Pangkalan Balai in the Banjuasin region had come across the carcasses of a mother animal and newborn infant in the forest. He tried to bring these back to his village, but decomposition was already too far advanced and (implicitly because of his contact with the creatures) the man died the following day.
4 'Rookmaker 1932' refers to a report attributed to 'v.d. [van den] Maaten'. Since subsequent articles make it clear that the author was O.J. Rookmaker, a colonial official stationed in the Rokan district, I employ Rookmaker's name throughout.
5 In a curious account tangentially linked with a possible orang pendek sighting, Pasteur (1960) reports a 'howling sound', described as 'ulu-ulu-lululu', attributed to a large number of creatures that reputedly surrounded a surveyors' base camp in Pulau Rimau in 1937.
6 Partly with regard to contrasts of temperment, Westenenk (1932) implies that different regional names—such as 'sedapa', 'orang pendek' and 'seguguh' (described as referring respectively to 'shy', 'gruff' and 'surly' and 'ill-tempered' beings)—label different representations, if not different creatures. Generally, however, the evidence does not substantiate any such regional pattern.
7 With tedious regularity (see e.g. Dammerman 1929, 1932), statements have appeared to the effect that Marco Polo referred to orang pendek in his travel account. In fact, what the Venetian described was a Sumatran practice of doctoring monkeys to look like tiny humans, pickling them and then exporting the product for sale as preserved 'pygmies'. Although this ancient practice acquired a particular significance in the context of a hoax perpetrated in 1932 (which I describe presently), it is difficult to see what real bearing this has on the orang pendek—unless it is simply to suggest that if Sumatrans can be duplicitous in one context, they can also be in others.
8 De Rienzi claimed also to have seen small people, belonging to a type he calls 'Aithalo-Pygmées', somewhere near Lampung Bay on Sumatra's southernmost tip (1836: 24). I am grateful to Robert Cribb (pers. comm., November 2005) for bringing the Lampung sighting to my attention.
9 Certainly, the officers of the Batavian Society (Bataviasch Genootschap)—including the noted Islamicist Snouck Hurgronje—were not impressed with the sergeant major's story, dismissing it, or at least the 'orang ratas', as no more than a fairy tale (Notulen 1899: 55–56).
10 Oostingh's observation occurred just after the appearance of the first published twentieth century reference to orang pendek. This was an article published in May 1917 by the naturalist Edward Jacobson, who reported on evidence also found in the vicinity of Mount Kabah. Nevertheless, Oostingh claimed never to have heard of the orang pendek before his reputed experience.
11 Herwaarden more exactly states that the head 'ends in more of a point' than in human beings. This, however, cannot refer to the lower part of the head, as he later refers to a receding chin. Heuvelmans, by contrast, has misinterpreted this feature as a pointed chin (1995: 133) and the mistake has further been perpetuated in popular books by Allen (1989) and Coleman and Huyghe (1999: 146–47).

12 Here, Herwaarden was evidently responding to earlier reports of short, wide footprints (Jacobson 1917, Westenenk 1918). He also remarks that the feet did not have the heels pointing forward, an obvious reference to the local notion that orang pendek have inverted feet, reported by Westenenk (1921) and Maier (1923).

13 Van Gulik (1967) describes gibbons as odourless. Nor is it clear whether body smells attributed to orang-utans would be noticeable to many Europeans (Herman Rijksen, pers. comm., February 2006; Ian Singleton, pers. comm., February 2006). Gorillas, by contrast, have a much stronger natural odour (Schaller 1963: 90–91).

14 Before encountering his sedapa on Pulau Rimau, Herwaarden had heard stories about the creature and had made enquiries among natives, both Malays and Kubu forest-dwellers. In this regard also, his account may be considered somewhat less credible than Oostingh's, as prior to his encounter, Oostingh was apparently unfamiliar with orang pendek and related his story only in response to Westenenk's question as to whether he had ever seen anything 'remarkable' in the jungles (Westenenk 1918). In his 1932 account, Westenenk further relates how Oostingh was reluctant to tell him of his experience earlier, as it had seemed so foolish. If either fabrication or a crucial influence of native representations is a possibility in Herwaarden's case, both are therefore much less likely in Oostingh's.

15 I am here reminded of a stereotype deriving from Indonesian government propaganda, often heard among eastern Indonesians, of Papuans of Irian Jaya as sleeping in trees like monkeys.

16 If I understand her argument correctly, the historian Frances Gouda (1995: 123–24, 138–42) claims that an ideology of racism caused Dutch colonials to perceive Indonesians and Sumatran apes as basically similar and that this accounts for observations like Herwaarden's of a subject that was in fact either fully simian or fully human. Support for this might be found in descriptions, such as that by Winter (1901: 219), of the Kubu as 'ugly representatives of the race "homo sapiens"'. Regardless of Dutch attitudes towards natives, however, I find it difficult to conceive how anyone could actually confuse an ape and a human, at least at a distance of a few metres. Winter, after all, does at least count Kubu as members of the human species.

17 According to Dammerman, the monkey was a member of the sub-family Semnopithecinae, genus *Semnopithecus*, but the species was impossible to identify as the specimen was too young (1932: 131). The genus in question now appears to be named *Presbytis*. Dance (1976: 66), however, says it was a 'lotong', *Trachypithecus pyrrhus*.

18 Apparently, the last article before the Second World War was a short piece by Van den Pijl (1938), mostly comparing orang pendek to the 'abominable snowman'. Unidentified newspaper articles contained in the Dubois archive are dated 1933 and 1935. I am grateful to Dr John de Vos of the Natural History Museum in Leiden for making this archive available to me. One of very few writings in English, the 1960 article by Pasteur (an employee of an oil company in Palembang) stands alone.

19 Although not a trained zoologist or anthropologist, in the course of her investigations Martyr has evidently gained sufficient knowledge of primate biology and behaviour to couch her findings in the appropriate language. Before beginning research, Martyr, who does not read Dutch, was not familiar with the colonial literature on orang pendek (Martyr, pers. comm., August 2004). She cites just one major secondary source (Heuvelmans 1958) in her 1990 article on the topic and was otherwise aware only of a journalistic account of the orang pendek by Benedict Allen (1989). Allen also cites some of the old Dutch sources, albeit rather inaccurately.

20 As most of the reports are unpaginated, page numbers are not always specified.

21 Martyr's article of 1990, her only substantial publication on the topic of orang pendek, is based entirely on interviews. Between 1993 and 1998, she conducted some 150 further interviews with informants who claimed to have seen orang pendek. Martyr herself says she observed orang pendek on three separate occasions, while Achmad Yanuar and Jeremy Holden claim one sighting each. A local Sumatran field assistant named Buya claimed to have seen orang pendek on no fewer than seven occasions. Also mentioned are another five 'inconclusive sightings' by the team. Never, it seems, were the creatures observed by more than one person simultaneously. Also, most sightings took place in 1994 or 1995 rather than during later field trips. In 1995, Holden set up infrared camera traps, but these failed to capture the image of an orang pendek. At the same time, the technique also failed to produce photographs of gibbons or orang-utans, even though the former are known definitely to be present in the area concerned (Holden 1997).

22 The latter assessment draws on conversations I had in October and November 2005 with European physical anthropologists, palaeontologists and primatologists, with whom I also subsequently corresponded by electronic mail. Casts of the prints are now held, unaccessioned, by the Natural History Museum in London. I am grateful to the museum's staff and particularly Dr Robert Kruszynski and Dr Chris Stringer, for allowing me to view the casts and for offering their views.

23 Language differences may also play a part in the representation. The word Europeans have translated as 'inverted' or 'backwards' is most likely Malay 'terbalik', a term generally translatable as 'turned the other way', but not always specifying a 180-degree inversion. A 'pigeon-toed' human might be described as having feet that are 'terbalik', as might other, equally empirical animals. It is in this sense that Nage describe the feet of the Water monitor (*Varanus salvator*) as 'terbalik' ('inverted').

24 The report that local people interpret nocturnal sounds resembling gong and drum orchestras (Malay 'gamelan') as a sign that orang pendek are in the vicinity (Westenenk 1918: 108) possibly reflects confusion with 'invisible beings', or 'brownies', called 'orang bunian'. These beings, according to East and Central Sumatrans, inhabit caves containing stalactites that produce musical tones (Moszkowski 1909: 996).

25 Evidently a perfunctory effort, the Pulau Rimau hunt was a largely nationalistic response by the Dutch to a proposal by the Royal Society of London to launch an expedition in search of the orang pendek (see Dammerman 1924: 182).

26 Regarding the blood and hair samples, Hill (1945: 254) suggests that 'these should have been sufficient for a decisive report, but seem to have been grossly mismanaged'. Referring especially to Dammerman's conclusion of 'faintly' human origins and the lack of a microscopic examination of the evidence, Hill further opines that 'the absence of a proper report seems flimsy evidence for denying the existence of the creature [the orang pendek] and submitting alternatively that the material was the result of an accident to the person who set the trap'.

27 Methods employed in analysing the sample are detailed in Brunner and Coman (1974).

28 In the Dutch newspaper *Het Vrije Volk* (14 March 1958), a retired Indonesian army captain is quoted to the effect that the Dutch had at one time offered a reward for the capture of a creature like this 'sindai', a clear reference to the orang pendek and bounties offered for a specimen in the 1920s (Anon. 1926c).

29 Although 'Kubu' has, broadly speaking, been applied to forest folk living as far north as the Rokan (LeBar 1972: I: 46–47), in the strictest sense the name specifies a population known historically only from Jambi southwards. Nevertheless, the Rokan district is home to similar non-Malay food-collectors and part cultivators known as 'Sakai' or 'Batin' (Loeb 1935: 290–93). Especially insofar as the Sakai

have been described as a 'Veddoid' type similar to some aboriginals of the Malay peninsula (Loeb ibid.), they would seem to be somewhat physically distinct from neighbouring Malays—or to have been so in the past. Thus, if Kubu have contributed to the image of the orang pendek elsewhere, it is equally possible that these people have played a similar part in representations of the letjo.

30 The Sumatran monkey with the shortest tail is the Pig-tailed macaque, *Macaca nemestrina*. But with a creamy brown pelage, a body length of just half a metre and a tail 18 centimetres long, this too is unlikely to be the source of the orang pendek.

31 In the same vein, Helfrich glosses 'goegoe' (= gugu) and 'gugo'ah' (= goego'ah), from the southern dialects of Besemah and Serawas respectively, with Malay terms for the orang-utan ('majas', Helfrich 1904; 'mawas' Helfrich 1927b). But this probably reflects no more than the supposition informing Schlegel and Müller's interpretation of 'orang panda' as a southwestern Sumatran name for the ape.

32 Contrary to what is sometimes suggested, Schlegel and Müller never observed orang-utans in southern Sumatra. Nor is it clear that either author ever visited the island. Their description is largely based on Müller's observation of Bornean orang-utans.

33 One source the authors seem not to cite is the colonial officer Luitjes (1925: 57), who thought he saw 'a pair of orang-utans fleeing in fairly low cover', in a forest on the east coast of Sumatra, near Lebuan Bilik.

34 The authors do not specify whether they refer to an animal belonging to the same genus as orang-utans (*Pongo*) or gibbons (*Hylobates*). A new genus, however, is suggested by the statement that, if related to the orang-utan, the species 'is likely to have evolved separately from the ancestral population [of *Pongo*] distributed all over South East Asia' (Martyr *et al.* n.d.: 11). Although yet to be fully documented scientifically (since an actual specimen has yet to be produced), the authors, who include Cambridge primatologist David Chivers, are nonetheless firm in their conviction that 'orang pendek really do exist' (ibid.).

6 Wildmen of western Indonesia and Mainland Southeast Asia

1 For example, in some versions of particular Toba stories, a homang can take the place of a monkey or ape (Voorhoeve 1927: 89).

2 The newspaper report goes on to state that 'four or six years ago [these creatures] were common near the river Batang Gadis, where they lived in a large forest named Siulang Aling. In recent times, orang pendek have seldom been encountered there. It is believed that they have moved to the boundary of the Rokan district.'

3 A possible cognate, 'manto', appears in Dairi Pakpak and is glossed as 'a kind of forest spirit' (Manik 1977: 195). Cognates seem not to occur in other north Sumatran languages, including Toba, Karo, Angkola, Mandailing and Simalungun.

4 More recently, MacKinnon has reiterated this view, suggesting a bear walking in such a way that the longer hind feet cover the broader fore feet (pers. comm., August 2004).

5 I am grateful to Paul Porodong, a doctoral student at the University of Kent, for making enquiries in his native Sabah concerning uses of 'batutut'. A search of published dictionaries of Sabah languages has failed to reveal words resembling 'batutut' that possess any relevant sense.

6 I am grateful to Clifford Sather (pers. comm., January 2007) for information on applications of Iban 'rambai'. Becarri says no more about the mayas rambei, other than to discount it as a third variety of Bornean orang-utan additional to another two that he does recognize as distinct forms (but not as separate species). Whatever its referent, the category is reminiscent of 'beruang rambai', described by the Earl of Cranbrook (cited in McNeely and Wachtel 1988: 259) as denoting a 'large unidentified primate', even though 'beruang' is the Malay term for

'bear'. Sather also mentions an Iban belief in spirits named 'antu rambai', credited with long, flowing hair on their arms and backs like an orang-utan.

7 Known in Malay as 'orang asli', or 'original people', ancestors of the aborigines, who include both 'negrito' and 'non-negrito' populations, inhabited the Malay Peninsula long before the arrival of the Malays. They speak languages classified as Austro-Asiatic rather than Austronesian (or Malayo-Polynesian), the family to which Malay and most Indonesian languages belong. The aboriginal languages have nevertheless adopted many words from Malay.

8 Howell does however refer to large malevolent beings with hair-covered bodies (1984: 104–5, 107–12, 274–75). These seem comparable to reports, some recently appearing in the Malaysian press, of hairy creatures leaving very large footprints and therefore dubbed with the American epithet 'bigfoot'. Among the peninsular Batek Negrito, 'Mawas' has apparently become completely spiritualized, being known only as the proper name of a thunder god (Endicott 1979: 163–65).

9 Another category of hairy manlike beings known to Malays are the 'bilian' (Anderson 1824: xxxii). These are described as anthropophagous creatures possessing extraordinarily long nails, but no mention is made of hairiness or a sharp forearm. Recalling the Sumatran idea that creatures called 'si bigau' (possibly a variant representation of the orang pendek) herd wild pigs, the bilian—a name curiously reminiscent of 'balian', a widespread Malay and western Indonesian term for 'shaman'—are said to 'tend' tigers, something which Sumatran shamans are also supposed to do. The notion further recalls the ownership of wild dogs by Labang's 'uyan' and indeed the widespread Southeast Asian representation of wild animals as the domestic animals of spirits. Bilian are further supposed to visit Malay habitations to demand fire, particularly on rainy nights.

10 In Malaya at the time was broadcaster Stewart Wavell (1958), who investigated the incidents and supplied additional details, but none that alter the overall story. Although Wavell provides no resolution of the mystery, he suggests that, once the story had broken, at least one encounter with a European—himself—facilitated its spread. Arguing against hoaxing Europeans as the instigators, Wavell points to the unlikelihood of a European female exposing her breasts. Yet none of the news reports explicitly described the 'woman' as bare-breasted; this is merely implied by the specification of a loin cloth as her only clothing. If the incidents were the work of pranksters, moreover, the 'woman' could have been a man suitably disguised.

11 Burchett employs the French verb 'gazouiller', which, in reference to birds, can mean 'to cheep' or 'warble'. Loofs-Wissowa (2001) has translated this term, perhaps tendentiously if not entirely inaccurately, as 'twittering', a word used to describe vocalizations of a Sri Lankan creature named 'nittaewo' (see Chapter 7).

12 Hypertrichosis, or more specifically congenital hypertrichosis lanuginosa, is a very rare disorder in which long hair grows on virtually every part of the body, including the upper cheeks, nose and ears. Other than defects of the teeth, hypertrichosis is not typically accompanied by any physical or mental disability that might suggest 'wildness'. There can be little doubt that Frederickson's illustration depicts a real subject. But when it was made is not specified; nor is the reference of the 'society at Calcutta' at all obvious. It is also unclear whether, in a comparative context, anything is to be made of the fact that, in Frederickson's picture, the 'wild man's' right foot appears to be turned the wrong way about!

13 A possible candidate is *Macaca arctoides*, the Stump-tailed macaque, found in Burma as well as Thailand, Malaysia, China and India. The male height, however, does not usually exceed 65 centimetres and the pelage is dark brown rather than red. Also the tail, although very short (0.32 to 6.9 centimetres), should nevertheless be detectable, especially in a dead specimen such as Bingham examined.

14 Potentially complicating the picture is a description, apparently drawn from rural Sundanese, of the 'aul', or 'ahool', as a monkey-faced, child-sized creature

with inverted feet which, however, also possesses wings, flies and feeds on large fish (Bartels n.d.). This the cryptozoologist Ivan Sanderson (1972) takes to be an extremely large bat. The representation of the aul as a flying creature, however, would appear to be a minority view.

15 This is not to say that the Javanese wewe, for example, is completely identical to the pontianak. Indeed, the Javanese instance, designated as 'poenthianak' or 'koenthianak' (Van Hien 1994: 129–30), is not only separately named but, rather than an old hag, appears as a beautiful woman who, instead of genitalia, possesses a cavity running through her whole body. This last feature the figure of course shares with the Nage 'logo lia' ('back hole') and comparable Florenese figures.

16 A figure reminiscent of both eastern Indonesian wildmen and malevolent female abductors occurs in a Balinese folktale recorded by Weck (1948: 31). Covered entirely in long hair and carrying on the hip a child that constantly cries from hunger but is occasionally fed with the flesh of human abductees, the evidently female monster is given the name 'dedengkeh'. However, not only does this creature lack the pendulous breasts of figures like Rangda, but the image appears specific to this particular tale, where indeed it is represented as the deliberate fabrication of a lecherous man deviously attempting to frighten a widow whose favours he seeks. Evidently another fabrication is the name itself. Although conceivably related to Balinese 'dengkek', 'seize, catch', 'be short in the body' (Barber 1979: 106), the monster is large-bodied rather than short. Interestingly, to realize his plan the deceiver disguises himself as a 'dedengkeh', a deception that recalls acts of fakery associated with the North American 'sasquatch' or 'bigfoot' and perhaps also the Trolak ape-men.

17 On the other hand, the rare endemic Mentawai gibbon (*Hylobates klossi*) is conceivably the source of the 'malevolent forest spirits' which the Sukkadei of the Mentawai Islands call 'silakkokoina' and describe as possessing long nails and shaggy hair. Among animals, the Mentawai gibbon is believed to possess an 'aura of superpersonal power' and is therefore 'sinister' and 'taboo' (Schefold 1988: 74, 327, 544).

18 Ridley (*Natural Science* vi, p. 23; cited in Beccari 1904) appears to infer the survival of *Pongo pygmaeus* in the peninsula entirely from the Malay name he transcribes as 'mowas'.

19 Similarly inconclusive is the occurrence of special terms for male and female orang-utans—'budak kelubi' and 'kudai'—among the Malays of Patani (southern Thailand; Wilkinson 1932: I: 158; II: 117), since these could conceivably reflect no more than historical links between Patani and Aceh, about 500 kilometres to the southwest. Distinct terms for male and female apes also occur in Bornean and Sumatran languages. Malay 'budak' otherwise denotes a young man, while 'kelubi' is the name of a species of palm (*Zalacca conferta*) related to the 'salak' (*Z. edulis*), which grows in marshy areas and on which orang-utans like to feed (Wilkinson 1932: I: 546).

20 I am indebted to Alexander Adelaar (pers. comm., 2006) for his insights into this matter.

21 That 'mawas' and cognates originally had a more inclusive primate reference might be inferred from 'mawa', used for a gibbon by Kedah and Penang Malays. Evidently a borrowing form Malay, 'mawak' (pronounced 'ma-wa') also refers to gibbons in Malaysian aboriginal languages (Means and Means 1987: 64). Like the wholly or partly synonymous 'wawa' or 'wa'wa" (Wilkinson 1932: II: 117), the application of these terms to gibbons may reflect onomatopoeia. On the other hand, the Penang Malay use of 'mawa' for a leaf-monkey (Winstedt 1964: 122) cannot be explained in this way.

22 With reference to apes, this sort of idea seems extremely widespread and has been reported among Africans in respect of chimpanzees (Reynolds 1967: 237).

Although Reynolds cites Richard Jobson in this connection, Jobson (1623: 185–86) actually refers more generally to primates ('monkeys' and 'babownes'). He also attributes the 'opinion' that primates are 'a race and kind of people', and that they refuse to speak to avoid work and 'subjection', not to Africans but to Spaniards.

23 Rijksen and Meijaard (1999: 422) state that Bontius's 'ourang outang', or a particular female specimen he observed closely, had come from 'the county of Sukadana in (west) Borneo'. But this seems incorrect and possibly reflects misunderstanding of his further report of having seen 'mountain-people with tails' (appendages Bontius does not attribute to the Javanese creatures) in the palace of the king of 'Succodana'.

7 Other Asian hominoids

1 Physical difference between Vedda and Sinhalese nowadays seems equivocal. Writing in the mid-nineteenth century, however, Bailey described the Vedda as shorter than Sinhalese and 'rather darker'. Vedda height he estimates as 1.37 to 1.55 metres for males and 1.32 to 1.42 metres for females (1863: 282, 283).

2 One of Lewis's informants, a man aged about 70 in 1914, had a grandfather living at the time, while another, probably younger man spoke of the extinction occurring 'four generations ago'. Yet another informant of Lewis's, a man of 48, described his grandfather as having actively participated in the extermination. This, then, suggests a rather more recent date, perhaps around 1825, or just 60 years before Nevill wrote. Reviewing the same literature, Hill (1945: 252) calculates the time of the event as 'not more than five generations ago, i.e., about 1795'.

3 While large game was plentiful in Lenama in the early nineteenth century, at the time Nevill wrote the tortoise—a food reputedly favoured by nittaewo—was nearing extinction in this region and 'may have been reduced to its present rarity by the nittaewo using it for food' (1886a: 68). According to Kadirgamar (1963: 282), the Lenama district was 'completely free of human habitation' when both he and Nevill wrote. This tends to suggest that the region was for a long time a wild place, considered unsuitable for cultivation but an environment suited to both nittaewo and human hunters.

4 Valentijn probably draws on Knox (1983[1681]: 26), who describes monkeys with head hair parted in the middle like men. The reference in both cases is probably to the previously-mentioned Purple-faced langur.

5 The only general work on the Vedda that mentions the nittaewo is Wijesekera (1964).

6 Sanderson (1961: 266) interprets this mention of 'moss' as probably a reference to vitamin-rich lichens.

7 This story may draw some inspiration from the history of an eastern Nepal war several centuries ago. Following the death of her husband, a Mangar (Nepali) princess, with apparent magnanimity, invited Sherpa warriors who had killed her husband to the prince's funeral. Desiring revenge, however, the princess 'mixed a deadly poison with the beer', thus killing 'some thousand Sherpas' (Nebesky-Wojkowitz 1956: 152–53).

8 As the source of this tale, both Sanderson (1961: 261) and Kirtley (1963: 318–20) cite a letter in an Indian newspaper by Prince Peter of Greece and Denmark. Sanderson appears to specify the date as 1939. Kirtley locates the capture in Sikkim, but also mentions two other very similar stories that place the alleged event elsewhere in the Himalayas.

9 *Gigantopithecus* was a huge ape known mostly from China, which apparently became extinct in the Middle Pleistocene. According to Ciochon *et al.* (1990:xv), *Homo erectus* in Mainland Asia would have been familiar with *Gigantopithecus*. This

encounter, the authors further suggest, could have given rise to legends of giant ape-like creatures or, as they express it, the 'Bigfoot myth'.

10 Sightings and other evidence have been reported from the north central and western provinces of Shanxi, Sichuan and Hubei and the southern and eastern provinces of Zhejiang, Fujian and Anhuei. Chinese researchers have also investigated reports from the autonomous regions of Tibet and Xinjiang. Referring to primatological field research they conducted in Sichuan, Greenwell and Poirier emphasize the ruggedness of regions in which wildmen are reputedly found, noting the serious impediment this poses for 'deep excursions' (1989: 50).

11 While the tone of Poirier *et al.* (1983) is generally sceptical with regard to wildmen reflecting unknown primates, in a later article Poirier wrote with Greenwell partly on the basis of primatological fieldwork conducted in Sichuan in the 1980s, the authors conclude that the probability of the representations reflecting some kind of primate or primates 'inconsistent with the known fauna of China' exceeds 50 per cent. Prior to this, Poirier had put the figure at just 5 per cent and Greenwell at 30 per cent (1989: 55).

12 Noting the relationship between size and climate, Von Koenigswald further suggested that orang-utans in tropical Sumatra and Borneo could have undergone reduction in size due to isolation on these islands.

13 Bacon specifies the homeland of these creatures as 'the land in which the emperor lives' and as 'not twenty days distant' from Cathay. Kirtley (1963: 320) suggests this might be Tibet.

14 Tchernine's book largely comprises an English summary of twentieth century investigations by Boris Porschnev and other Russian scientists into wildmen reported from various parts of the former Soviet Union. Some of this material is also discussed by Sanderson (1961). A reputed sighting of an almas by a Mongolian scholar is critically reviewed by Heaney (1983).

15 Tchernine does not mention a body odour. The story of Zana is attributed to Professor A.A. Mashkovtsev, a colleague of Boris Porshnev's, who heard it while collecting reports of Caucasian wildmen in 1962 (Tchernine 1971: 155).

8 Outside Asia

1 'Euro-American' denotes residents of North America descended from Europeans and therefore includes 'Euro-Canadians'. To be sure, people reporting encounters with sasquatch in recent decades have included people of Amerindian descent. But this does not contradict the fundamentally Euro-American character of the category as it has evolved over the last 100 years.

2 Of eleven reports of reputed nineteenth-century sightings (Green 1973: 5–8), as many as five appear to relate to creatures smaller than humans; another two were of human size, whereas just four were apparently larger. Instances of small size, where height is specified with reasonable clarity, include 1.5 metres, 1.2 metres, 'like a child of 7 or 8' and 1.4 metres (the specimen named 'Jacko', which I discuss below). In another instance, the subject is simply described as 'squatty'. With regard to creatures apparently larger than humans, in only one case is a height specified (over two metres). These figures should be compared to those for 1900 to 1925, where of 18 cases just two (reported from 1907 and 1909) concern creatures smaller than humans, while 13 are larger. Another three reports do not mention size (Green ibid.: 1973: 9–13). The impression that sasquatch have grown in size over time and especially during the twentieth century, is also borne out by putative eyewitness accounts reviewed chronologically by Bord and Bord (1982).

3 Possibly the earliest reference to a hairy hominoid recognized by northwestern American natives is found in a 1793 report by the Spanish naturalist José Mariano Mozino (Mozino 1970). Mozino describes a figure the Nootka Indians of Vancouver

Island call 'matlox', a creature covered in 'stiff black bristles', with a head similar to a human's but with 'fangs' larger, sharper and stronger than a bear and with very long arms and long 'claws' on all four limbs. Although physically powerful and much feared by traditional Nootka, the creature's size is not mentioned.

4 After being discovered by the railway tracks, Jacko recovered and scrambled up a bluff, with the railwaymen in pursuit. His capture was accomplished by dropping a rock on him, which rendered him senseless; he was then bound, placed on the train and taken to Yale. Possibly, he was already wounded when found lying beside the railway tracks. Either way, these details would seem to explain how it was possible for a group of men to overcome such a powerful creature.

5 A man named William Roe, who claimed to have observed a sasquatch in 1955 near Tete Jaune Cache, just west of the Rocky Mountains, similarly defines 'sasquatch' as referring to 'giant hairy Indians' (Green 1978: 55). The notion may not be exclusive to Euro-Americans. In addition to and evidently separate from the boqs, the Bella Coola speak of a group of Indians who, following an enemy attack, fled to the interior of King Island, where they 'grew hairy and lost their human characteristics' (McIlwraith 1992: 63). Apparently, then, like Europeans the Bella Coola entertain a belief in, as it were, feral humans, for whom profuse body hair is *acquired* from conversion to a wild or uncultured state.

6 The idea that Jacko was sold to Barnum seems to be based on an incorrect identification with 'Jo-Jo the Dog-Faced Boy'. Exhibited in 1884, Jo-Jo was actually a youth afflicted with Hypertrichosis lanuginosa who was born in St. Petersburg in 1868. Krantz speculates that Jacko was Barnum's original Jo-Jo but in this role did not live long and by 1885 was replaced by 'a hairy faced young man of very ordinary-looking strength and body build' (1992: 204). This, however, is based only on an unattributed and unsubstantiated report of an article appearing in 1884 in an unnamed newspaper in Duluth, Minnesota, which is supposed to have described a 'strange animal arriving from the west'.

7 According to a popular magazine article from the 1950s, Alexander Caulfield Anderson, a sometime official of the Hudson's Bay Company, 'mentioned wild men hurling rocks down the mountains in 1864' when he was travelling in the Fraser Canyon in the vicinity of Yale (Franklin 1959: 4). The source of this report, however, is nowhere made clear and I have been unable to locate any reference to 'wild men', resembling Jacko or otherwise, in any of Anderson's several writings on the history of British Columbia. Nor does the alleged incident receive any mention in John Green's numerous published records of sasquatch sightings (see Green 1970, 1973 and 1978.). There is, however, one detail hinting at a connection between the Jacko story and what Anderson is supposed to have reported in 1864: whereas Anderson's wild men hurled rocks, Jacko himself was reputedly struck by a rock, hurled down a mountainside by a railwayman involved in his capture. Shackley (1983: 36) states that a 'consistent stream of other sightings' were reported by the *Daily Colonist* subsequent to Jacko's capture, but these too are not registered by Green.

8 The 'smelly cave' recalls Gibbs's description of foul-smelling hominoids recognized by the Chehalis. Indeed, an unpleasant odour has sometimes been attributed to the modern sasquatch (Bord and Bord 1982). According to Green (1978: 444), however, a large majority of putative encounters do not include any mention of smell and he suggests that, where they do, this is extraneous and not a natural body odour.

9 As a possible source of sasquatch lore, Franklin (1959: 33) refers to unspecified 'reports of Spanish explorers with caged apes on their ships getting drunk in British Columbia waters and leaving the cages unlocked'. However, regardless of what these 'reports' might be, it is virtually impossible that Jacko, who appeared nearly a century after Spaniards were visiting western Canada, was descended from any such escaped 'apes'.

10 Owing to lack of space, South and Central American wildman images cannot be accommodated in the present work. Suffice it to say that images of hair-covered hominoids, some relatively small, are also reported from these regions (see Sanderson 1965, Heuvelmans 1995). Certain figures, it has been suggested, may reflect the influence of the European wildman through the medium of Spanish or Portuguese colonialism (Robe 1972). However, other possibilities are raised by the natural presence in Central and South America of New World monkeys, some of which—like the bald-headed 'uakari'—are tailless or short-tailed.

11 Joyner (1977) cites 29 reports from the nineteenth and early twentieth centuries, many of them involving claimed sightings. His 1980 addendum comprises a further four. The material has been analysed by Brisbane writer Malcolm Smith (1989) and primatologist Colin Groves (1988), who calculates that Joyner's 33 reports comprise no more than 13 independent cases. I am grateful to Mr. Joyner for kindly sending me copies of his articles.

12 Reference to a small Australian hairy hominoid can be found in an internet article by Gary Opit ('Yowies—fact or fiction? 2000; http://www.n2.net/prey/bigfoot/articles/opit.html). Known as 'Brown Jacks' (a name reminiscent of 'Jacko') and identified as the referent of names in several Aboriginal languages, this is described as an apelike creature with dark hair, standing 1.25 metres and sometimes moving on all fours. Curiously, none of the published authors mentions this 'Brown Jacks' and Graham Joyner, the leading authority on the yahoo, informs me he has never heard of it (pers. comm., March 2005).

13 I am grateful to Martin Walsh of Cambridge University for alerting me to Hichens' 1928 article. As is general in Bantu languages, the initial /a/ in 'agogwe' would appear to be the general noun class prefix for humans, or humanlike beings. But since this is also applied to animals in some Bantu languages, it is not certain that the word denotes something locally classified as human, or even hominoid (Walsh, pers. comm., September 2006). On the other hand, the initial /n/ in 'ngogwe' is the usual Bantu prefix for nouns denoting animals; and in this connection, Walsh has suggested that, if 'agogwe' is not a misprint, Hichens may have substituted this for 'ngogwe', perhaps by way of correction, in his 1937 piece. Referring to the Tanzanian Ihanzu people, Todd Sanders of the University of Toronto mentions as a possible comparison the term 'ahing'wi', denoting aboriginal but now invisible 'bush-dwelling creatures' that Ihanzu describe as having bodies divided laterally between a human half and a wooden half consisting of a log. The figure thus suggests the widespread image of the 'half-man' (see Needham 1980) and the partly vegetal 'green man', often considered a variant of the European wildman. Sanders further remarks that, as the Ihanzu word for 'log' is 'igogo', this image could conceivably be the source of Hichens' 'agogwe'. Another peculiarity of ahign'wi is their ability to produce porridge, milk, or meat from rocks, which they then leave in the bush for humans they favour (Sanders, pers. comm., September 2006)—an idea that suggests an inversion of the agogwe practice of receiving beer from people for whom they perform agricultural labours.

14 Although Burgoyne entitles his article 'little furry men', this simply replicates Hichens' usage and Burgoyne does not actually describe the figures he saw as hairy.

15 As the English name should suggest, other colonials likened the animal to a bear (no species of which is found in Africa). Heuvelmans reviews evidence suggesting that the Nandi Bear may indeed be a member of the *Cynocephala* (baboons and allies) of an unknown species (1995: 473–75); but in the end he favours an old and rather large ratel, or 'honey-badger' (family Mustelidae), as the main source of the representation.

16 Reported only by Cordier (1963: 179) and most likely recorded in a less than accurate transcription, other names for the kakundakari include: Kumu

'amajúngi', Tembo 'ambátcha', Topeke (=Beeke?) 'lisisingo' and Lega 'niaka ambuguza'. Other names for the kikomba include: Kumu 'apamándi' or 'apamaánji', Tembo 'tshingómbe' and Lega 'zaluzúgu' or 'zalazúgu'. (The same names are given but in some instances with a slightly different transcription in Cordier 1973.) I am most grateful to Dr Robert Botne of Indiana University, a specialist in Congo languages and especially Lega, for commenting on these various African terms.

17 Although the derivation of 'kikomba' (from Konzo, a Bantu language) is unknown, the reference to wrestling might suggest a connection with the proto-Bantu roots '*-kombat' and '*-kumbat', meaning 'to embrace' and 'hold in the arms' (Botne 1994: 89, 90). In Swahili (a non-Bantu language), 'kikomba' means 'ravenous hunger' and is also the name of a small primate, a galago or 'bushbaby' (Johnson 1939).

18 A more circumstantial yet equally striking parallel, more particularly to Van Herwaarden's story of his encounter with an orang pendek, is evident from the account of a European named Pessina, recorded by Cordier (1973: 188) in 1963. While out hunting one afternoon in the Masisi forest (near Lake Kivu), Pessina saw coming towards him what he first thought was a chimpanzee. Noticing something human in its form and behaviour, however, he hesitated before taking a shot, thus giving the beast sufficient time to escape.

19 In the African case, the only hint of abduction is a local idea that, despite its small size, the kakundakari is capable of carrying off a 14 year old child (Cordier 1963: 179). This absence of abduction stories may seem odd in view both of African claims that chimpanzees will abduct human children (Richards 1995: 267, regarding Mende ideas) and incidents where youngsters have actually been taken (Migeod 1926, cited in Richards 1995). Indeed, a chimpanzee practice of attacking, killing and even consuming young humans is primatologically well attested (Goodall 1986: 282–83; see also Wrangham and Peterson 1996).

20 Heuvelmans (1990a: 27) attributes 'pendulous breasts' to the anonymous Kenyan creature Roumeguère interprets as an archaic hominin. In fact, the only indication of such appendages appears in an account by a Maasai man who claimed to have come across the corpse of a female specimen and who inferred from the breasts, which he described as smooth and hairless and stretched to a length of 20 centimetres, that the creature had recently given birth (Roumeguère 1990: 176).

21 Photographic evidence is available from Duffy (1996: 112) and Kleiweg's Plate 11, which shows Paul Schebesta with two Congo pygmies.

22 Although scarcely relevant to the identification of African hairy hominoids, which generally appear to lack language, in a broader comparative context it is interesting that Johnston, probably following taller African informants, described pygmies as speaking the languages of neighbouring peoples 'imperfectly' and moreover as 'replacing certain consonants by little gasps' (1902: 180). Since the characterization is remarkably similar to speech peculiarities ascribed to both the Sumbanese mili mongga and the northwestern North American zonokwa, this could conceivably reflect a widespread tendency to represent certain kinds of human outsiders in quite specific ways and to that extent point to a possible source of non-African images as well.

23 Heuvelman's 'red' and 'black' types appear to be merged in the 'niankonkla' of the Toulepleu region and the Guéré country of the Ivory Coast. Occurring in either colour, these small beings, identified as the earliest inhabitants of the land, appear quite human but are distinguished by large heads, long head hair, beards and hairy bodies. They also go naked and—in a way recalling the Central African kikomba—are given to releasing animals caught in hunters' snares. More fantastic attributes include their ability to cause death or madness in anyone who speaks to them, but also a former provision of valuable amulets to favoured hunters. Yet the niankonkla are evidently not the most fantastic, or spiritually powerful,

of small humanlike beings recognized by Ivory Coast peoples (Schnell 1948, cited in Heuvelmans 1980: 493, 495). In Ghana, the Ashanti speak of 'fairies' called 'mmoatia', which they represent in wooden statues resembling pygmies. Ashanti claim these beings occur in three colours, 'black, red and white'. Like the niankonkla, light-coloured mmoatia provide humans with charms in a silent trade (Rattray 1927: 25–26).

24 The resemblance between kalanoro and wildman figures elsewhere was first brought to my attention by Genese Sodikoff of Rutgers University, who generously provided much information on the category from her own ethnographic fieldwork on Madagascar. How the name 'kalanoro' might be analyzed is not entirely clear. 'Kala' seems comparable to components of terms in a variety of Malayo-Polynesian languages denoting entities credited with spiritual power (cf. Blust 1983), as does 'kaka', which replaces 'kala' in 'kakanoro' or 'kakanoru', the variant employed in Mayotte. I have yet to find any relevant sense of 'noro'.

25 Betsileo describe angalapono, or some creatures thus named, as living in underwater caves where they do not get wet (Sibree 1896: 235). Similarly, 'kalanoro' elsewhere denotes female beings which dwell at the bottom of lakes and which Decary (1950: 207) designates as 'naiads' or 'mermaids'. But it is not clear how these are related to other creatures bearing the same name. Other Betsileo categories sharing one or more features with kalanoro include the 'kinoly', who are credited with red eyes and long nails and thieving 'elves' called 'siona', whose bodies are covered in lichen and who, like Osborn's wildmen, like to warm themselves by fires left burning by sleeping foresters (Sibree 1896: 230, 233, 236).

9 Pacific images

1 'Negrito' (literally 'small black') has been applied generally to pygmy, or pygmoid, peoples of Southeast Asia and Oceania. 'Pygmy' refers to groups with an average male height of 1.5 metres or less; 'pygmoid' has been used for short populations somewhat taller than this.

2 Oppenheimer's field assistant, whom he describes as a 'very down-to-earth midwife', added that she felt 'very privileged and excited' to see the dwarfs, apparently as they are rarely encountered. She reported the sighting to him with some embarrassment, because she thought that, as a westerner, he would not believe her. Oppenheimer states that he has 'absolutely no reason to disbelieve that she was describing a real event'; at the same time, he can offer no opinion as to what she and her husband may have seen, adding that 'ideally the sighting should have been more fully investigated and recorded at the time'.

3 A possible exception to their lack of material culture is a 'sharp-edged stone club' described in a story. In the same tale, a 'clan' of maero are described as living in a 'bush fortress' (Cowan 1977: 64–65).

4 According to one story, a female escapee was 'unable to relate her experiences owing to fright and collapse' (Beattie 1919: 212). Interestingly, in both this story and another related by Beattie, a maero appeared immediately after someone had shot and killed a bird, which the hairy being carried off as well. Describing apparently identical creatures the Tuhoe tribe call 'nanakia', Best states that 'their principal food supply was birds, which they speared with their hands'; he also describes these nanakia—a rather general epithet referring to anything harmful or bothersome—as living in trees (1972: 998, 996). Noteworthy here is the Maori reliance on birds in mammal-poor New Zealand (Bellwood 1987: 132).

5 Gudgeon relates this episode to a general dislike of fire by 'gods'. Cowan (1930: 10) says that maero as well as 'fairies' ('patu-paiarehe') dislike both fire and steam from cooking-ovens.

6 'Patu-paiarehe' may sometimes subsume 'maero' (see Cowan 1977). In one place, Gudgeon (1906) refers to maero (or 'mohoao') as 'atua', a term often glossed as 'god' or 'spirit'. As Best (1972) points out, however, 'atua' has quite various referents and can refer to anything considered strange or extraordinary.

7 Among these, Luomala includes the kakamora of the Solomon Islands. But this is not clearly correct, for the Solomons builders are alternatively identified as a quite different category of beings named 'masi' (Fox 1924: 345–46).

8 Of course, the theme of nocturnal assistance provided by mysterious or elusive beings occurs in European representations of elves and fairies—indeed, in folk traditions the world over. A further instance is the nocturnal hoeing attributed to the East African agogwe (Chapter 8). Where heavy work is credited to small creatures, an attribution of disproportionate physical strength might follow as an obvious inference. But, conversely, assistance with construction could be inferred from physical evidence of powerful creatures. Thus, in the ancient Mediterranean, similar feats of construction were attributed to Cyclops giants, belief in the existence of which may have derived from local familiarity with the bones of large extinct mammals (see Mayor 2000: 6–7, 201). Whatever the case, the origin of such ideas most obviously lies in attempts to explain feats of ancient engineering which seem impossible in relation to known technology.

9 Although Erdland refers to the Marshall Island beings as 'spirits', he otherwise describes them in completely prosaic terms. Christian's account alternates between a representation of the chokalai as disappeared or remnant 'negritos' and as 'woodland elves' or 'fairies' whom he explicitly compares to the Maori patu-paiarehe (1899: 11). The term 'dwarf', employed frequently by Christian, naturally encompasses both possibilities.

10 In eastern Indonesia and New Guinea, local people ascribe the anusless state to fully zoologically attested animals, namely, fruit bats. In South America it is sloths that are thought to lack an anus (Forth 2004a: 123–24).

11 The version featuring the log bridge includes a motif also found in the previously mentioned story from the Sangir Islands. In the Sangir tale, humans destroy an anthropophagous and apparently hairy giant by cutting a log bridge halfway through; the bridge then collapses when the giant attempts to cross and it falls and drowns in a lake (De Vries 1925: 75–78).

12 'Pas-ta'ai' means 'ceremony of the ta'ai', the name Saisiat give to the little people, which is distinct from names employed in other aboriginal languages. Suggesting a particular derivation of 'ta'ai' as a generic term for the dwarfs, Ta'ai occurs as the name of a specific male dwarf who, in another variant of the Saisiat myth, was found singing in a cave and who, with his wife—named To'ai—introduced the pas-ta'ai ceremony to the Saisiat (Pache 1964: 39 Ferrell 1968: 66). According to Pache, it was this named pair, the dwarf leader and his mate, who managed to escape the general extermination. Some Saisiat regard Ta'ai as a male agricultural 'deity' (Ferrell 1969a: 50) and apparently not a dwarf at all. The neighbouring Taokas people designate their 'ancestor ceremony' as 'paatai', thus with a name similar to that of the Saisiat ceremony (Ferrell 1969b: 183).

13 Giants, for example the 'kapre' (from 'kaffir'), huge hominoids that may well be based on African slaves brought by the Spanish, are another matter.

14 I am indebted to Tessa Minter, a doctoral student at Leiden University, for alerting me to this lowland stereotype. Although Worcester does not mention long breasts, photographs included in his article (1912: 840) might seem to bear out the attribution. It is noteworthy, as well, that Worcester, at the time Secretary of the Interior in the American administration of the Philippines, denies Agta nakedness and tree-dwelling and also the idea that negritos engage in obscene dancing.

15 In the early sixteenth century, Pigafetta (1975: 113) recorded local reports from Mindanao of 'hairy men', apparently residing in northern Mindanao, who were

'great fighters and good archers' and who fed on 'the raw hearts of men'. The obscure statement that they carried 'swords a palm in length' could suggest small size. The reference may well have been to the Mamanua negritos of northeastern Mindanao who, although evidently not cannibals, hunted with bows and poisoned arrows until the end of the nineteenth century (Maceda 1964: 17, 39–40).

16 Another possibility is more recent contact between Indonesians and Melanesians. Bellwood (1997: 224–27) discusses archaeological evidence indicating maritime trade links, dating to 2000 to 3000 years ago, between New Britain and southeastern Sabah (in northeast Borneo), where orang-utans still occur. On the other hand, it is not clear whether hominoidal images maintained by inhabitants of New Britain, such as the dwarfs mentioned by Brown (1910), concern creatures that are hairy or in any way apelike.

10 Conclusion: what were the ebu gogo?

1 One thinks, for example, of Victor Turner's argument regarding the deployment of monstrous figures and masks depicting these, in African initiation ceremonies (1967: 103–06).

2 Nage, it will be recalled, cite the absence of a tail in considering whether the ebu gogo should be classified as human or animal. Since 'men with tails' (by no means an anatomical impossibility) are reported from various parts of the world (see for example Chapters 5 and 6), it is remarkable how regularly wildmen are described as lacking these appendages and how tailed humans are represented independently of hairy hominoids.

3 As shown in connection with the Sumatran orang pendek, explanations of hairy hominoids as descendants of human-animal mating do occasionally occur elsewhere. According to Keith Thomas (1983: 134–35), the European wildman, too, was a hybrid figure and as such attested to a continuity seventeenth-entury Europeans perceived between human and non-human beings. Obviously, this view differs from those of Bernheimer, White and others and especially from interpretations of the mediaeval wildman as a feral human, or an image symbolic of such a condition.

4 In the Nage case, a specific resemblance between the ebu gogo and spirits ('nitu') may be found in details of their legendary extermination, which followed the wildmen's attendance at a feast and their participation in circle-dancing. Nage also tell stories of beautiful female spirits appearing as uninvited guests and dancers at nocturnal feasts—and similar tales are told of female 'pontianak' in more easterly parts of Flores (Arndt 1932: 148–50). However, whereas the surreptitious attendance of spirits at celebrations has disastrous consequences for the human hosts, typically leading to the break-up of settlements (for which reason such stories form part of the oral histories of clans and villages; Forth 1998a: 74–75), quite the opposite outcome followed in the case of the ebu gogo.

5 None of the numerous differences between wildmen and spirits entails that, in addition to experience of human society, spirits can have no grounding in observation of non-human animals. As shown elsewhere (see Forth 2004a), experience of natural species may to a large extent inform spiritual images, as should be expected from the frequent association of particular spirits with particular species. Ironically, however, this implicit epistemological commonality between wildmen and spiritual beings obtains only to the extent that neither wildmen nor spirits are completely imaginary.

6 This idea was first suggested to me by Derek Freeman's (1968) interpretation of Southeast Asian thunder deities.

7 The contradiction is perhaps resolvable, but only by adopting a Freudian analysis of the extermination legend as a symbolic representation of the ultimate replacement of senior by junior generations.

8 Prominent but apparently smaller breasts, that is, ones not so long they can be hung over the shoulders, have sometimes been described for recently observed Chinese wildmen and also the sasquatch.
9 A striking example is provided by Nubian river spirits called 'dogri', said to live under the waters of the Nile. In addition to over-the-shoulder breasts, the spirits possess long tails, donkey-like legs and large ears. Their spiritual nature is confirmed by their ability to assume the further forms of humans and various animals. Interestingly, as well as long head hair, the beings are 'sometimes' described as having hair 'all over the body'. Buried in a plethora of fantastic details, however, this ambiguous reference to hirsutism does not compromise their classification as spirits (Grauer and Kennedy 1978: 122). On Sumba, a malevolent 'river spirit' ('patau luku'), taking the form of a dog-headed but not expressly hirsute female with extremely long breasts, will squirt milk at people it encounters; anyone struck by the milk will shortly die from drowning.
10 There is one report of such feet for the Australian 'yahoo' (Chapter 8). In other instances, such as the Chinese 'fei-fei' and the 'wokolo' of West Africa (Chapters 7 and 8), it is difficult to decide whether these are instances of wildmen or another kind of category. In any case, recently described Chinese wildmen seem not to possess inverted feet.
11 While one might expect body odour to be linked with accusations of dirtiness, the latter quality is far less frequently mentioned for wildmen. Instances include one report of the African kakundakari and the ape-like creature encountered by Oostingh (Chapter 5).
12 As noted in Chapters 7 and 8, capture tales concerning yeti, Chinese and Central Asian wildmen, as well as the wildman of Europe, sometimes describe the victims as being rendered susceptible by first making them drunk. Intoxication also figures in the legend of the ebu gogo, but this relates to extermination rather than capture of wildmen. Indeed, the theme is relatively restricted, evidently being confined to adjacent regions of Asia and Europe. In view of the numerous other humanlike features of wildmen, it is hardly surprising that a liking for alcohol and susceptibility to intoxication should be attributed to them. It is known that non-human primates, also, can acquire a taste for alcohol. If Beeckman (1973 [1718]: 38) is to be believed, the young orang-utan he kept for seven months not only loved 'strong liquors', but would steal punch and brandy from his ship's stores.
13 For examples from Flores see Chapter 3 (regarding Florenese 'pontianak') and Forth (1998a: 74–75); for eastern and central Sumatra see Moszkowski (1909).
14 Writing on the image of the witch, Needham employs this term to refer to a polythetic class whose components are often, but not necessarily, conjoined.
15 Lack of appreciable culture is therefore a completely general attribute of wildman figures, notwithstanding stories suggesting a minority of them (including Sumbanese, Central Asian, Malagasy and North American exemplars) are not afraid of human fires and will sometimes seek their warmth.
16 This commonly voiced opinion, often expressed as 'people seeing what they want too see', obviously ignores the question of where wildman images come from in the first place.
17 Other than primates, the only relatively large wild mammals present on Flores are Palm civets, wild (apparently feral) cats, feral dogs, wild pigs, deer and porcupines. It may readily be agreed that none of these is likely to be the basis of representations of hairy hominoids.
18 When *Homo sapiens* reached Flores is not known. The earliest skeletal evidence for modern humans on the island, excavated at Liang Bua (also the discovery site of the remains recently interpreted as *Homo floresiensis*), is about 10,500 years old (Brumm *et al.* 2006). However, since humans inhabited the neighbouring island of Timor by 40,000 years ago (O'Connor 2007), 15,000 years ago is a reasonable

and probably a conservative, estimate. Human palaeontology and archaeology on Flores, it should be noted, are still in their infancy.

19 See, for example, Mathijsen (1915), Kleiweg de Zwaan (1915) and Klokke *et al.* (1988). In some tales, monkeys appear as helpmeets to humans (see Wielenga 1913, Kleiweg 1915: 41–43). Revealing Christian influence, a Manggarai myth depicts monkeys as informing God about man-eating spirits called 'poti', whom God then condemns to the fires of hell (Verheijen 1951: 152).

20 Nage sometimes ascribe monkey intelligence to their derivation from humans, as described in variants of a myth known throughout Flores and elsewhere in Indonesia. But this pseudo-Darwinian notion does not apply to the ebu gogo. Nor, of course, are the wildmen considered ancestral to humans (Forth 2006).

21 Jacob (1967) gives the height of the Liang Toge specimen as 148.8 centimetres, which is within the range for present day Florenese females. Physical anthropologist David Bulbeck has remarked that 'the so-called Negritoid from Liang Toge does not qualify as a Negrito on any criteria—the individual is too large and its morphological features do not match those of any of the groups conventionally assigned to Negritos' (pers. comm., April 2005). As noted previously, 'negrito', a term virtually synonymous with 'pygmy', is usually taken to refer to populations with an average male height of 1.5 metres or less. In a Southeast Asian context, 'negrito' normally refers to extant populations (generally known as Semang) inhabiting the Malay Peninsula and southern Thailand; certain Philippines' groups, mostly in Luzon (Chapter 9); and the Andaman Islanders. Nowhere else in the region have distinct and currently extant populations classifiable as negritos been found, even in Sumatra and Borneo where, on comparative grounds, there has been some expectation of finding them.

22 'Cline' refers to a gradual change in physical features detectible across the distributional range of a population, usually correlated with environmental or geographic transitions.

23 In her master's thesis on the orang pendek, Rintjema (2001: 93) similarly raises the possibility of Sumatran stories of 'ape-man' deriving from ancient Mongoloid denigrations of Australoids. As a rule of thumb, 'Mongoloid' can be taken as a reference to a physical appearance like that of Chinese and many Mainland Southeast Asians and 'Australoid' to people resembling Melanesians and Australian Aborigines.

24 Such breasts cannot link Kubu with the orang pendek, as the Sumatran hominoids are *not* described as long-breasted.

25 The genealogy of the Nage clan Mudi includes an ancestor who was reputedly hirsute, as well as extremely strong and an incorrigible thief and adulterer (1998a: 81–2); he too was not identified with the ebu gogo. Otherwise, I never heard anything that might suggest cases of abnormal hair growth in the Nage region. Nor is there any reason to believe that such conditions, possibly linked to chemical deficiency (Elias and Gwinup 1983), might be more common in some parts of Flores than others, or on Flores more than other islands. Central Florenese are familiar with people born with 'tails', apparently an elongated coccyx. The fact that these individuals, also, are not linked with local hominoids accords with the usual description of wildmen as tailless.

26 Unfortunately, this putative eastern Ngadha lithic technology has not been investigated archaeologically. Molnar's mention of a 'spear point' (2000: 30), however, suggests that the reference is not simply to the polished stone adzes and chisels that have frequently been found throughout Flores (Verhoeven 1968). Called 'lightning teeth', these artefacts Nage and other Florenese construe not as products of human manufacture but as material manifestations of lighting strikes (Forth 1998a: 212).

27 The single exception was an informant who speculatively invoked this idea with reference to the forebears of a reclusive ‘Ua man who wore a long beard, but was also relatively tall. I should also mention that the ‘Ua have generally impressed me as being physically no different from other Nage.

28 A similar case is suggested by the Philippine negrito (Agta) category ‘ebuked’, applied pejoratively by sedentized Agta and non-negrito lowlanders alike to notionally ‘wild’ (that is, non-sedentized) Agta. The negative qualities associated with ‘ebuked’ are partly ones lowlanders ascribe to negritos in general, but no Agta groups will identify themselves by this term (Estioko and Griffin 1975).

29 The figure of eight derives from Morwood’s 2005 article. In his recent book (Morwood and Van Oosterzee 2007: 106, 228), he says at least twelve individuals and in another place thirteen.

30 There is disagreement over how robust *H. floresiensis* may have been, or have appeared (Jacob *et al.* 2006). Several commentators, however, speak of relatively short legs, differently aligned thigh bones—thus affecting the manner of walking—and very wide pelvic bones, giving ‘a different overall body shape’ from modern humans (Mirazon Lahr and Foley 2004: 1043).

31 This capacity, called ‘humeral torsion’, is significantly less in *Homo floresiensis* than in all other hominins and even larger apes and compares with that found in gibbons and quadrupedal monkeys (like the *Macaca*). By itself, the feature would seem to counter the suggestion that the hominin used tools; however, the slope of the shoulders apparently compensates in this respect (Culotta ibid.).

32 Gert van den Bergh and Bert Roberts, two members of the *Homo floresiensis* discovery team who visited the Nage region in 2004, were told that ebu gogo had long arms and fingers, pot-bellies and slightly protruding ears (Forth 2005: 14; Morwood and Van Oosterzee 2007: 154). Curiously, these details do not appear in the ethnographic record I compiled over a period of twenty years, beginning in 1984. On the other hand, the being with the crooked lower arms observed in the 1970s had a belly that appeared distended. Morwood (ibid.) claims that local descriptions of ebu gogo fit *Homo floresiensis* ‘to a T’. But if he refers to descriptions I recorded, as his reference to my 1998 book might suggest, this is less than accurate.

33 Amerindian myth may echo geological history over similarly long periods. To cite one of many possible examples, Moody and Catchpole (1992) have quite conclusively shown how Athapaskan oral traditions reflect volcanic eruptions that occurred in the White River basin of Alaska about 1300 years and possibly even 2000 years ago. In a more general vein, it is noteworthy how, in recent decades, archaeologists and anthropologists have increasingly recognized the value of oral narrative as a potential source of information on past events (see e.g. Gazin-Schwartz and Holtorf 1999).

34 The main difference is that the abductee in the northern Sumatran and Rotinese stories is male, whereas in the Bornean story she is female. The first story was related to me in March 2006 by Juara Ginting, himself a member of the north Sumatran Karo people. The Rotinese and Bornean narratives are recorded in Jonker (1911: 71–72) and Rutter (1999: 60–65).

References

Abel, Clarke. 1825. 'Some account of an Orang Outang of remarkable height found on the Island of Sumatra'. *Asiatic Researches* 15: 489–98.

Adam, Tassilo. 1944. 'The Kubus: Sumatra forest-dwellers'. *Knickerbocker Weekly* 4(14): 21–23.

Adriani, N. and A.C. Kruyt. 1898. 'Van Posso naar Parigi, Sigi en Lindoe'. *Mededeelingen van wege het Nederlandsch Zendingsgenootschap* 42: 369–536.

Allen, Benedict. 1989. *Hunting the gugu: In search of the lost ape-men of Sumatra.* London: Macmillan.

Allport, Gordon W. and Leo Postman. 1947. *The psychology of rumor*. New York: Henry Holt and Company.

Anderson, Atholl. 1990. 'The beast without: The moa as a colonial frontier myth in New Zealand'. In R.G. Willis, ed. *Signifying animals: Human meaning in the natural world.* London: Unwin Hyman, pp. 236–45.

Anderson, John. 1824. *Physical and commercial considerations relative to the Malayan peninsula and the British settlements in the Straits of Malacca.* Prince of Wales Island: William Cox.

Anon. 1924a. 'De orang pendek op Sumatra'. *De Locomotief*, 1 August 1924.

Anon. 1924b. 'De "Orang Pendek" op Sumatra'. *Deli Courant*, 3 September 1924. (A letter to the editor from an unnamed administrator from Palembang).

Anon. 1924c. 'Aapmenschen op Sumatra'. *Algemeen Landbouwweekblad voor Nederlandsch-Indië* 8: 890.

Anon. 1926a. 'Aardrijkskundig Nieuws: Orang sedapa?' *Tijdschrift van het Aardrijkskundig Genootschap* (2e serie) 43: 117–20.

Anon. 1926b. 'De orang pendek weer ter sprake gebracht'. *Het Koloniaal Weekblad* 26(43): 10.

Anon. 1926c. 'Een levenrancier van orang pendek's!' *Het Koloniaal Weekblad* 26(9): 8.

Anon. 1927a. 'Orang pendek'. *Het Handelsblad*, 20 September 1927.

Anon. 1927b. 'De "Orang Pendek"'. *Bataviaasch Nieuwsblad*, 5 September 1927.

Anon. 1932. 'Nog eens de orang pendek'. *De Sumatra Post*, 15 June 1932.

Anon. 1933. 'De radja en de orang pendek. Waaraan Amat uit Rokan zijn misvormde hand dankt. Hoe de rijstsnoeper gevangen werd en wist te ontvluchten'. (Unattributed newspaper article, Dubois Collection, Natural History Museum, Leiden, The Netherlands; annotation gives date of 4th May 1933).

Anon. 1935. 'Onderhoud met een ouden tijgerjager. In Atjeh's bosschen had hij den orang pendek meermalen gezien'. (Article in an unidentified Dutch language newspaper, Dubois Collection, Natural History Museum, Leiden, The Netherlands; annotation gives date of 26 May 1935).

Anon. 1995. 'Orang Pendek Kerinci, Antara Mitos dan Realitas'. *Warta Massa* (weekly edition), 1 September 1995, pp. 1, 8.

Anon. 1996. '"Mutant" chimp gets gene check'. *Science* 274(5288): 727.

Appell, George. 1991. 'Individuation of the drives of sex and aggression in the linguistic and behavioural repertoire of the Rungus'. In Vinson H. Sutlive, Jr., ed. *Female and male in Borneo: Contributions and challenges to gender studies*. Borneo Research Council Monograph Series, Volume 1. Williamsburg, VA: The Borneo Research Council, Inc, pp. 57–120.

Argue, Debbie, Denise Donlon, Colin Groves and Richard Wright. 2006. '*Homo floresiensis*: Microcephalic, pygmoid, *Australopithecus*, or *Homo*?'. *Journal of Human Evolution* 51: 360–74.

Arndt, Paul. 1930. 'Die Religion der Ngad'a'. *Anthropos* 24: 817–61.

—— 1932. *Mythologie, Religion und Magie im Sikagebiet (östl. Mittelflores)*. Ende-Flores: Arnoldus-Druckerei.

—— 1933. *Li'onesisch-Deutsches Wörterbuch*. Ende-Flores: Arnoldus-Druckerei.

—— 1951. *Religion auf Ost-flores, Adonare und Solor*. Studia Instituti Anthropos, 1. Wien-Mödling: Missionsdruckerei St. Gabriel.

—— 1961. *Wörterbuch der Ngadhasprache*. Studia Instituti Anthropos, 15. Posieux, Fribourg, Suisse: Anthropos-Institut.

Arne, Antti. 1973. *The types of the folktale: A classification and bibliography*. Translated and enlarged by Stith Thompson. Second Revision. Helsinki: Academia Scientiarum Fennica.

Astuti, Rita. 1995. '"The Vezo are not a kind of people": Identity, difference and "ethnicity" among a fishing people of western Madagascar'. *American Ethnologist* 22: 464–82.

Astuti, Rita, Gregg Solomon and Susan Carey. 2004. *Constraints on conceptual development: A case study of the acquisition of folkbiological and folksociological knowledge in Madagascar*. Oxford: Blackwell Publishing.

Atran, Scott. 1990. *The cognitive foundations of natural history: Towards an anthropology of science*. Cambridge: Cambridge University Press.

Auffenberg, Walter. 1981. *The behavioral ecology of the Komodo Monitor*. Gainesville, FL: University Presses of Florida.

Bacon, Roger. 1928. *The Opus majus of Roger Bacon*. 2 volumes. Translated from the Latin by Robert Belle Burke. Philadelphia, PA: University of Pennsylvania Press. (Originally published in 1268.)

Bader, Hermann. 1953. *Die Reifefeiern bei den Ngada (Mittelflores, Indonesien)*. St.-Gabrieler Studien 14. Moedling bei Wien: St.-Gabriel.

Bailey, John. 1863. 'An account of the wild tribes of the Veddahs of Ceylon: Their habits, customs and superstitions'. *Transactions of the Ethnological Society of London* 2: 278–320.

Bakels, Jet. 1995. 'Zo onmenselijk, zo aan een mens gelijk'. *NRC Handelsblad*, 18 January.

—— 2000. *Het verbond met de tijger: Visies op mensenetende dieren in Kerinci, Sumatra*. Leiden: Research School of Asian, African and Amerindia Indian Studies, Leiden University.

Barber, C.C. 1979. *Dictionary of Balinese-English*. 2 volumes. Aberdeen: Aberdeen University Library (Occasional Publications No. 2).

Barendregt, Bart. 2002. 'Representing the ancient other'. *Indonesia and the Malay World* 30(88): 277–307.
Barnes, R.H. 1974. *Kédang: A study of the collective thought of an eastern Indonesian people*. Oxford: Clarendon Press.
Barone, Jennifer. 2007. 'Bondo mystery ape proves to be a chimpanzee with unusual habits'. *Discover* 28(3): 20.
Bartels, Ernest. n.d. 'The aũl'. Mss in files of the Society for the Investigation of the Unexplained, Columbia, New Jersey (cited in Sanderson 1972).
Bartlett, Harley Harris. 1973. *The labors of the datoe and other essays on the Bataks of Asahan (North Sumatra)*. Michigan Papers on South and Southeast Asia 5. Ann Arbor: Center for South and Southeast Asian Studies, The University of Michigan.
Bayanov, Dmitri. 1982. 'A note on folklore in hominology'. *Cryptozoology* 1: 46–48.
Bean, Robert Bennet. 1913. 'Notes on the hairy men of the Philippine islands and elsewhere'. *American Anthropologist* 15(3): 415–24.
Beattie, H. 1919. 'Traditions and legends collected from the natives of Murihiku (Southland, New Zealand)'. *Journal of the Polynesian Society* 28: 212–25.
Beccari, Odoardo. 1904. *Wanderings in the great forests of Borneo: Travels and researches of a naturalist in Sarawak*. Translated from the Italian by E.H. Giglioli. London: Archibald Constable & Co Ltd.
Beeckman, Daniel. 1973. *A voyage to and from the Island of Borneo*. Folkestone, England: Dawsons of Pall Mall. (Originally published in 1718.)
Begbie, P.J. 1834. *The Malayan Peninsula embracing its history, manners and customs of the inhabitants, politics, natural history etc. from its earliest records*. Madras: Vepery Mission Press.
Bellwood, Peter. 1979. *Man's conquest of the Pacific: The prehistory of Southeast Asia and Oceania*. New York: Oxford University Press.
—— 1987. *The Polynesians: Prehistory of an island people*. Revised edition. London: Thames and Hudson.
—— 1997. *Prehistory of the Indo-Malaysian archipelago*. Revised edition. Honolulu: University of Hawai'i Press.
Bellwood, Peter, Goenadi Nitihaminoto, Geoffrey Irvin, Gunadi, Agus Waluyo and Daud Tanudirjo. 1998. '35,000 years of prehistory in the northern Moluccas'. *Modern Quaternary Research in Southeast Asia* 15: 233–75.
Belo, Jane. 1949. *Bali: Rangda and Barong*. Monographs of the American Ethnological Society, XVI. New York: J.J. Augustin.
Berlin, Brent. 1992. *Ethnobiological classification: Principles of categorization of plants and animals in traditional societies*. Princeton, NJ: Princeton University Press.
Bernheimer, Richard. 1952. *Wild men in the Middle Ages: A study in art, sentiment and demonology*. Cambridge, MA: Harvard University Press.
Best, Elsdon. 1972. *Tuhoe, the children of the mist: A sketch of the origin, history, myths and beliefs of the Tuhoe tribe of the Maori of New Zealand, with some account of other early tribes of the Bay of Plenty district*. 2nd edition, Volume 1. Wellington: A.H. & A.W. Reed (for the Polynesian Society). (Originally published in 1925.)
Bickmore, Albert S. 1868. *Travels in the East Indian archipelago*. London: Murray.
Birdsell, Joseph B. 1975. *Human evolution: An introduction to the new physical anthropology*. Chicago: Rand McNally College Publishing Company.
—— 1993. *Microevolutionary patterns in Aboriginal Australia: A gradient analysis of clines*. New York: Oxford University Press.

Blanford, W.T. 1891. 'Mammalia'. In W.T. Blanford, ed. *The fauna of British India including Ceylon and Burma*. London: Taylor and Francis.
Blench, Roger and Matthew Spriggs, eds. 1997. *Archaeology and language I: Theoretical and methodological orientations*. London: Routledge.
Bloch, Maurice. 1986. *From blessing to violence: History and ideology in the circumcision ritual of the Merina of Madagascar*. Cambridge: Cambridge University Press.
Blomberg, Rolf. 1948. *Sydvart*. Stockholm: Gebers.
Blust, Robert. 1983. *A linguistic key to the early Austronesian spirit world*. Unpublished manuscript.
Boas, F. 1935. *Kwakiutl culture as reflected in mythology*. Memoirs of the American Folklore Society, Volume 28. New York: American Folklore Society.
Bondeson, J. and A.E.W. Miles. 1996. 'The hairy family of Burma: A four generation pedigree of congenital hypertrichosis lanuginosa'. *Journal of the Royal Society of Medicine* 89(7): 403–8.
Bontius, Jacobus. 1931. *On tropical medicine*. Opuscula Selecta Neerlandicorum de Arte Medica X. Amsterdam: Nederlandsch Tijdscrift voor Geneeskunde.
Bord, Janet and Colin Bord. 1982. *The Bigfoot casebook*. Harrisburg, PA: Stackpole Books.
Botne, Robert. 1994. *A Lega and English dictionary with an index to Proto-Bantu roots*. (Compiled by R. Botne with Kisanga Salama-Gray). East African Languages and Dialects, Volume 3. Köln: Rüdiger Köppe Verlag.
Boyer, Pascal. 1994. *The naturalness of religious ideas: A cognitive theory of religion*. Berkeley, CA: University of California Press.
—— 2001. 'Cultural inheritance tracks and cognitive predispositions: The example of religious concepts'. In Harvey Whitehouse, ed. *The debated mind: Evolutionary psychology versus ethnography*. Oxford: Berg, pp. 57–89.
Brasser, J.G. 1926. *Jacht op groot wild in Nederlandsch Oost-Indië*. Zutphen: Thieme.
Brewster, A.B. 1922. *The hill tribes of Fiji*. London: Seeley, Service & Co. Limited.
Brooke, Charles. 1866. *Ten years in Sarawak*. 2 volumes. London: Tinsley Brothers.
Brow, James. 1978. *Vedda villages of Anuradhapura: The historical anthropology of a community in Sri Lanka*. Seattle: University of Washington Press.
Brown, George. 1910. *Melanesians and Polynesians: Their life-histories described and compared*. London: Macmillan and Co. Ltd.
Brown, P.T., M.J., Morwood, R.P., Soejono, Jatmiko, E. Wayhu Saptomo and Rokus Awe Due. 2004. 'A new small-bodied hominin from the late Pleistocene of Flores, Indonesia'. *Nature* 431: 1055–61.
Brumm, Adam, Fachroel Aziz, Gert D. van den Bergh, Michael J. Morwood, Mark W. Moore, Iwan Kurniawan, Douglas R. Hobbs and Richard Fullagar. 2006. 'Early stone technology of Flores and its implications for *Homo floresiensis*'. *Nature* 441: 624–28.
Brunner, Hans. 2003. 'Final report'. (Unpublished report on unidentified mammalian hair, mailed electronically to A. Sanderson by H. Brunner, 20 February 2003 and to the author, 3 November 2004.)
Brunner, Hans and Brian J. Coman. 1974. *The identification of mammalian hair*. Melbourne: Inkata Press.
Brunvand, Jan Harold. 2001. *Encyclopedia of urban legends*. Santa Barbara, CA: ABC-Clio.
Burchett, Wilfred G. 1965. *La seconde résistance: Vietnam 1965*. Translated from English by Michel Deutsch. Paris: Gallimard.
Burgoyne, Cuthbert. 1938. 'Little furry men'. *Discovery* 19: 51.

Burney, David A. and Ramilisonina. 1998. 'The *kilopilopitsofy*, *kidoky* and *bokyboky*: Accounts of strange animals from Belo-sur-mer, Madagascar and the Megafaunal "extinction window"'. *American Anthropologist* 100(4): 957–66.

Burns, J.W. 1929. 'Introducing B.C.'s hairy giants'. *Maclean's*, 1 April 1929, p. 9.

Büttikofer, Johann. 1893. 'Notes and news'. *American Anthropologist* 9(3): 337–9.

Byrne, Peter. 1958. 'Searching for the abominable snowman'. *New York Journal-American*, June 15.

—— 1975. *The search for Big Foot: Monster, myth or man*? Washington, DC: Acropolis Books.

Cartmill, Matt. 2001. 'Taxonomic revolutions and the animal-human boundary'. In Raymond Corbey and Wil Roebroeks, eds. *Studying human origins: Disciplinary history and epistemology*. Amsterdam Archaeological Studies 6. Amersterdam: Amsterdam University Press, pp. 97–106.

Christian, F.W. 1899. *The Caroline Islands: Travel in the sea of the little lands*. London: Methuen & Co.

Chuan Kun Ho, Guo Xing Zhou and David R. Swindler. 1995. 'Dental evolution of the orang-utan in China'. *Human Evolution* 10(4): 249–64.

Ciochon, Russell, John Olsen and Jamie James. 1990. *Other origins: The search for the giant ape in human prehistory*. New York: Bantam Books.

Clark, Ella E., ed. 1955. 'George Gibbs' account of Indian mythology in Oregon and Washington territories'. *Oregon Historical Quarterly* 56(3): 293–325.

Classen, Constance. 1992. 'The odor of the other: olfactory symbolism and cultural categories'. *Ethos* 20: 133–66.

Codrington, R.H. 1891. *The Melanesians: Studies in their anthropology and folklore*. Oxford: The Clarendon Press.

Colarusso, J. 1980. 'Ethnographic information on a wild man of the Caucasus'. In Halpin and Ames, eds, pp. 255–64.

Coleman, Loren and Patrick Huyghe. 1999. *The field guide to the bigfoot, yeti and other mysteries worldwide*. New York: Avon Books.

Cook, S.V. 1924. 'The leprachauns of Kwa Ngombe'. *Journal of the East African and Ugandan History Society* 20: 24.

Coolsma, S. 1910. *Hollandsch-Soendaneesch Woordenbook*. Leiden: A.W. Sijthoff.

—— 1930. *Soendaneesch-Hollandsch Woordenboek*. Leiden: A.W. Sijthoff.

Coomans de Ruiter, L. 1929. 'De stand van het orang pendek vraagstuk'. Collectie L.C. Westenenk, nummer toegang 2.21.205.71. Nationaal Archief, The Hague.

Corbey, Raymond and Bert Theunissen, eds. 1995. *Ape, man, apeman: Changing views since 1600*. Leiden: Department of Prehistory, Leiden University.

Cordier, Charles. 1963. 'Deux anthropoïdes inconnus marchant debout au Congo ex-Belge'. *Genus* 19: 175–82.

—— 1973. 'Animaux inconnus au Congo'. *Zoo* 38(4): 185–91.

Courtney, Roger. 1940. *A greenhorn in Africa*. London: Herbert Jenkins Limited.

Covarrubias, M. 1937. *Island of Bali*. New York: Alfred A. Knopf.

Cowan, James. 1930. *Fairy folk tales of the Maori*. Second edition. Auckland: Whitcombe & Tombs Ltd.

—— 1977. *Legends of the Maori*. Volume 1. Wellington: Harry H. Tombs Ltd. (Originally published 1930).

Cramer, A.H.W. 1927. Letter in *Algemeen Handelsblad*, 5 July 1927.

Culotta, Elizabeth. 2006. 'How the hobbit shrugged: Tiny hominid's story takes new turn'. *Science* 312: 983–4.

Curtis, Garniss, Carl Swisher and Roger Lewin. 2000. *Java man: How geologists changed the history of human evolution*. New York: Scribner.
Daegling, David. 2004. *Bigfoot exposed: An anthropologist examines America's enduring legend*. Walnut Creek, CA: Altamira Press.
Dammerman, K.W. 1924. 'De orang pandak van Sumatra'. *De Tropische Natuur* 13: 177–82.
—— 1926. 'Geheimzinnige dieren'. *Algemeen Landbouwweekblad voor Nederlandsch-Indie* 10: 194–8.
—— 1929. 'The orang pendek or ape-man of Sumatra'. *Proceedings of the Fourth Pacific Science Congress* 3: 121–6.
—— 1932. 'De nieuwe-ontdekte orang pendek'. *De Tropische Natuur* 21: 123–31.
Dance, Peter. 1976. *Animal fakes and frauds*. Maidenhead: Sampson Low.
Darwin, Charles. 1998. *The expression of the emotions in man and animals*. Third edition. New York and Oxford: Oxford University Press.
Davis, Phillip W. and Ross Saunders. 1980. *Bella Coola Texts*. British Columbia Provincial Museum, Heritage Record No. 10. Victoria: British Columbia Provincial Museum.
Decary, Raymond. 1950. *La faune malgache, son rôle dans les croyances et les usages indigénes*. Paris: Payot.
Dennell, Robin and Wil Roebroeks. 2005. 'An Asian perspective on early human dispersal from Africa'. *Nature* 438: 1099–104.
Deraniyagala, S.U. 1992. *The prehistory of Sri Lanka: An ecological perspective*. Memoir Volume 8, Part II. Colombo: Department of Archaeological Survey, Government of Sri Lanka.
De Rienzi, G.L. Domeny. 1836. *Océanie ou cinquième partie du monde: Revue géographique et ethnographique de la Malaisie, de la Micronésie, de la Polynésie et de la Mélanésie*. Paris: Fermin Didot Frères.
De Roo van Alderwerelt, J. 1890. 'Eenige mededeelingen over Soemba'. *Tijdschrift voor Indische Taal-, Land- en Volkenkunde* 33: 565–95.
De Santy, H.W. Th. 1925. 'Bijdragen over het voorkomen van de orang pandak op Sumatra VI. De Sedapak op Sumatra'. *Het Koloniaal Weekblad* 25(2): 3–5.
De Vries, Jan. 1925/1928. *Volksverhalen uit Oost-Indië*. 2 volumes (1925 and 1928 respectively). Zutphen: W.J. Thieme & Cie.
De Waal, Frans. 1997. *Bonobos: the forgotten ape*. Berkeley, Los Angeles: The University of California Press.
De Wals, Hans. 1937. *Jan en ik: De jagers van de brandsavannen*. Batavia & Amersterdam: G. Kolff & Co.
De Zylva Mudliyar, A. 1886. 'The Nittaweo: Note by A. de Zylva Mudliyar'. *The Taprobanian*. February issue, p. 68. Bombay: Education Society's Press.
Dickason, Olive. 1980. 'The concept of *l'homme sauvage*'. In Halpin and Ames, eds, pp. 65–82.
Dietrich, Stefan. 1995. 'Tjeritera Patigolo Arkian: Struktur und Variation in der Gründungsmythe der Fürstenhauses von Larantuka (Ostindonesien)'. *Tribus: Jahrbuch des Linden-Museums* 44 (October 1995). Stuttgart: Linden-Museum.
DiFonzo, Nicholas and Prashant Bordia. 2007. *Rumor psychology: Social and organizational approaches*. Washington, DC: American Psychological Association.
Dixon, Roland B. 1920. 'A new theory of Polynesian origins'. *Proceedings of the American Philosophical Society* 59: 261–67.
Djajadinigrat, Hoesein. 1934. *Atjèhsch-Nederlandsch woordenboek*. Batavia: Landsdrukkerij.

Dobson, Jessie. 1953–54. 'James Hunter and the early knowledge of the anthropoid apes'. *Proceedings of the Zoological Society of London* 123: 1–12.

Dosedla, H.C. 2006. 'Oral lore concerning "wildman" figures from the central highlands of Papua New Guinea'. (Unpublished paper, presented at the 'Masterclass' conference entitled *Images of the wildman in Southeast Asia*, University of Leiden, February 2006.)

Drake, Richard A. 1989. 'Construction sacrifice and kidnapping rumour panics in Borneo'. *Oceania* 59: 269–79.

Dubois, Eugene. 1908. 'On the correlation of the black and the orange-coloured pigments and its bearing upon the interpretation of red-hairedness'. *Man* 8(46): 87–89.

Du Chaillu, Paul B. 1861. *Explorations and adventures in equatorial Africa*. London: J. Murray.

Dudley, Edward J. and Maximillian E. Novak, eds. 1972. *The wild man within: An image in western thought from the Renaissance to Romanticism*. Pittsburgh: University of Pittsburgh Press.

Duff, Wilson. 1952. *The Upper Stalo Indians of the Fraser Valley, British Columbia*. Victoria: British Columbia Provincial Museum.

Duffy, Kevin. 1996. *Children of the forest: Africa's Mbuti Pygmies*. Prospect Heights, IL: Waveland Press.

Echols, John M. & Hassan Shadily. 1989. *An Indonesia-English dictionary*. Third edition. Revised and edited by John U. Wolff and James T. Collins, in cooperation with Hassan Shadily. Ithaca, NY: Cornell University Press.

Elias, Alan N. and Grant Gwinup. 1983. *Hirsutism*. New York: Praeger Publishers.

Elwes, H.J. 1915. (Untitled note 'on the possible existence of a large Ape, unknown to science, in Sikkim', dated 27 April , 1915.) *Proceedings of the General Meetings for Scientific Business of the Zoological Society of London*. London: Longmans, Green and Co, p. 294.

Ely, John J., M. Leland, M. Martino, W. Swett and C.M. Moore. 1998. 'Technical report: Chromosomal and mtDNA analysis of Oliver'. *American Journal of Physical Anthropology* 105(3): 395–403.

Endicott, Kirk M. 1979. *Batek negrito religion: The world-view and rituals of a hunting and gathering people of peninsular Malaysia*. Oxford: Clarendon Press.

Erb, Maribeth. 1987. 'When rocks were young and earth was soft: Ritual and mythology in Northeastern Manggarai'. PhD thesis, State University of New York at Stony Brook.

—— 1991. 'Construction sacrifice, rumors and kidnapping scares in Manggarai : further comparative notes from Flores'. *Oceania* 62(2): 114–26.

—— 1999. *The Manggaraians: A guide to traditional lifestyles*. Singapore: Times Editions.

Erdland, August. 1914. *Die Marshall-Insulaner: Leben und Sitte, Sinn und Religion eines Südsee-Volkes*. Münster: Aschendorff.

Erlinga, F.S. 1984. *Soendaas-Nederlands woordenboek*. Dordrecht: Foris.

Estioko, Agnes A. and P. Bion Griffin. 1975. 'The Ebuked Agta of northeastern Luzon'. *Philippine Quarterly of Culture and Society* 3: 237–44.

Evans, I.H.N. 1953. *The religion of the Tempasuk Dusuns of North Borneo*. Cambridge: Cambridge University Press.

Falk, Dean, Charles Hildebolt, Kirk Smith, Mike Morwood, Thomas Sutikna, Peter Brown, Jatmiko, E. Wayhu Saptomo, Barry Brunsden and Fred Prior. 2005. 'The brain of Homo floresiensis'. *Science* 308: 242–45.

Ferrell, Raleigh. 1968. '"Negrito" ritual and traditions of small people on Taiwan'. In N. Matsumoto and T. Mabuchi, eds. *Folk religion and the worldview in the southwestern Pacific*. Tokyo: Keio Institute of Cultural and Linguistic Studies, pp. 63–72.

—— 1969a. *Taiwan aboriginal groups: Problems in cultural and linguistic classification*. Nankang, Taipei, Taiwan: Institute of Ethnology Academia Sinica.

—— 1969b. 'Paiwanic ethno-linguistic groups of the west-central Taiwan "black pottery" culture area'. *Bulletin of the Institute of Ethnology* (Academia Sinica, Taipei) 28: 159–96.

Fleagle, John G. 1999. *Primate adaptation and evolution*. Second edition. San Diego and London: Academic Press.

Flood, Josephine. 1983. *Archaeology of the dreamtime*. Sydney and London: Collins.

Forbes, Henry O. 1885. *A naturalist's wanderings in the eastern archipelago: A narrative of travel and exploration from 1878 to 1883*. London: Sampson Low, Marston, Searle & Rivington.

Forth, Gregory. 1981. *Rindi: An ethnographic study of a traditional domain in eastern Sumba*. Verhandelingen van het Koninklijk Instituut voor Taal-, Land-, en Volkenkunde 93. The Hague: Martinus Nijhoff.

—— 1989. 'Animals, witches and wind: Eastern Indonesian variations on the "thunder complex"'. *Anthropos* 84: 89–106.

—— 1991. 'Construction sacrifice and head-hunting rumours in central Flores (eastern Indonesia)'. *Oceania* 61(3): 257–66.

—— 1995. 'Ethnozoological classification and classificatory language among the Nage of eastern Indonesia'. *Journal of Ethnobiology* 15(1): 45–69.

—— 1998a. *Beneath the volcano: Religion, cosmology and spirit classification among the Nage of eastern Indonesia*. Verhandelingen van het Koninklijk Instituut voor Taal-, Land-, en Volkenkunde 177. Leiden: KITLV Press.

—— 1998b. 'On deer and dolphins: Nage ideas regarding animal transformation'. *Oceania* 68(4): 271–93.

—— 2001. *Dualism and hierarchy: Processes of binary combination in Keo society*. Oxford: Oxford University Press.

—— 2004a. *Nage birds: Classification and symbolism among an eastern Indonesian people*. London and New York: Routledge.

—— 2004b. 'A "cock" and "bull" story: Nage sex terms and their implications for ethnozoological classification'. *Anthropological Linguistics* 46(4): 427–49.

—— 2005. 'Hominids, hairy hominoids and the science of humanity'. *Anthropology Today* 21(3): 13–17.

—— 2006. 'Flores after floresiensis: Implications of local reaction to recent palaeoanthropological discoveries on an eastern Indonesian island'. *Bijdragen tot de Taal-, Land- en Volkenkunde* 162 (2/3): 332–45.

—— 2007. 'Images of the wildman inside and outside Europe'. *Folklore* 118: 261–82.

Fox, Charles E. 1924. *The threshold of the Pacific: An account of the social organization, magic and religion of the people of San Cristoval in the Solomon Islands*. London: Kegan Paul, Trench, Trubner & Co. Ltd.

Franklin, Stephen. 1959. 'The trail of the Sasquatch' *Weekend Magazine* Vol. 9, No. 14, 11 April 1959, pp. 2–33. (A weekly supplement issued with several Canadian newspapers.)

Frederickson, A.D. 1889. *Ad orientem*. London: W.H. Allen & Co.

Freeman, Derek. 1968. 'Thunder, blood and the nicknaming of God's creatures'. *The Psychoanalytical Quarterly* 37(3): 353–399.

Fürer-Haimendorf, C. von. 1954. 'The "scalp" of an "abominable snowman"? A yeti-hide head-dress'. *The Illustrated London News* 224: 477 (March 27, 1954).
Galdikas, Birute. 1995. *Reflections of Eden: My years with the orangutans of Borneo.* Boston: Little, Brown.
Gaudebout, P. and L. Molet. 1957. Coutumes et texts Tanala. *Mémoires de l'Institut Scientifique de Madagascar*. Série C. Volume 4, pp. 35–96.
Gazin-Schwartz, Amy and Cornelius J. Holtorf, eds. 1999. *Archaeology and folklore.* London: Routledge.
Gervaise, Nicolas. 1701. *An historical description of the Kingdom of Macasar*. (In Three Books). London: Tho. Leigh and D. Midwinter. (Original French edition *Description historique du royaume de Macacar*, published in 1688.)
Gibbons, Ann. 2007. 'Hobbit's status as a new species gets a hand up'. *Science* 316 (6 April): 34.
Gibson, Thomas. 1986. *Sacrifice and sharing in the Philippine highlands: Religion and society among the Buid of Mindoro*. LSE Monographs on Social Anthropology 57. London and Dover, New Hampshire: The Athlone Press.
Gibson, Walter M. 1856. *The prison of Weltevreden; and a glance at the East Indian Archipelago*. London: Sampson Low, Son and Co.; New York: J.C. Riker.
Goldenberg, Linda. 2007. *Little people and a lost world: An anthropological mystery*. Minneapolis: Twenty-first Century Books.
Goodall, Jane. 1986. *The chimpanzees of Gombe: Patterns of behavior*. Cambridge, MA: The Belknap Press of Harvard University Press.
Gouda, Frances. 1995. *Dutch culture overseas: Colonial practice in the Netherlands Indies, 1900–1942*. Amsterdam: Amsterdam University Press.
Grauer, Armand and John G. Kennedy. 1978. 'The dogri: Evil beings of the Nile'. In John G. Kennedy, ed. *Nubian ceremonial life: Studies in Islamic syncretism and cultural change*. Cairo: The American University of Cairo Press, pp. 114–24.
Green, John. 1970. *Year of the Sasquatch*. Second edition. Agassiz, British Columbia: Cheam.
—— 1973. *The Sasquatch File*. Agassiz, British Columbia: Cheam.
—— 1978. *Sasquatch: The apes among us*. North Vancouver, British Columbia: Hancock House.
—— 1980. 'What is the Sasquatch?'. In Halpin and Ames, eds, pp. 237–244.
Greenwell, J. Richard and Frank E. Poirier. 1989. 'Further investigations into the reported *yeren*—the wildman of China'. *Cryptozoology* 8: 47–57.
Groves, Colin P. 1986. 'The yahoo, the yowie and reports of Australian hairy bipeds'. *Cryptozoology* 5: 47–54.
—— 1987. 'Noise, signal and evidence (response to Bayanov and Joyner)'. *Cryptozoology* 6: 128–29.
—— 1988. 'On type I and type II errors in cryptozoology; or, was Proteus a yahoo?' *Cryptozoology* 7: 123–4.
Gudgeon, T.W. 1906. 'Tipau-kura and manifestations of the spirit world'. *Journal of the Polynesian Society* 15: 27–57.
Gurung, K.K. and Raj Singh. 1996. *Field guide to the mammals of the Indian subcontinent*. San Diego: Academic Press.
Hagen, B. 1883. 'Beiträge zur Kenntnis der Battareligion'. *Tijdschrift voor Indische Taal-, Land- en Volkenkunde* 28: 498–545.
—— 1908. *Die Orang Kubu auf Sumatra*. Frankfurt am Main: Joseph Baer & Co.
Hallpike, C.R. 1969. 'Social hair'. *Man* (n.s.) 4: 256–64.

Halpin, Marjorie. 1980a. 'Investigating the goblin universe'. In Halpin and Ames, eds, pp. 3–26.

—— 1980b. 'The Tsimshian monkey mask and sasquatch'. In Halpin and Ames, eds, pp. 211–28.

Halpin, Marjorie and Michael Ames, eds. 1980. *Manlike monsters on trial: Early records and modern evidence*. Vancouver: University of British Columbia Press.

Hambruch, Paul. 1932, 1936. 'Ponape'. In Georg Thilenius, ed. *Ergebnisse der Suedsee-Expedition 1908–1910. II. Ethnographie: B. Micronesien*, vol. 7, part 1 (1932) and part 3 (1936). Hamburg.

Harrisson, Tom. 1955. 'Maias v. Man'. *Sarawak Museum Journal*. New series, Vol. 6 (no. 6): 600–32.

Hazeu, Godard Arend Johannes. 1907. *Gajosch-Nederlandsch woordenboek met Nederlandsch-Gajosch register*. Batavia: Landsdrukkerij.

Headland, Thomas N. 1992. 'The Tasaday: A hoax or not?'. In T.N. Headland, ed. *The Tasaday controversy: Assessing the evidence*. Special Publication No. 28. Washington D.C.: American Anthropological Association, pp. 215–24.

Heaney, Michael. 1983. 'The Mongolian almas: A historical re-evaluation of the sighting by Baradiin'. *Cryptozoology* 2: 40–52.

Helfrich, O.L. 1904. 'Midden Maleische woordenlijst'. *Verhandelingen van het Bataviaasch Genootschap van Kunsten en Wetenschappen* 53.

—— 1927a. 'Uit de folklore van Zuid-Sumatra'. *Bijdragen tot de Taal-, Land- en Volkenkunde* 83: 193–315.

—— 1927b. 'Nadere bijdragen to de kennis van het Midden Maleisch (Besemahsch & Serawajsch dialect)'. *Verhandelingen van het Bataviaasch Genootschap van Kunsten en Wetenschappen* 68, part 3: 1–128.

Hendriks, Hendrik. 1897. *Het Burusch van Masarete*. 's-Gravenhage: Martinus Nijhoff.

Heuvelmans, Bernard. 1958. *On the track of unknown animals*. Translated from the French by Richard Garnett. London: Rupert Hart-Davis.

—— 1980. *Les bêtes humaines d'Afrique*. Paris: Plon.

—— 1990a. Preface to Roumeguère-Eberhardt.

—— 1990b. 'The metamorphosis of unknown animals into fabulous beasts and of fabulous beasts into known animals'. *Cryptozoology* 9: 1–12.

—— 1995. *On the track of unknown animals*. Revised third English edition. London and New York: Kegan Paul International.

Hewkin, James A. 1993–96. 'Sasquatch investigations in the Pacific Northwest, 1993'. *Cryptozoology* 12: 72–75.

Hichens, William. 1928. 'Africa's mystery beasts'. *The World Wide Magazine* 62 (no. 369): 171–76.

—— 1937. 'African mystery beasts'. *Discovery* 18: 369–73.

Hidding, K.A.H. 1935. *Gebruiken en godsdienst der Soendaneezen*. Batavia: G. Kolff & Co.

Hill, Catherine, Pedro Soares, Maru Mormina, Vincent Macaulay, Dougie Clarke, Petya B. Blumbach, *et al.* 2007. 'A mitochondrial stratigraphy for Island Southeast Asia'. *The American Journal of Human Genetics* 80: 29–43.

Hill, W.C. Osman. 1945. 'Nittaewo—an unsolved problem of Ceylon'. *Loris: A Journal of Ceylon Wild Life*. Vol. IV, No. 1 (December): 251–62.

Hobbs, Joseph J. 2001. 'People and caves in Madagascar'. *Focus* 46(3): 1–7.

Hobbs, J.J. and Christopher L. Salter. 2006. *Essentials of world regional geography*. Fifth edition. Southbank, Victoria, Australia and Belmont, CA: Brooks/Cole Thomson Learning.

Holden, Jeremy. 1997. 'Project Orang Pendek: Report on photo-trapping April 1996–March 1997'. Flora and Fauna International and Research and Development Centre for Biology (LIPI). Unpublished manuscript.

Howell, Signe. 1984. *Society and cosmos: Chewong of peninsular Malaysia.* Singapore: Oxford University Press.

Hung, Hsiao-chun, Yoshiyuki Iizuka and Peter Bellwood. 2006. 'Taiwan jade in the context of Southeast Asian archaeology'. In E. Bacus, I. Glover and V. Pigott, eds. *Uncovering Southeast Asia's past.* London: British Museum, pp. 203–15.

Hunter, Monica. 1961. *Reaction to conquest: Effects of contact with Europeans on the Pondo of South Africa.* London: Oxford University Press (for the International African Institute). (Originally published in 1936.)

Ivens, Walter G. 1927. *Melanesians of the south-east Solomon Islands.* London: Kegan Paul, Trench, Trubner.

—— 1930. *The island builders of the Pacific; how & why the people of Mala construct their artificial islands, the antiquity & doubtful origin of the practice, with a description of the social organization, magic & religion of their inhabitants.* Philadelphia: Lippincott.

Izzard, Ralph. 1955. *The abominable snowman adventure.* London: Hodder and Stoughton.

Jacob, Teuku. 1967. *Some problems pertaining to the racial history of the Indonesian region: A study of human skeletal and dental remains from several prehistoric sites in Indonesia and Malaysia.* Utrecht: Drukkerij Neerlandia.

Jacob, T., E. Indriati, R.P. Soejono, K. Hsu, D.W. Frayer, R.B. Eckhardt, A.J. Kuperavage, A. Thorne and M. Henneberg. 2006. 'Pygmoid Australomelanesian *Homo sapiens* skeletal remains from Liang Bua, Flores: Population affinities and pathological abnormalities'. *Proceedings of the National Academy of Sciences* 103(36): 13421–426.

Jacobson, Edward. 1917. 'Rimboeleven in Sumatra'. *De Tropische Natuur* 6(5): 69–72.

—— 1918. 'Nog eens de "orang pandak"'. *De Tropische Natuur* 6(11): 173–76.

Jaivin, Linda. 1985. 'Is there a wildman?' *Asiaweek*, May 24, 1985, pp. 26–39.

Jardine, Sir William. 1833. *The natural history of monkeys.* Edinburgh: W.H. Lizars.

Jobson, Richard. 1632. *The golden trade, or a discovery of the river Gambra and the golden trade of the Aethiopians.* London: Nicholas Okes.

Johnson, Frederick. 1939. *A standard Swahili English dictionary (founded on Madan's English-Swahili dictionary).* London: Oxford University Press.

Johnston, Sir Harry. 1902. 'The pygmies and ape-like men of the Uganda borderland'. *The Pall Mall Magazine* 26: 173–84.

Jolly, Clifford J. 1970. 'Hadropithecus: A lemuroid small-object feeder'. *Man* (n.s.) 5(4): 619–26.

Jonker, J.C.G. 1911. *Rottineesche teksten met vertaling.* Leiden: Brill.

Joulian, Frédéric. 1995. 'Représentations traditionnelles du chimpanzé en Côte d'Ivoire: Etude préliminaire de quelques critères d'Humanité'. In Corbey and Theunissen, eds, pp. 275–87.

Joustra, M. 1904. 'Karo-Bataksche vertellingen'. *Verhandelingen van het Bataviaasch Genootschap van Kunsten en Wetenschappen* 56(1): 1–123.

—— 1907. *Karo-Bataksch woordenboek.* Leiden: Brill.

—— 1912. *Nederlandsch-Karosche woordenlijst.* Leiden: Van Doesburgh.

Joyner, Graham C. 1977. *The hairy man of south eastern Australia.* Canberra: National Library of Australia.

—— 1980. 'More historical evidence for the yahoo, hairy man, wild man or Australian "gorilla"'. Unpublished manuscript provided by the author.
—— 1984. 'The orang-utan in England: An explanation for the use of *yahoo* as a name for the Australian hairy man'. *Cryptozoology* 3: 55–57.
Kadirgamar, S.J. 1963. 'Comments on Rambukwella'. *Journal of the Royal Asiatic Society Ceylon Branch* 8 (part 2, new series): 277–90.
Kalff, S. 1925. 'Indische monstruositeiten'. *De Tropische Natuur* 14: 113–21.
Kapferer, Jean-Noel. 1990. *Rumors: uses, interpretations, images*. New Brunswick, NJ: Transaction Publishers. (Translated from the French original, 1987, by Bruce Fink.)
Kapita, Oemboe Hina. 1976. *Sumba di dalam jangkauan jaman*. Waingapu: Panitia Penerbit Naskah-Naskah Kebudayaan Daerah Sumba, Dewan Penata Layanan Gereja Kristen Sumba.
—— 1982. *Kamus Sumba/Kambera-Indonesia*. Waingapu: Panitia Penerbit Naskah-Naskah Kebudayaan Daerah Sumba, Dewan Penata Layanan Gereja Kristen Sumba.
Keers, W. 1948. *An anthropological survey of the eastern Lesser Sunda Islands*. Amsterdam: Koninklijke Vereeniging Indisch Instituut.
Kirlin, R. Lynn and Lasse Hertel. 1980. 'Estimates of pitch and vocal tract length from recorded vocalizations of purported bigfoot'. In Halpin and Ames, eds., pp. 274–90.
Kirtley, Bacil F. 1963. 'Notes upon a Central Asian legend'. *Folklore* 74: 318–20.
—— 1964. 'Unknown hominids and New World legends'. *Western Folklore* 23(2): 77–90.
Kleiweg de Zwaan, J.P. 1915. 'De aap in het volksgeloof der inlanders van den Indischen Archipel'. *Tijdschrift van het Koninklijk Nederlandsch Aardrijkskundig Genootschap* 32: 35–55.
—— 1942. *De dwergvolken (anthropologisch beschouwd)*. Den Haag: N.V. Servire.
Klokke, A.H., A. Klokke-Coster and M. Saha. 1988. *Ngaju-Dayak dierverhalen: Orale literatuur uit Midden-Kalimantan*. Koninklijk Instituut voor Taal-, Land- en Volkenkunde Werkdocumenten 2. Dordrecht, Holland/Providence, U.S.A.: Foris Publications.
Knibbs, Stanley George Curthoys. 1929. *The savage Solomons as they were & are*. London: Seeley, Service & Co.
Knox, Robert. 1983. *An historical relation of the island Ceylon in the East Indies*. New Delhi: Navrang. (First published in 1681 by Richard Chiswell, London.)
Kopstein, Felix. 1930. *Een zoölogische reis door de tropen met camera en verrekijker door de Indo-Australische dierenwereld*. Batavia: Kolff.
Krantz, Grover S. 1992. *Big footprints: A scientific inquiry into the reality of sasquatch*. Boulder, CO: Johnson Books.
Kreemer, J. 1910. 'De Loeboe's in Mandailing'. *Bijdragen tot de Taal-, Land- en Volkenkunde* 66 (3): 303–336.
Lambek, Michael. 1981. *Human spirits: A cultural account of trance in Mayotte*. Cambridge: Cambridge University Press.
Lamberton, C. 1937. 'Contribution à la connaissance de la faune subfossile de Madagascar. Note III: Les Hadropithèques'. *Bulletin de l'Academie Malgache (nouvelle série)* 20: 127–69.
La Roncière, Charles de. 1925. *La découverte de l'Afrique au Moyen-âge, cartographes et explorateurs*. Le Caire and Paris: Champion.
Laubscher, B.J.F. 1937. *Sex, custom and psychopathology: A study of South African pagan natives*. London: Routledge & Kegan Paul Ltd.

Leach, E.R. 1958. 'Magical hair'. *Journal of the Royal Anthropological Institute* 88(2): 147–64.
LeBar, Frank M. 1972. *Ethnic groups of Insular Southeast Asia. Volume 1: Indonesia, Andaman Islands and Madagascar*. New Haven, CT: Human Relations Area Files Press.
Leguat, François. 1891. *The voyage of François Leguat of Bresse to Rodriguez, Mauritius, Java and the Cape of Good Hope*. 2 volumes. (Transcribed from the first English edition; edited and annotated by Pasfield Oliver.) London: Hakluyt Society. (Original French title: *Voyage et aventures de François Leguat et de ses compagnons en deux îles désertes* (1690–1698).)
Lekkerkerker, Cornelis. 1916. *Land en volk van Sumatra*. Leiden: E.J. Brill.
Lessa, William. 1961. *Tales from Ulithi Atoll: A comparative study of Oceanic folklore*. Berkeley, CA: University of California Press.
—— 1980. *More tales from Ulithi Atoll: A content analysis*. Berkeley and Los Angeles: University of California Press.
Lévi-Strauss, Claude. 1982. *The way of the masks*. Translated from the French by Sylvia Modelski. Vancouver: Douglas & McIntyre.
Lewis, Frederick. 1914. 'Notes on an exploration in Eastern Uva and Southern Panama Pattu'. *Journal of the Royal Asiatic Society Ceylon Branch* 23 (67): 276–93.
—— 1915. 'Notes on animal and plant life in the Vedda country'. *Spolia Zeylanica* 10: 119–65.
Lin, Hen-li. 1956. 'Songs of the Pas-Taai'. Bulletin 2, Institute of Ethnology, Academia Sinica, Taipei. (In Chinese, with an English summary.)
Little, Henry W. 1884. *Madagascar: Its history and people*. Edinburgh and London: William Blackwood and Sons.
Loeb, Edward M. 1935. *Sumatra: Its history and people*. Wiener Beiträge zur Kulturgeschichte und Linguistik, Vol. 3. Vienna: Verlag des Institutes für Völkerkunde der Universitaet Wien.
Long, Barney. 1995. 'Project Orang Pendek support expedition to Sumatra, Indonesia, 26 June–23 August 1995'. Unpublished manuscript.
Loofs-Wissowa, Helmut. 2001. 'Would the real orang-utan please stand up and be counted: In search of unidentified relic hominoids in Southeast Asia'. In Paul Cropper, ed. *Myths and monsters*. (Proceedings of a conference held in 2001, available at http://www.strangenation.com.au/mm2001.htm)
Luitjes, A. 1925. 'De orang pandak. . . . een beer?' *De Tropische Natuur* 14: 56–58.
Luomala, Katharine. 1951. *The Menehune of Polynesia and other mythical little people of Oceania*. Honolulu: Bernice Pauahi Bishop Museum.
Mabuchi, Toichi. 1974. *Ethnology of the southwestern Pacific: The Ryukyus-Taiwan-Insular Southeast Asia*. Asian Folklore and Social Life Monographs Vol. 59. Taipei: The Orient Cultural Service.
Maceda, Marcelino N. 1964. *The culture of the Mamanua (northeast Mindanao) as compared with that of other negritos of Southeast Asia*. Manila: Catholic Trade School.
MacKinnon, John Ramsay. 1974. *In search of the red ape*. London: Collins.
Maier, R. 1923. 'De orang pandak of orang pendek'. *De Tropische Natuur* 12: 154–56.
Maître, Henri. 1912. *Les jungles moï: Exploration et histoire des hinterlands moï du Cambodge, de la Cochinchine, de l'Annam et du bas Laos*. Paris: Larose.
Maple, Terry L. 1980. *Orang-utan behavior*. New York. Van Nostrand Reinhold Company.

Marsden, William. 1783. *The history of Sumatra, containing an account of the government, laws, customs and manners of the native inhabitants, with a description of the natural productions and a relation of the ancient political state of that island.* London: Payne.

Manik, Tindi Radja. 1977. *Kamus bahasa Dairi Pakpak-Indonesia.* Jakarta: Pusat Pembinaan dan Pengembangan Bahasa, Departemen Pendidikan dan Kebudayaan.

Martyr, Deborah. 1990. 'An investigation of the orang-pendek, the "short man" of Sumatra'. *Cryptozoology* 9: 57–65.

—— 1994. 'The orang pendek of west Sumatra: report on fieldwork in Taman Nasional Kerinci Seblat, Sumatra, Indonesia'. Unpublished manuscript.

—— 1995a. 'Project Orang Pendek 1995, a Flora and Fauna Preservation Society project: Quarterly report April–June 1995'. Unpublished manuscript.

—— 1995b. 'Project Orang Pendek 1995, a Flora and Fauna Preservation Society project: progress report June–August 1995'. Unpublished manuscript.

—— 1995c. 'Project Orang Pendek 1995: A report on fieldwork conducted in Taman Nasional Kerinci Seblat, Sumatra, Indonesia and elsewhere April–November 1995'. A Flora and Fauna International Project. Unpublished manuscript.

—— 1995d. 'Project Orang Pandak 1995'. A Flora and Fauna Preservation Society Project. Unpublished manuscript.

—— 1996. 'Project Orang Pendek: Annual report 1996'. Flora and Fauna International in partnership with the Indonesian Institute of Sciences (LIPI). Unpublished manuscript.

—— 1998. 'Quarterly report on field work: Kerinci-Seblat National Park Sumatra and surrounding forests January–March 1998'. Indonesian Institute of Sciences (LIPI), PHPA, Flora and Fauna International. Unpublished manuscript.

—— 1999a. 'Project Orang Pendek Flora and Fauna International: Report on Field Work Kerinci-Seblat National Park, Sumatra and surrounding forest Jan.–July 1999'. Unpublished manuscript.

—— 1999b. 'The Sipurak "hook": Interim report on fieldwork north of Gunung Sumbing and Gunung Nilo, Jambi province, Sumatra'. Unpublished manuscript.

Martyr, Deborah, Achmad Yanuar, Jeremy Holden and David Chivers. n.d. 'An assessment of the status of orangpendek in Sumatra'. Unpublished manuscript.

Mathijsen, A. 1915. 'Eenige fabels en volkslegenden van de onderafdeeling Beloe op het eiland Timor'. *Verhandelingen van het Bataviaasch Genootschap van Kunsten en Wetenschappen* 61(2): 1–96.

Mayor, Adrienne. 2000. *The first fossil hunters: Paleontology in Greek and Roman times.* Princeton, NJ: Princeton University Press.

Mazur, Oleh. 1980. *The wild man in the Spanish Renaissance and Golden Age theatre.* Ann Arbor, MI: University Microfilms International (for Villanova University).

McDougall, Harriette. 1992. *Sketches of our life at Sarawak.* Singapore: Oxford University Press. (First published in 1882 by the Society for Promoting Christian Knowledge.)

McGrew, William. 2004. *The cultured chimpanzee: Reflections on cultural primatology.* Cambridge: Cambridge University Press.

McHugh, J.N. 1955. *Hantu hantu: An account of ghost belief in modern Malaya.* Singapore: Donald Moore.

McIlwraith, T.F. 1992. *The Bella Coola Indians.* Volume 1. Toronto: University of Toronto Press.

McNeely, Jeffrey A. and Paul Spencer Wachtel. 1988. *Soul of the tiger: Searching for nature's answers in exotic Southeast Asia.* New York: Doubleday.

Means, Nathalie and Paul B. Means. 1987. *Senoi-English, English-Senoi dictionary*. Edited by Gordon P. Means. Toronto: The Joint Centre on Modern East Asia, University of Toronto and York University.

Melalatoa, M. Junus and S.R.H. Sitanggang. 1985. *Kamus bahasa Gayo-Indonesia*. Jakarta: Pusat Pembinaan dan Pengembangan Bahasa, Departemen Pendidikan dan Kebudayaan.

Messner, Reinhold. 2000. *My quest for the yeti: Confronting the Himalaya's deepest mystery*. Translated from the German by Peter Constantine. New York: St. Martin's Press.

Migeod, F.W.H. 1926. *A view of Sierra Leone*. London: Kegan Paul, Trench, Trubner.

Mirazon Lahr, M. and R. Foley. 2004. 'Human evolution writ small'. *Nature* 431: 1043–44.

Mivart, St. George Jackson. 1873. *Man and apes: An exposition of structural resemblances and differences bearing upon questions of affinity and origin*. London: Robert Hardwicke.

Molnar, Andrea K. 2000. *Grandchildren of the Ga'e ancestors: Social organization and cosmology among the Hoga Sara of Flores*. Verhandelingen van het Koninklijk Instituut voor Taal-, Land- en Volkenkunde 185. Leiden: KITLV Press.

Mommersteeg, Adriaan and Margaretha H. Dirkzwager, eds. 1999. *Punu nangé: Ceritera dari So'a, Flores*. Jakarta: Yayasan Obor Indonesia dan Yayasan Asosiasi Tradisi Lisan.

Moncrieff, R.W. 1966. *Odour preferences*. London: Leonard Hill.

Monro, D. 1898. 'Notes on a journey through a part of the middle island of New Zealand'. Appendix C in Thomas M. Hocken, *Contributions to the early history of New Zealand (Settlement of Otago)*. London: Sampson, Low, Marston & Co. Ltd, pp. 230–63.

Moodie, D.W. and A.J.W. Catchpole. 1992. 'Northern Athapaskan oral traditions and the White River volcano'. *Ethnohistory* 39(2): 148–71.

Morwood, M.J., P. Brown, Jatmiko, T. Sutikna, E. Wahyu Saptomo, K.E. Westaway, *et al.* 2005. 'Further evidence for small-bodied hominins from the Late Pleistocene of Flores, Indonesia'. *Nature* 437: 1012–17.

Morwood, M.J., R.P. Soejono, R.G. Roberts, T. Sutikna, C.S.M. Turney, K.E. Westaway, *et al.* 2004. 'Archaeology and age of a new hominin from Flores in eastern Indonesia'. *Nature* 431: 1087–91.

Morwood, Mike and Penny Van Oosterzee. 2007. *A new human: The startling discovery and strange story of the "Hobbits" of Flores, Indonesia*. New York: HarperCollins.

Moszkowski, Max. 1909. 'Sagen und Fabeln aus Ost- und Zentralsumatra'. *Anthropos* 4: 989–97.

Moussay, Gerard. 1995. *Dictionnaire Mingangkabau-Indonesien-Français*. 2 volumes. Paris: Editions l'Harmattan & Association Archipel.

Moyne, Lord and Kathleen Haddon. 1936. 'The pygmies of the Aiome Mountains, Mandated Territory of New Guinea'. *The Journal of the Royal Anthropological Institute of Great Britain and Ireland* 66: 269–90.

Mozino, José Mariano. 1970. *Noticias de Nutka: An account of Nootka Sound in 1792*. Translated and edited by I.H. Wilson. Seattle and London: University of Washington Press. (Originally published in 1793.)

Napier, John. 1972. *Bigfoot: The yeti and sasquatch in myth and reality*. New York: E.P. Dutton & Co.

Nebesky-Wojkowitz, Rene von. 1956. *Where the gods are mountains: Three years among the people of the Himalayas.* Translated from the German by Michael Bullock. London: Weidenfeld and Nicholson.

Needham, Rodney. 1956. 'The Siwang of central Malaya'. *Journal of the Malayan Branch of the Royal Asiatic Society* 29(1): 49–69.

—— 1978. *Primordial characters.* Charlottesville, VA: University of Virginia Press.

—— 1980. *Reconnaissances.* Toronto: University of Toronto Press.

Neumann, J.H. 1926–27. 'Bijdrage tot de geschiedenis de Karo-Batakstammen' (parts 1 and 2). *Bijdragen tot de Taal-, Land- en Volkenkunde* 82: 1–36; 83: 162–80.

—— 1951. *Karo-Bataks—Nederlands Woordenboek.* Lembaga Kebudajaan Indonesia ('Koninklijk Bataviaasch Genootschap van Kunsten en Wetenschappen'). Medan (Sumatra): v/h Varekamp & Co.

Nevill, Hugh. 1886a. 'The Nittawewo of Ceylon'. *The Taprobanian.* February issue, pp. 66–68. Bombay: Education Society's Press.

—— 1886b. 'Brown bear of Ceylon'. *The Taprobanian.* January issue, pp. 1–2. Bombay: Education Society's Press.

Noble, Charles Frederick. 1762. *A voyage to the East Indies in 1747 and 1748. Containing an account of the islands of St. Helena and Java. Of the city of Batavia. Of the government and political conduct of the Dutch. Of the empire of China, with a particular description of Canton.* London, T. Becket and P.A. Dehondt.

Noone, H.V.V. 1945. 'Stone age relics at Bandarawela'. *Loris: A Journal of Ceylon Wild Life.* Vol. IV No. 1 (December 1945): 263–267.

Notulen. 1899. *Notulen van de Algemeene en Bestuurs-Vergaderingen van het Bataviaasch Genootschap van Kunsten en Wetenschappen.* Deel XXXVII. 6 March, 1899 Batavia: Albrecht & Co, pp. 55–56.

O'Connor, Sue. 2007. 'New evidence from East Timor contributes to our understanding of earliest modern human colonisation east of the Sunda Shelf'. *Antiquity* 81: 523–35.

Onvlee, Louis. 1925. *Eeenige Soembasche vertellingen.* Leiden: Drukkerij A. Vros.

Oppitz, Michael. 1968. *Geschichte und Sozialordnung der Sherpa.* Innsbruck: Universitätsverlag Wagner.

Osborn, Chase Salmon. 1925. *Madagascar: Land of the man-eating tree.* London: Hutchinson & Co.

Osman, Mohd. Taib. 1989. *Malay folk beliefs: An integration of disparate elements.* Kuala Lumpur: Dewan Bahasa dan Pustaka.

Pache, Alois. 1964. *Die religiösen Vorstellungen in den Mythen der formosanischen Bergstämme.* Studia Instituti Anthropos, Volume 19. Wien-Mödling: Missionsdruckerei St. Gabriel.

Pan, Chiu-rong. 1999. 'A study of the paSta'ay (small people ceremony) and heavenly god ceremony of the Saisiat'. Master's Thesis for the Institute of Ethnology, National Chengchi University. (In Chinese, passages translated for the author by F. Hsu)

Pareira, M. Mandalangi and E. Douglas Lewis. 1998. *Kamus Sara Sikka Bahasa Indonesia.* Ende, Flores: Penerbit Nusa Indah.

Pasteur, J.D. 1960. 'The "Orang Pendek" of Sumatra'. *Panorama Stanvac* 5 (no. 2): 16–21.

Pelras, Christian. 1996. *The Bugis.* Oxford: Blackwell.

Persoon, Gerard. 1989. 'The Kubu and the outside world (South Sumatra, Indonesia)'. *Anthropos* 84: 507–19.

Pigafetta, Antonio. 1975. *Magellan's voyage: A narrative account of the first navigation.* Translated and edited by R.A. Skelton. London: The Folio Society. (Originally published in 1528, Paris.)

Pleyte, C.M. 1894. *Batakse vertellingen.* Utrecht: H. Honig.

Poirier, Frank E. and J. Richard Greenwall. 1992. 'Is there a large, unknown primate in China? The Chinese *yeren* or wildman'. *Cryptozoology* 11: 70–82.

Poirier, Frank E., Hu Hongxing and Chung-Min Chen. 1983. 'The evidence for wildman in Hubei Province, People's Republic of China'. *Cryptozoology* 2: 25–39.

Raffles, Thomas Stamford. 1817. *The history of Java.* 2 volumes. London: Murray.

Rambukwella, Captain A.T. 1963. 'The Nittaewo: The legendary pygmies of Ceylon.' *Journal of the Royal Asiatic Society Ceylon Branch* 8 (part 2, new series): 265–70.

Randerson, James. 2007. 'Found: The giant lion-eating chimps of the magic forest'. *The Guardian*, Saturday July 14, 2007.

Ransome, L.H. 1889. 'The river Antanambalana, Madagascar'. *Proceedings of the Royal Geographical Society and Monthly Record of Geography* 11(5): 295–305.

Rappaport, Roy. 1974. *Pigs for the ancestors: Ritual in the ecology of a New Guinea people.* New Haven, CT: Yale University Press.

Rattray, Robert Sutherland. 1927. *Religion and art in Ashanti.* London: Oxford University Press.

Ravenstein, E.G., ed. 1901. *The strange adventures of Andrew Battell of Leigh, in Angola and the adjoining regions.* London: Hakluyt Society.

Rawling, C.G. 1913. *The land of the New Guinea pygmies: An account of the story of a pioneer journey of exploration into the heart of New Guinea.* Philadelphia: J.B. Lippincott Company.

Reynolds, Vernon. 1967. *The apes: The gorilla, chimpanzee, orangutan and gibbon—their history and their world.* New York: E.P. Dutton & Co.

Richards, Anthony. 1981. *An Iban-English Dictionary.* Oxford: Clarendon Press.

Richards, Paul. 1995. 'Local understandings of primates and evolution: Some Mende beliefs concerning chimpanzees'. In Corbey and Theunissen, eds, pp. 265–73.

Ridderhof, W.H.R. 1929. 'Over orang pendek, santeng en kentoes'. *Het Nederlandsche Java-Instituut* 5: 36–38.

Riedel, J.G.F. 1871. 'De volksoverleveringen betreffende de voormalige gedaante van Noord-Selebes en den oorsprong zijner bewoners'. *Tijdschrift voor Nederlandsch Indië* (3rd series, 5th volume, part 1) 5(1): 288–303.

Rigg, Jonathan. 1862. *A dictionary of the Sunda language.* Batavia: Lange & Co.

Rigsby, Bruce. 1971. 'Some Pacific Northwest native language names for the sasquatch phenomenon'. *Northwest Anthropological Research Notes* 5(2): 153–56.

Rijksen, Herman D. and Erik Meijaard. 1999. *Our vanishing relative: The status of wild orang-utans at the close of the twentieth century.* Dordrecht: Kluwer Academic Publishers.

Rintjema, Annelieke. 2001. 'The unidentified small hominoids of Lempur: A search for cultural evidence for the existence of the ape-men of Sumatra'. MA thesis, Cultural Anthropology, State University of Utrecht.

Rivers, W.H.R. 1914. *The history of Melanesian society.* 2 Volumes. Cambridge: Cambridge University Press.

Robe, Stanley R. 1972. 'Wild men and Spain's Brave New World'. In E.J. Dudley and M.E. Novak eds, *The wild man within: An image in western thought from the Renaissance to Romanticism*, pp. 39–53. Pittsburgh, PA: University of Pittsburgh Press.

Rookmaker, O.J. 1932. 'De "Orang Letjo" en de "Orang Pendek"'. *Het Koloniaal Weekblad* 32(19): 219–21.

Rosaldo, Renato. 1982. 'Utter savages of scientific value'. In Eleanor Leacock and Richard Lee, eds. *Politics and history in band societies.* Cambridge: Cambridge University Press, pp. 309–325.

Roth, Henry Ling. 1896. *The natives of Sarawak and British North Borneo.* 2 volumes. London: Truslove and Hanson.

Roumeguère-Eberhardt, Jacqueline. 1990. *Le dossier X: Les hominidés non identifiés des forêts d'Afrique.* Paris: Editions Robert Laffont.

Rütimeyer, L. 1903. Die Nilgalaweddas in Ceylon. *Globus* 83 (13): 201–07, 220–23.

Rutter, Owen. 1999. *The dragon of Kinabalu and other Borneo stories.* Kota Kinabalu: Natural History Publications (Borneo).

Sanders, Andrew. 1995. *A deed without a name: The witch in society and history.* Oxford: Berg.

Sanderson, Ivan T. 1961. *Abominable snowmen: Legend come to life; the story of sub-humans on five continents from the early ice age until today.* Philadelphia and New York: Chilton Company.

—— 1965. *Ivan Sanderson's book of great jungles.* (with David Loth.) New York: J. Messner.

—— 1967. 'The Wudewasa or Hairy Primitives of Ancient Europe'. *Genus* 23: 109–40.

—— 1972. *Investigating the unexplained: A compendium of disquieting mysteries of the natural world.* Englewood Cliffs, NJ: Prentice-Hall.

Saragih, Jatias Edison. 1989. *Kamus Simalungun Indonesia.* Pematangsiantar (North Sumatra, Indonesia): Gereja Kristen Protestan Simalungun.

Sarasin, Fritz. 1906. *Versuch einer Anthropologie der Insel Celebes, zweiter teil: Die Varietäten des Menschen auf Celebes.* Wiesbaden: CW Kreidel's Verlag.

Sarasin, Paul and Fritz Sarasin. 1903. 'Über die Toala von Süd-Celebes'. *Globus* 83(18): 277–280.

Satjadibrata, R. 1954. *Kamus basa Sunda: Katut k°etjap-k°etjap asing nu geus ilahar.* Djakarta: Perpustakaan Perguruan, Kementerian P.P. dan K.

Schadeberg, Thilo. 1999. 'Batwa: The Bantu name for the invisible people'. In Karen Biesbrouck, Stefan Elders and Gerda Rossel, eds. *Central African hunter-gatherers in a multidisciplinary perspective: Challenging elusiveness.* Leiden: Research School CNWS, pp. 21–39.

Schaller, George B. 1963. *The mountain gorilla: Ecology and behavior.* Chicago: The University of Chicago Press.

—— 1964. *The year of the gorilla.* Chicago and London: The University of Chicago Press.

Schebesta, Paul. 1928, *Orang-Utan: Bei den Urwaldmenschen Malayas und Sumatras.* Leipzig: F.A. Brockhaus.

—— 1931. *Among Congo pigmies.* Translated from the German by Gerald Griffin. London: Hutchinson & Co.

Schebesta, Paul and Victor Lebzelter. 1933. *Anthropology of the Central African pygmies in the Belgian Congo.* Prague: Czech Academy of Sciences and Arts.

Schefold, Reimar. 1988. *Lia: Das grosse Ritual auf den Mentawai-Inseln (Indonesien).* Berlin: Dietrich Reimer.

Schlegel, Hermann and Saloman Müller. 1839–1844. 'Bijdragen tot de natuurlijke historie van den Orang-oetan (Simia satyrus)'. In *Verhandelingen over de natuur-*

lijke geschiedenis der Nederlandsche Overzeesche Bezittingen, door de Leden der Natuurkundige Commissie in Indië en andere Schrijvers. Vol. 2, Zoologie, pp. 1–28. Leiden: S. en J. Luchtmans en C.C. van der Hoek.

Schnell, Raymond. 1948. 'A propos de l'hypothèse d'un peuplement négrille ancient de l'Afrique occidentale'. *L'Anthropologie* 52(3–4): 229–42.

Schwarz, J.A.T. 1907. *Tontemboansche Teksten: Vertaling*. Leiden: E.J. Brill.

—— 1908. *Tontemboansch-Nederlandsch Woordenboek*. Leiden: E.J. Brill.

Schweinfurth, Georg. 1873. *The heart of Africa*. 2 volumes. London: S. Low, Marston, Low and Searle.

Seltmann, Friederich. 1987. *Die Kalang: Eine Volksgruppe auf Java und ihre Stamm-Myth*. Stuttgart: Franz Steiner Verlag Wiesbaden.

Shackley, Moira. 1983. *Wildmen: Yeti, Sasquatch and the Neanderthal enigma*. London: Thames and Hudson.

Sharp, Lesley. 1993. *The possessed and the dispossessed: Spirits, identity and power in a Madagascar migrant town*. Berkeley: University of California Press.

Sharpe, Margaret and Dorothy Tunbridge. 1997. 'Traditions of extinct animals, changing sea-levels and volcanoes among Australian Aboriginals: Evidence from linguistic and ethnographic research'. In Blench and Spriggs, eds, pp. 345–61.

Shea, Brian T. 1984. 'Between the gorilla and the chimpanzee: A history of debate concerning the existence of the *kooloo-kamba* or gorilla-like chimpanzee'. *Journal of Ethnobiology* 4(1): 1–13.

Shipman, Pat. 2001. *The man who found the missing link: Eugène Dubois and his life-long quest to prove Darwin right*. New York: Simon & Schuster.

Sibree, James.1896. *Madagascar before the conquest: The island, the country and the people*. London: T. Fisher Unwin.

—— 1915. *A naturalist in Madagascar*. London: Seeley, Service and Co.

Skeat, Walter William. 1900. *Malay magic: An introduction to the folklore and popular religion of the Malay Peninsular* [sic]. London: Macmillan and Co. Ltd.

Skeat, Walter William and Charles Otto Blagden. 1906. *Pagan races of the Malay Peninsula*. 2 volumes. London: Macmillan.

Smith, Malcolm. 1989. 'Analysis of the Australian "hairy man" (yahoo) data'. *Cryptozoology* 8: 27–36.

Snouck Hurgronje, Christiaan. 1906. *The Achehnese*. Translated from the Dutch by A.W.S. O' Sullivan. Leyden: Brill.

Sody. 1925. 'Nachschrift'. *De Tropische Natuur* 14: 59–60.

Speiser, Felix. 1913. *Two years with the natives in the western Pacific*. London: Mills & Boon, Ltd.

Sperber, Dan. 1975. *Rethinking symbolism*. Translated from the French by Alice L. Morton. Cambridge: Cambridge University Press.

Spittel, Richard L. 1933. *Far-off things: Treating of the history, aborigines, myths and jungle mysteries of Ceylon*. Ceylon: Colombo Apothecaries.

—— 1936, 'Leanama—Land of the Nittaewo'. *Loris* 1: 37–46.

—— 1963 'Comments on Rambukwella'. *Journal of the Royal Asiatic Society Ceylon Branch* 8 (part 2, new series): 271–75.

Sprague, Roderick. 1980. 'Carved stone heads on the Columbia River and sasquatch'. In Halpin and Ames, eds, pp. 229–34.

Spriggs, Matthew. 1997. *The island Melanesians*. Oxford: Blackwell Publishers.

Spriggs, Matthew and Atholl Anderson. 1993. 'Late colonization of East Polynesia'. *Antiquity* 67: 200–217.

Steedly, Mary M. 1993. *Hanging without a rope: Narrative experience in colonial and postcolonial Karoland.* Princeton, NJ: Princeton University Press.

St. John, Spenser. 1862. *Life in the forests of the Far East.* 2 volumes. London: Smith, Elder and Co.

Stoczkowski, Wiktor. 2002. *Explaining human origins: Myth, imagination and conjecture.* Cambridge: Cambridge University Press.

Stonor, Charles. 1955. *The sherpa and the snowman.* London: Hollis & Carter.

Sugathapala de Silva, M.W. 1972. *Vedda language of Ceylon: Texts and lexicon.* München: R. Kitizinger.

Supanov, Valentin B. 1988. 'A mathematical analysis of "snowman" (wildman) eyewitness reports'. *Cryptozoology* 7: 61–65.

Sutlive, Vinson H. and Joanne Sutlive. 1994. *Handy reference dictionary of Iban and English.* Kuala Lumpur: Ampang Press.

Suttles, Wayne. 1972. 'On the cultural track of the sasquatch'. *Northwest Anthropological Research Notes* 6(1): 65–90.

—— 1980. 'Sasquatch: The testimony of tradition'. In Halpin and Ames, eds, pp. 245–54.

Taft, Michael. 1980. 'Sasquatch-like creatures in Newfoundland: A study in the problem of belief, perception and reportage'. In Halpin and Ames, eds, pp. 83–96.

Taki, J. and D. Tryon. 1997. 'The lost languages of Erromango (Vanuatu)'. In Blench and Spriggs, eds, pp. 362–70.

Tchernine, Odette. 1971. *In pursuit of the abominable snowman.* New York: Taplinger Publishing Co.

Ter Haar, Barend J. 2006. *Telling stories: Witchcraft and scapegoating in Chinese history.* Leiden and Boston: Brill.

Thomas, Keith. 1983. *Man and the natural world: Changing attitudes in England 1500–1800.* London: Allen Lane.

Thompson, Stith. 1946. *The folktale.* New York: The Dryden Press.

—— 1975. *Motif-index of folk-literature: A classification of narrative elements in folktales, ballads, myths, fables, mediaeval romances, exempla, fabliaux, jestbooks and local legends.* Revised and enlarged edition. 5 volumes. Bloomington, IN: Indiana University Press.

Tishkov, Arkady. 1993–96. 'Observation of a Yeti in the Himalayas of Tibet'. *Cryptozoology* 12: 58–65.

Tocheri, M., C.M. Orr, S.G. Larson, T. Sutikna, Jatmiko, E. Wahyu Saptomo, Rokus Awe Due, T. Djubiantono, M.J. Morwood and W.L. Jungers. 2007. 'The primitive wrist of *Homo floresiensis* and its implications for hominin evolution'. *Science* 317 (21 September): 1743–45.

Tressol, James. 1961. *Strange patterns in the South Seas.* Garden City, New York: Doubleday.

Trinkaus, Erik and Pat Shipman. 1993. *The Neandertals: Changing the image of mankind.* New York: Alfred A. Knopf.

Tukang Hikayat. 1932. 'The orang pendek?'. *The Java Gazette: The Official Organ of the British Chamber of Commerce for the Netherlands East Indies* 1(5): 253.

Turnbull, Colin. 1961. *The forest people: A study of the pygmies of the Congo.* New York: Simon and Schuster.

Turner, Victor. 1967. *The forest of symbols.* Ithaca, NY: Cornell University Press.

Valentijn, Francois. 1978. *Francois Valentijn's Description of Ceylon.* Translated and edited by Sinnappah Arasaratnam. London: Hakluyt Society. (First published in 1726.)

Valeri, Valerio. 2000. *The forest of taboos: Morality, hunting and identity among the Huaulu of the Moluccas*. Madison, Wisconsin: The University of Wisconsin Press.

Van Bekkum, Willem. 1946. 'Geschiedenis van Manggarai (West-Flores) II. Todo en Pongkor'. *Cultureel Indië* 8: 65–75

Van Beurden, Marcel. 2002. 'Jacht op de wildeman'. *Skepter* 15 (3): 12–18.

Van den Pijl, L. 1938. 'De orang pendek als sneeuwman'. *De Tropische Natuur* 27: 55–56.

Van Dongen, G.J. 1906. 'Bijdrage tot de kennis van de Ridan-Koeboes'. *Tijdschrift voor het Binnenlandsch Bestuur* 30: 225–53.

—— 1931. 'Een en ander over de Koeboes, naar aanleiding van Prof. P.P. Schebesta's publicities over dit volk'. *Bijdragen tot de Taal-, Land- en Volkenkunde* 88: 519–618.

Van Gennep, Arnold. 1910. *La formation des légendes*. Paris: E. Flammarion.

Van Gulik, Robert H. 1967. *The gibbon in China: An essay in Chinese animal lore*. Leiden: Brill.

Van Herwaarden, J. 1924. 'Een ontmoeting met een aapmensch'. *De Tropische Natuur* 13: 103–06.

Van Hien. H.A. 1912. *De Javaansche geestenwereld en de betrekking, die tusschen de geesten en de zinnelijke wereld bestaat, verduidelijkt door petangan's of tellingen, bij de Javanen en Soendaneezen in gebruik. Deel I: De geschiedenis der godsdiensten en der tijdrekenkunde op Java.* Bandoeng: Fortuna.

—— 1994. *Wali Sanga: De Javaanse geheime leer.* Breda: Warung Bambu.

Vansina, Jan. 1965. *Oral tradition: A study in historical methodology*. Chicago: Aldine.

Verheijen, J.A.J. 1950. 'De stem der dieren in de Manggaraise folklore'. *Bijdragen tot de Taal-, Land- en Volkenkunde* 106: 55–78.

—— 1951. *Het hoogste wezen bij de Manggaraiers*. Studia Instituti Anthropos 4. Wien-Mödling: St. Gabriel.

—— 1967. *Kamus Manggarai I: Manggarai-Indonesia*. 's-Gravenhage: Martinus Nijhoff.

—— 1982. *Komodo: Het eiland, het volk en de taal*. Verhandelingen van het Koninklijk Instituut voor Taal-, Land-, en Volkenkunde 96. The Hague: Martinus Nijhoff.

Verhoeven, Theodor. 1958. 'Proto-Negrito in den Grotten auf Flores. *Anthropos* 53: 229–32.

—— 1968. 'Vorgeschichtliche Forschungen auf Flores, Timor und Sumba'. In *Anthropica*. Studia Instituti Anthropos Vol. 21. St. Augustin bei Bonn: Verlag des Anthropos-Instituts, pp. 393–403.

Vicedom, G.F. and H. Tischner. 1943. *Die Mbowamb: Die Kultur der Hagenberg-Stämme im östlichen Zentral-Neuguinea*. 3 volumes. Hamburg: Cram de Gruyter.

Viveiros de Castro, E. 1998. 'Cosmological deixis and Amerindian perspectivism'. *The Journal of the Royal Anthropological Institute* 4: 469–88.

Vogelpoel, E.W.P. 1930. 'De "orang pendek"'. (Letter to the editor of an unidentified Dutch language newspaper, Dubois Collection, Natural History Museum, Leiden, The Netherlands; annotation gives date of 11th June 1930).

Volz, Wilhelm. 1922. *Im Dämmer des Rimba: Sumatras Urwald und Urmensch.* Breslau: F. Hirt.

Von Koenigswald, R. 1981. 'Gibt es in China noch Orang Utans?'. *Natur und Museum* 111(8): 260–61.

Voorhoeve, P. 1927. *Overzicht van de volksverhalen der Bataks*. Vlissingen: Boekdrukkerij Firma F. van de Velde Jr.

Warneck, Johannes. 1909. *Die Religion der Batak: Ein Paradigma für die animistischen Religionen des Indischen Archipels.* Religions-Urkunden der Völker, Abteilung IV, Band I. Leipzig: Dieterisch'sche Verlagsbuchhandlung, Theodor Weicher.
Wavell, Stewart. 1958. *The lost world of the east: An adventurous quest in the Malayan hinterland.* London: Souvenir Press.
Weck, Wolfgang. 1948. *Märchen von Bali.* Leipzig: Die Humboldt-Bücherei, Gerhard Mindt.
Werner, E.T.C. 1922. *Myths and legends of China.* London: George G. Harrap & Co.
Westenberg, C.J. 1892. 'Aanteekeningen omtrent de godsdienstige begrippen der Karo-Bataks'. *Bijdragen tot de Taal-, Land- en Volkenkunde* 41: 208–53.
Westenenk, L.C. 1918. 'Orang pandak (boschmenschen) op Soematra'. *De Tropische Natuur* 12: 108–110.
—— 1921. 'Sedapak . . . the missing link?'. *De Tropische Natuur* 10: 120–21.
—— 1932. *Het rijk van Bittertong.* 's-Gravenhage: Leopold.
White, Hayden V. 1978. *Tropics of discourse: Essays in cultural criticism.* Baltimore, MD: Johns Hopkins University Press.
Whitten, Tony, Muslimin Mustafa and Gregory S. Henderson. 2000. *The ecology of Sumatra.* Hong Kong: Periplus.
Wielenga, D.K. 1910. 'Soemba: In angst en vreezen'. *Macedoniër* 14: 302–12.
—— 1913. 'Soembaneesche verhalen. In't dialect van Kambera, met vertalingen en aanteekeningen'. *Bijdragen tot de Taal-, Land- en Volkenkunde* 68: 1–290.
Wieringa, Edwin. 1998. 'Who are the Kalang?: An unknown minority group on Java and their so-called myth of origin'. *Anthropos* 93: 19–30.
Wijesekera, Nandadeva. 1964. *Veddas in transition.* Colombo: M. D. Gunasena.
Wilkinson, R.J. 1932. *A Malay-English dictionary (romanised).* 2 Volumes. Mytilene, Greece: Salavopoulos and Kinderlis.
Wilson, A.H. 1950. 'Guadalcanal's undiscovered people just another tall tale'. *Pacific Islands Monthly* 20(8): 7.
Winstedt, Richard Olaf. 1964. *A practical modern Malay-English dictionary.* Fourth edition. Kuala Lumpur: Marican.
Winter, H.R. 1901. 'Ook onderdanen onzer Koningin, een bezoek aan de tamme Koeboes'. *De Indische Gids* 23(1): 208–47.
Wolpert, Lewis. 1992. *The unnatural nature of science.* London: Faber and Faber.
Worcester, Dean C. 1912. 'Head-hunters of northern Luzon'. *National Geographic* 23: 833–930.
Wrangham, Richard W. and Dale Peterson. 1996. *Demonic males: Apes and the origins of human violence.* Boston: Houghton Mifflin.
Yanuar, Achmad. 1998. 'Quarterly report on field work: Kerinci-Seblat National Park Sumatra and surrounding forests January–March 1998'. Indonesian Institute of Sciences (LIPI), PHPA, Flora and Fauna International. Unpublished manuscript.
—— 1999. 'Project Orang Pendek Flora and Fauna International: Report on field work April–July 1999'. Unpublished manuscript.
Yanuar, Achmad and Deborah Martyr. 1995. 'Project Orang Pendek: Tentative final report on field work May–November 1995'. Flora and Fauna International. Unpublished manuscript.
Yerkes, Robert M. and Ada W. Yerkes. 1929. *The great apes: A study of anthropoid life.* New Haven, CT: Yale University Press.
Young, Emma. 2004. 'The beast with no name'. *New Scientist* 184(2468): 32–35.

Yuan Zhenxin and Huang Wanpo. 1981. 'A challenge to science: the mystery of the wildman'. In Steve Moore, ed. *Wildman: China's yeti*. Fortean Times Occasional Papers No. 1. London, pp. 5–14.

Yule, Henry. 1903. *Hobson-Jobson: A glossary of colloquial Anglo-Indian words and phrases and of kindred terms, etymological, historical, geographical and discursive*. New edition edited by William Crooke. London: John Murray.

Zhou Guoxing. 1982. 'The state of wildman research in China'. *Cryptozoology* 1: 13–23.

—— 1984. 'Morphological analysis of the Jiulong mountain "manbear" (wildman) hand and foot specimens'. *Cryptozoology* 3: 58–70.

Zollinger, Heinrich. 1850. 'Verslag van eene reis naar Bima en Sumbawa, en naar eenige plaatsen op Celebes, Saleijar en Flores, gedurende de maanden Mei tot December 1847'. *Verhandelingen van het Bataviaasch Genootschap van Kunsten en Wetenschappen* 23 (part 4).

Index

For Product Safety Concerns and Information please contact our EU representative GPSR@taylorandfrancis.com
Taylor & Francis Verlag GmbH, Kaufingerstraße 24, 80331 München, Germany

www.ingramcontent.com/pod-product-compliance
Ingram Content Group UK Ltd.
Pitfield, Milton Keynes, MK11 3LW, UK
UKHW020021200926
475702UK00014B/1996
* 9 7 8 0 4 1 5 5 3 3 4 8 5 *